Monographs on Statistics and Applied Probability 103

Stereology for Statisticians

MONOGRAPHS ON STATISTICS AND APPLIED PROBABILITY

General Editors

1 Stochastic Population Models in Ecology and Epidemiology *M.S. Barlett* (1960)
2 Queues *D.R. Cox and W.L. Smith* (1961)
3 Monte Carlo Methods *J.M. Hammersley and D.C. Handscomb* (1964)
4 The Statistical Analysis of Series of Events *D.R. Cox and P.A.W. Lewis* (1966)
5 Population Genetics *W.J. Ewens* (1969)
6 Probability, Statistics and Time *M.S. Barlett* (1975)
7 Statistical Inference *S.D. Silvey* (1975)
8 The Analysis of Contingency Tables *B.S. Everitt* (1977)
9 Multivariate Analysis in Behavioural Research *A.E. Maxwell* (1977)
10 Stochastic Abundance Models *S. Engen* (1978)
11 Some Basic Theory for Statistical Inference *E.J.G. Pitman* (1979)
12 Point Processes *D.R. Cox and V. Isham* (1980)
13 Identification of Outliers *D.M. Hawkins* (1980)
14 Optimal Design *S.D. Silvey* (1980)
15 Finite Mixture Distributions *B.S. Everitt and D.J. Hand* (1981)
16 Classification *A.D. Gordon* (1981)
17 Distribution-Free Statistical Methods, 2nd edition *J.S. Maritz* (1995)
18 Residuals and Influence in Regression *R.D. Cook and S. Weisberg* (1982)
19 Applications of Queueing Theory, 2nd edition *G.F. Newell* (1982)
20 Risk Theory, 3rd edition *R.E. Beard, T. Pentikäinen and E. Pesonen* (1984)
21 Analysis of Survival Data *D.R. Cox and D. Oakes* (1984)
22 An Introduction to Latent Variable Models *B.S. Everitt* (1984)
23 Bandit Problems *D.A. Berry and B. Fristedt* (1985)
24 Stochastic Modelling and Control *M.H.A. Davis and R. Vinter* (1985)
25 The Statistical Analysis of Composition Data *J. Aitchison* (1986)
26 Density Estimation for Statistics and Data Analysis *B.W. Silverman* (1986)
27 Regression Analysis with Applications *G.B. Wetherill* (1986)
28 Sequential Methods in Statistics, 3rd edition
G.B. Wetherill and K.D. Glazebrook (1986)
29 Tensor Methods in Statistics *P. McCullagh* (1987)
30 Transformation and Weighting in Regression
R.J. Carroll and D. Ruppert (1988)
31 Asymptotic Techniques for Use in Statistics
O.E. Bandorff-Nielsen and D.R. Cox (1989)
32 Analysis of Binary Data, 2nd edition *D.R. Cox and E.J. Snell* (1989)
33 Analysis of Infectious Disease Data *N.G. Becker* (1989)
34 Design and Analysis of Cross-Over Trials *B. Jones and M.G. Kenward* (1989)
35 Empirical Bayes Methods, 2nd edition *J.S. Maritz and T. Lwin* (1989)
36 Symmetric Multivariate and Related Distributions
K.T. Fang, S. Kotz and K.W. Ng (1990)

37 Generalized Linear Models, 2nd edition *P. McCullagh and J.A. Nelder* (1989)

38 Cyclic and Computer Generated Designs, 2nd edition
J.A. John and E.R. Williams (1995)

39 Analog Estimation Methods in Econometrics *C.F. Manski* (1988)

40 Subset Selection in Regression *A.J. Miller* (1990)

41 Analysis of Repeated Measures *M.J. Crowder and D.J. Hand* (1990)

42 Statistical Reasoning with Imprecise Probabilities *P. Walley* (1991)

43 Generalized Additive Models *T.J. Hastie and R.J. Tibshirani* (1990)

44 Inspection Errors for Attributes in Quality Control
N.L. Johnson, S. Kotz and X. Wu (1991)

45 The Analysis of Contingency Tables, 2nd edition *B.S. Everitt* (1992)

46 The Analysis of Quantal Response Data *B.J.T. Morgan* (1992)

47 Longitudinal Data with Serial Correlation—A State-Space Approach
R.H. Jones (1993)

48 Differential Geometry and Statistics *M.K. Murray and J.W. Rice* (1993)

49 Markov Models and Optimization *M.H.A. Davis* (1993)

50 Networks and Chaos—Statistical and Probabilistic Aspects
O.E. Barndorff-Nielsen, J.L. Jensen and W.S. Kendall (1993)

51 Number-Theoretic Methods in Statistics *K.-T. Fang and Y. Wang* (1994)

52 Inference and Asymptotics *O.E. Barndorff-Nielsen and D.R. Cox* (1994)

53 Practical Risk Theory for Actuaries
C.D. Daykin, T. Pentikäinen and M. Pesonen (1994)

54 Biplots *J.C. Gower and D.J. Hand* (1996)

55 Predictive Inference—An Introduction *S. Geisser* (1993)

56 Model-Free Curve Estimation *M.E. Tarter and M.D. Lock* (1993)

57 An Introduction to the Bootstrap *B. Efron and R.J. Tibshirani* (1993)

58 Nonparametric Regression and Generalized Linear Models
P.J. Green and B.W. Silverman (1994)

59 Multidimensional Scaling *T.F. Cox and M.A.A. Cox* (1994)

60 Kernel Smoothing *M.P. Wand and M.C. Jones* (1995)

61 Statistics for Long Memory Processes *J. Beran* (1995)

62 Nonlinear Models for Repeated Measurement Data
M. Davidian and D.M. Giltinan (1995)

63 Measurement Error in Nonlinear Models
R.J. Carroll, D. Rupert and L.A. Stefanski (1995)

64 Analyzing and Modeling Rank Data *J.J. Marden* (1995)

65 Time Series Models—In Econometrics, Finance and Other Fields
D.R. Cox, D.V. Hinkley and O.E. Barndorff-Nielsen (1996)

66 Local Polynomial Modeling and its Applications *J. Fan and I. Gijbels* (1996)

67 Multivariate Dependencies—Models, Analysis and Interpretation
D.R. Cox and N. Wermuth (1996)

68 Statistical Inference—Based on the Likelihood *A. Azzalini* (1996)

69 Bayes and Empirical Bayes Methods for Data Analysis
B.P. Carlin and T.A Louis (1996)

70 Hidden Markov and Other Models for Discrete-Valued Time Series
I.L. Macdonald and W. Zucchini (1997)

71 Statistical Evidence—A Likelihood Paradigm *R. Royall* (1997)
72 Analysis of Incomplete Multivariate Data *J.L. Schafer* (1997)
73 Multivariate Models and Dependence Concepts *H. Joe* (1997)
74 Theory of Sample Surveys *M.E. Thompson* (1997)
75 Retrial Queues *G. Falin and J.G.C. Templeton* (1997)
76 Theory of Dispersion Models *B. Jørgensen* (1997)
77 Mixed Poisson Processes *J. Grandell* (1997)
78 Variance Components Estimation—Mixed Models, Methodologies and Applications *P.S.R.S. Rao* (1997)
79 Bayesian Methods for Finite Population Sampling *G. Meeden and M. Ghosh* (1997)
80 Stochastic Geometry—Likelihood and computation *O.E. Barndorff-Nielsen, W.S. Kendall and M.N.M. van Lieshout* (1998)
81 Computer-Assisted Analysis of Mixtures and Applications—Meta-analysis, Disease Mapping and Others *D. Böhning* (1999)
82 Classification, 2nd edition *A.D. Gordon* (1999)
83 Semimartingales and their Statistical Inference *B.L.S. Prakasa Rao* (1999)
84 Statistical Aspects of BSE and vCJD—Models for Epidemics *C.A. Donnelly and N.M. Ferguson* (1999)
85 Set-Indexed Martingales *G. Ivanoff and E. Merzbach* (2000)
86 The Theory of the Design of Experiments *D.R. Cox and N. Reid* (2000)
87 Complex Stochastic Systems *O.E. Barndorff-Nielsen, D.R. Cox and C. Klüppelberg* (2001)
88 Multidimensional Scaling, 2nd edition *T.F. Cox and M.A.A. Cox* (2001)
89 Algebraic Statistics—Computational Commutative Algebra in Statistics *G. Pistone, E. Riccomagno and H.P. Wynn* (2001)
90 Analysis of Time Series Structure—SSA and Related Techniques *N. Golyandina, V. Nekrutkin and A.A. Zhigljavsky* (2001)
91 Subjective Probability Models for Lifetimes *Fabio Spizzichino* (2001)
92 Empirical Likelihood *Art B. Owen* (2001)
93 Statistics in the 21st Century *Adrian E. Raftery, Martin A. Tanner, and Martin T. Wells* (2001)
94 Accelerated Life Models: Modeling and Statistical Analysis *Vilijandas Bagdonavicius and Mikhail Nikulin* (2001)
95 Subset Selection in Regression, Second Edition *Alan Miller* (2002)
96 Topics in Modelling of Clustered Data *Marc Aerts, Helena Geys, Geert Molenberghs, and Louise M. Ryan* (2002)
97 Components of Variance *D.R. Cox and P.J. Solomon* (2002)
98 Design and Analysis of Cross-Over Trials, 2nd Edition *Byron Jones and Michael G. Kenward* (2003)
99 Extreme Values in Finance, Telecommunications, and the Environment *Bärbel Finkenstädt and Holger Rootzén* (2003)
100 Statistical Inference and Simulation for Spatial Point Processes *Jesper Møller and Rasmus Plenge Waagepetersen* (2004)
101 Hierarchical Modeling and Analysis for Spatial Data *Sudipto Banerjee, Bradley P. Carlin, and Alan E. Gelfand* (2004)
102 Diagnostic Checks in Time Series *Wai Keung Li* (2004)
103 Stereology for Statisticians *Adrian Baddeley and Eva B. Vedel Jensen* (2004)

Monographs on Statistics and Applied Probability 103

Stereology for Statisticians

Adrian Baddeley
Eva B. Vedel Jensen

A CRC Press Company
Boca Raton London New York Washington, D.C.

Library of Congress Cataloging-in-Publication Data

Catalog record is available from the Library of Congress

International Standard Book Number 1-58488-405-3
Printed in the United States of America 1 2 3 4 5 6 7 8 9 0
Printed on acid-free paper

Mais il se trompait sur les apparences: ce qui n'arrive que trop, soit qu'on se serve ou non de microscopes.*

Voltaire, *Micromégas* (1752)

* "But he was deceived by appearances: which happens only too often, whether or not we use microscopes." [571, chap. 5]

Contents

1 Introduction **1**

2 Classical stereology **9**

2.1 Volume fraction 9

2.2 Surface area density and length density 17

2.3 Particles 25

2.4 Particle size 32

2.5 Thick sections 38

2.6 Counting profiles in two dimensions 41

2.7 Projections 42

2.8 Statistical critique of classical stereology 46

2.9 Advice to consultants 48

2.10 Exercises 50

2.11 Bibliographic notes 52

3 Overview of modern stereology **55**

3.1 Motivation 55

3.2 Statistical foundations 56

3.3 Random sampling designs 63

3.4 Geometrical identities 71

3.5 Geometrical probability 75

3.6 Advice to consultants 78

3.7 Exercises 82

3.8 Bibliographic notes 83

4 Geometrical identities **87**

4.1 Plane sections and line transects 88

4.2 Quadrats 97

4.3 Curvature 104

4.4 Exercises 107

4.5 Bibliographic notes 108

5 Geometrical probability **111**

5.1 UR points 112

5.2 IUR and FUR lines in the plane 114

5.3 FUR planes 117

5.4 FUR quadrats in the plane 119

5.5 Isotropic directions 123

5.6 IUR planes 126

5.7 IUR quadrats 129

5.8 Advice to consultants 129

5.9 Exercises 130

5.10 Bibliographic notes 133

6 Statistical formulations of stereology **135**

6.1 Terminology 135

6.2 Generic models 137

6.3 Parameters 139

6.4 Reference spaces 142

6.5 Estimation in the *Random* case 144

6.6 Estimation in the *Restricted* case 147

6.7 Estimation in the *Extended* case 149

6.8 Advice to consultants 151

6.9 Exercises 152

6.10 Bibliographic notes 153

7 Uniform and isotropic uniform designs **155**

7.1 Estimation of volume and area 155
7.2 Length and surface area estimation 162
7.3 Exercises 168
7.4 Bibliographic notes 172

8 Vertical and local designs **175**

8.1 Vertical sections 176
8.2 Vertical slices and vertical projections 188
8.3 Local lines 192
8.4 Local vertical sections 199
8.5 Local isotropic sections 202
8.6 Variance comparisons of local estimators 204
8.7 Exercises 205
8.8 Bibliographic notes 206

9 Ratio estimation **209**

9.1 Ratio estimation in finite populations 209
9.2 Stereological estimation *via* ratios 213
9.3 Estimation of stereological ratios 216
9.4 Data modelling 224
9.5 Exercises 227
9.6 Bibliographic notes 228

10 Discrete sampling and counting **229**

10.1 Basic principles 229
10.2 Unbiased sampling rules for particles 233
10.3 Decomposition 242
10.4 Reweighting 244
10.5 Variance reduction 248
10.6 Fractionator 249
10.7 Exercises 250
10.8 Bibliographic notes 252

11 Inference for particle populations **255**

11.1 Strategies for particle stereology 255

11.2 The design-based case 257

11.3 Model-based case 269

11.4 Exercises 275

11.5 Bibliographic notes 276

12 Design of stereological experiments **277**

12.1 Design constraints 277

12.2 Parameters and estimators 278

12.3 Design strategies 281

12.4 Implementation 284

12.5 Sources of non-sampling error 289

12.6 Optimal design 290

13 Variance of stereological estimators **295**

13.1 Variances in model-based stereology 296

13.2 Variances in design-based stereology 302

13.3 Variance comparisons and Rao-Blackwell 314

13.4 Prediction variance 317

13.5 Nested ANOVA 318

13.6 Particle number 321

13.7 Advice to consultants 324

13.8 Exercises 326

13.9 Bibliographic notes 327

14 Frontiers and open problems **329**

14.1 Variances 329

14.2 Inhomogeneous materials and tissues 330

14.3 Anisotropy and digital stereology 330

14.4 Spatial arrangement 331
14.5 Stereology of extremes 331
14.6 Fractal behaviour of surfaces 331
14.7 Non-recognition artefacts 331
14.8 Reconciliation with other areas 332
14.9 Shape models 332

A Sampling theory 333
A.1 Population and sample 333
A.2 Estimation from a sample 335
A.3 Sampling designs 336
A.4 Nonuniform sampling designs 338
A.5 Variances of estimators 339

List of notation 341

References 345

Index 377

Preface

Above all, we have written this book to help statisticians to give effective advice to researchers in microscopy.

Stereology (or 'quantitative microscopy') is a valuable research tool in biological science, materials science, geological science and in many technologies. It rests heavily on statistical principles, especially random sampling and sampling inference. It bears a strong resemblance to survey sampling.

Like survey sampling, stereology has many traps — potential errors of methodology, such as sampling bias — into which we may easily fall if they are not pointed out. Important scientific findings (the structure of human liver; the structure of quenched steel; age-related deterioration of the human brain) have been overturned or cast into doubt by the discovery of stereological flaws.

Stereological methods are now much more flexible and powerful than they were twenty years ago. Major advances have been achieved by applying statistical ideas. Yet, despite its scientific importance and its strong reliance on statistics, modern stereology remains largely unknown to the statistical community. The essential literature is widely dispersed in applications journals. Many of the major recent advances in stereology have been direct applications of statistical ideas by non-statisticians.

The time is ripe for a book on stereology for statistical readers. We have usually recommended the excellent treatise of Weibel [580, 581] to our statistical colleagues, but it has been overtaken by developments since 1980. Athough there are several recent textbooks on stereology for applications [49, 175, 272, 328, 444, 427, 489] our book differs by emphasising statistical issues such as the precise sampling conditions, the nature of sampling inference, and the performance of estimators. Existing books on the probabilistic foundations of stereology [543, Chap. 10], [262, 503] give a good foundation for 'model-based' stereological inference. Our book is unique in that it also covers 'design-based' inference (based on random sampling designs) and emphasises statistical issues.

Statisticians could be making a far greater and far more useful contribution to stereology than they are making at present. We hope this book will help them to do so.

Acknowledgements

We owe an enormous debt to Roger Miles, pioneer of modern stereology, who has been an inspiration to us both.

The influence of our good friends and collaborators Hans Jørgen Gundersen and Luis Cruz-Orive is present throughout this book. Their work is a major part of modern stereology. We have shared countless lecture halls and forums with them, and this book inevitably bears the imprint of their views on stereology and on how it can be communicated.

The text has benefited from the valuable comments of the stereological community, our statistical colleagues, our students in statistics, and participants in stereology courses run by the International Society for Stereology. We thank especially Luis Cruz-Orive, Karl-Anton Dorph-Petersen, Hans Jørgen Gundersen, Ian James, Nazim Khan, Michael Levitan, Antonietta Mira, Kevin Murray, Berwin Turlach, Rolf Turner, and an anonymous CRC reviewer.

Images were generously provided by Jenny Bevan, Joanne Chia, Luis Cruz-Orive, Karl-Anton Dorph-Petersen, Arun Gokhale, Hans Jørgen Gundersen, Ian Harper, Charter Mathison, Paul McCormick, Niels Marcussen, Jens Randel Nyengaard, Joachim Ohser, John Robinson and D.C. Sterio. The excellent illustrations were drawn by Kaj Vedel, and the others were drawn by AB.

The International Society for Stereology has played an important supporting role through its officers, its activities, and its membership. This work was partially supported by MaPhySto — Centre for Mathematical Physics and Stochastics — funded by a grant from the Danish National Research Foundation; and by grants from the Australian Research Council and the Danish Natural Science Research Council. AB also acknowledges support from CWI (Amsterdam, Netherlands), CSIRO (Australia) and the Universities of Bath, Cambridge, Leiden and Western Australia.

CHAPTER 1

Introduction

Aims

The main practical purpose of stereology is to extract quantitative information from microscope images. For example, Figure 1.1 shows a microscope image of a type of rock. Stereology makes it possible to infer, *from this image alone*, properties of the rock such as its percentage composition, the average volume of the crystalline mineral grains, the area of grain boundary surfaces per unit volume of rock, the length of grain edges per unit volume of rock, and the mean solid angle between facets of a grain.

Figure 1.1 *Optical microscope image of a polished plane section of granuloblastic rock. Image width about 3 mm. Reproduced by kind permission of Dr C. Mathison, University of Western Australia. Copyright © C. Mathison.*

This is a form of statistical inference. The microscope image is a sample of the material of interest. Our goal is to draw statistical inferences about parameters of the material overall (such as percentage composition) from observations in the sample. A stereological analysis of Figure 1.1 yields *estimates* of the percentage composition and other parameters of the rock.

Yet this may seem impossible, because the image is a *two-dimensional* plane section of the *three-dimensional* rock specimen. The rock was cut, polished, and viewed under the optical microscope in transmitted polarised light to produce this image. The shapes seen in the image are plane sections of individual solid grains of crystalline mineral. We cannot observe the three-dimensional extent of each grain, so we cannot directly measure its volume or surface area. Even simply counting the shapes in the section is not equivalent to counting three-dimensional grains.

Thus, a key problem is *how to extrapolate from two-dimensional plane sections to three-dimensional reality.* This is the main objective of stereology, a discipline defined in 1963 as "the spatial interpretation of sections" [169], its name coined from the Greek *stereos* (solid). Weibel [580, Introduction] writes:

Stereology
is
(1) *a body of mathematical methods*
relating
(2) *three-dimensional parameters defining the structure*
to
(3) *two-dimensional measurements obtainable on sections of the structure.*

Stereology is not tomography. The aim of stereology is not to reconstruct the three-dimensional geometry of the material, but only to estimate certain parameters such as its percentage composition. Three-dimensional reconstruction and imaging techniques, such as computed tomography [436], serial section reconstruction [348, 384, 469] and confocal microscopy [461, 608], effectively reassemble the solid geometry of the material from information in densely-spaced plane sections or projections. Stereology is a completely different activity, in which statistical inferences are drawn from a *single* plane section, or a handful of sections, taken at random locations scattered sparsely through the material. It is a form of sampling inference for plane sections of solid materials.

The first stereological method was devised by a geologist, A.E. Delesse, in 1847. Until that time, in order to measure the mineral composition of a rock, one would have crushed the rock to fine pieces and assayed the constituents, or inferred the composition from physical properties. Instead, Delesse reasoned that a polished plane section of rock could be regarded as 'representative' of the composition of the entire rock. The constituent minerals should be present in a plane section in roughly the same proportions in which they occur throughout the entire rock. That is, the percentage of *area* occupied by a mineral in the plane section should reflect its percentage of *volume* in the solid rock. To determine mineral composition we have only to measure the relative areas of each mineral visible in a polished plane section of the rock [155, 156]. This technique is still used in quantitative microscopy today.

Delesse's key geometrical insight was that **three-dimensional reconstruction is not necessary** in order to measure volume. By a principle known to Archimedes, the volume of a three-dimensional solid — of arbitrary shape —

is the ‘sum’ of the areas of its plane sections. This means that volumes of solid objects can be estimated from the areas of plane sections, under the right conditions, without reassembling the sections into a three-dimensional solid.

The greatest advantage of stereological methods is their efficiency. The introduction of Delesse’s technique vastly reduced the amount of effort required to get a reasonably accurate estimate of the mineral composition of a rock. Apart from reducing the amount of effort, Delesse’s technique is also surprisingly accurate in absolute terms in many applications. By the widespread adoption of Delesse’s method, and of subsequent stereological methods, a pragmatic scientific community showed its gradual acceptance of the idea that random sampling can produce accurate results.

Scope

Modern stereological techniques are not confined to the interpretation of two-dimensional plane sections of three-dimensional materials like Figure 1.1. The same principles apply equally to other types of microscopy, in which the image may be a two-dimensional projection of the material, a three-dimensional volume image, a cylindrical core sample, a one-dimensional transect, etc. These may all be regarded as samples, and are amenable to stereological methods. The material under investigation could itself be two-dimensional, for example, a skin graft, the retina, or a sheet of plastic. What these methods share in common is the use of statistical sampling principles in a geometrical context. Modern stereology may be defined as “sampling inference for geometrical objects” [28, 147].

Stereological techniques offer a highly efficient way of estimating *some* parameters of a solid material from samples such as plane sections. They are closely related to the methods of survey sampling [83, 555] and share the same advantages and disadvantages. The main advantages of stereology are efficiency, simplicity, general validity, and broad applicability. Consequently, each new development in stereological methods has become a powerful new tool in many fields of science. The main disadvantage of stereology is that only certain parameters are accessible from a given kind of sample.

Virtually all microscopy involves sampling, because of the magnification involved. Hence there is no escape from sampling issues, such as sampling design, sampling bias and efficient estimation. Stereology is therefore a useful guide to some of the fundamental issues of scientific method in microscopy.

Applications

Stereological techniques are used widely in science and engineering, especially in the life sciences (anatomy, histology, neuroscience, pathology, plant biology, soil science, veterinary science), in geology (mineralogy, petroleum geology), in materials science and engineering (metallurgy, composites, concrete, ceramics), industry (food science) and clinical medicine (dermatology, nephrology, oncology). They apply to many different materials, specimen preparations, and modalities of microscopy (notably optical transmission and reflected light microscopy, transmission electron microscopy (EM), back-scattered EM, acoustic microscopy, and 2D/3D confocal optical microscopy). They also apply to non-microscopical imagery and spatial data.

Modern stereology is a truly interdisciplinary science in which theory and applications must work hand-in-hand. Key theoretical advances have been made by researchers in geology, materials science, biological science, clinical medicine, statistics, image analysis and mathematics. This activity is supported by the International Society for Stereology (ISS), founded in 1961, which unites researchers in all these fields. The official journals of the Society are the *Journal of Microscopy* (Royal Microscopical Society, Oxford, UK) and *Image Analysis & Stereology* (formerly *Acta Stereologica*, Ljubljana). These journals, and the proceedings of the 4-yearly International Congresses and European Congresses for Stereology, publish new research in stereology. Innumerable specialist journals also carry stereological research.

Stereological thinking

Although the main thrust of stereological techniques is to get quantitative information about solid materials, stereology also plays a very important role in the *qualitative* interpretation of microscope images.

For example, as emphasised above, the shapes seen in a plane section should not be misinterpreted as three-dimensional objects. Mistakes of this kind have been responsible for a litany of scientific errors, including a misunderstanding of the organisation of the mammalian liver which persisted for a century, and an important misunderstanding of the microstructure of quenched steel. To detect and avoid such errors, the experimenter needs to think in three dimensions, and to think about sampling effects — what we might call "stereological thinking" [171, 495].

One important methodological problem in microscopy is *sampling bias*. Consider a solid material that contains discrete 'particles', such as the individual grains in a metal, or the individual cells in a biological tissue. A plane section of this material introduces a sampling bias: larger particles are more likely to be represented on the plane section, since they are more likely to be hit by the section plane. Thus, for example, the relative abundance of two types of cells in the tissue is not equal to the relative frequency with which these cells will be observed in a plane section.

Modern stereology, like modern statistics [9], retains a strong sense of the need to examine the validity of scientific methods and conclusions. Methodological errors in the quantitative use of the microscope are quite common in our experience. In 1985 the stereologist and anatomist H. Haug identified a flaw in all existing experimental evidence for the statement that the human brain progressively loses neurons with age [250, 251]. Methods for counting neurons are still controversial, judging by some piquant correspondence in the journal *Trends in Neurosciences* [23, 46, 84, 195]. There is a long-running discussion over the industry standards for measuring grain size in metallography (cf. [437]).

Statistical theory

There are essentially two ways of formulating sampling inference in stereology, depending on whether the sampling variation arises from the natural variability of the material (a *model-based* approach) or from the random positioning of the section plane (a *design-based* approach).

The classical theory of stereology was model-based. The material or tissue being sectioned was assumed to be 'spatially homogeneous'. This is a stochastic modelling assumption, which treats the contents of the material as a realisation of a random process, and assumes the process is stationary in some sense. Stereological relations, such as Delesse's principle, can then be derived. Unbiased estimators of three-dimensional quantities are obtained, under the assumption of homogeneity, from an arbitrary plane section of the material.

The design-based approach originates from the work of Australian statisticians R.E. Miles and P.J. Davy [146, 147, 408]. They pointed out the strong analogy between stereology and sample surveys [83, 555]. The three-dimensional material can be regarded as a population of interest, and the plane section as a random sample of it. Instead of assuming a spatially homogeneous material, we may alternatively conduct the experiment so that our sampling of the material is spatially homogeneous. Recall that, in "randomised design based" survey sampling, estimators of population parameters are unbiased by virtue of the randomisation of the sample, without any need for assumptions about the population structure (cf. [555, p. 3]). By analogy, estimators of three-dimensional quantities in stereology are unbiased, under appropriate randomisation of the section plane, regardless of the spatial arrangement of the material.

The advent of design-based stereological methods vastly increased the power, flexibility, adaptability and scope of applications of stereology. It also stimulated the development of some new sampling techniques which do not have an analogue in the model-based viewpoint. These developments also established a much closer connection between stereology and statistics.

About this book

This book sets out the basic principles of stereology from a statistical viewpoint. Its main aim is to help biostatisticians and statistical consultants to give advice to their research collaborators and clients about experiments that involve stereology. To aid the consultant, we focus on practical implications as well as basic theory, discuss many examples, give literature references to applications as well as statistical theory, draw attention to common methodological errors, and suggest effective ways of communicating statistical issues to the client.

The book could also serve as the text for a course on stereology for statistics students, or for an extension of a course on survey sampling theory. Students and teachers of statistics will find that stereology is an interesting twist on the usual diet of survey sampling theory, with the extra ingredient of spatial reasoning. We think this book would fit naturally into a sampling theory course at undergraduate or postgraduate level.

The book may also be used as a reference and bibliography on stereological sampling theory for research in statistical science.

Basic undergraduate statistical concepts will be assumed (such as expected value, variance, parameter estimation, unbiased estimation, and sources of variability). We will also need the key ideas of survey sampling theory [83, 555] (such as random sampling designs, estimation from samples, and sampling bias). These are summarised in Appendix A.

This book focuses mainly on **design-based** stereology. There is currently no textbook on the design-based approach, and we think there is an urgent need for one. The essential literature on design-based inference is widely scattered and inaccessible to the newcomer. Yet this is a valuable scientific methodology, depending heavily on statistical concepts, of which statisticians should be aware, and to which statisticians can make substantial contributions. By focusing on design-based stereology in this book, we are able to highlight the statistical connections more easily.

The **model-based** approach is also treated (especially Chapters 3, 6 and 11) but is not our primary focus. For a full-length and rigorous account of stereology from this standpoint, see [543, Chap. 10] and [510].

Plan of the book

Chapters 2 and 3 give a general survey of the science of stereology, emphasising applications. Chapter 2 covers classical stereology while Chapter 3 is an overview of 'modern' stereology. These two chapters should help the reader to see the overall perspective and to navigate through the rest of the book.

A more methodical treatment of stereological theory starts in Chapter 4. Chapters 4–6 cover the essential mathematical and statistical theory needed for stereology. They deal with geometry, probability, and statistical inference, respectively.

Chapters 7–11 present the core techniques of stereology: the estimation of 'absolute' geometrical quantities, such as volume and surface area, using isotropic uniform designs (Chapter 7) or vertical or local designs (Chapter 8); the estimation of 'relative' quantities such as volume fraction (Chapter 9); and statistical inference for a population of discrete objects, such as cells or mineral grains (Chapters 10 and 11).

The subsequent chapters discuss implementation and problems. The implementation of these techniques in practical sampling designs is discussed in Chapter 12. Chapter 13 summarises our current understanding of the variance of stereological estimators. Chapter 14 describes open problems for further research. An Appendix covers essential background on Sampling theory.

CHAPTER 2

Classical stereology

This chapter introduces the 'classical' concepts and methods of stereology. It may serve as a self-contained introduction to basic stereological ideas.

2.1 Volume fraction

2.1.1 Delesse's principle

From the early 19th century, microscopes were used in mineralogy and metallurgy to study the internal structure of materials. Rocks and metals had to be cut with a saw, polished to a flat surface, and viewed under the microscope in reflected light. This technique of 'metallography' was used mainly for the qualitative study of microscopic structure; but in 1847 the young French geologist A.E. Delesse* proposed that it could also be used for quantitative purposes.

Delesse reasoned that under suitable conditions, a polished plane section of rock could be regarded as 'representative' of the entire rock. The constituent minerals should be present in a plane section in the same proportions in which they occur throughout the entire rock. Assuming the rock is "spatially homogeneous", Delesse [155, 156] proved that

$$V_V = A_A \tag{2.1}$$

where V_V is the average fraction of *volume* in the solid rock that is occupied by the mineral of interest, and A_A is the average fraction of *area* on a plane section that is occupied by the same mineral. In brief, a mineral's percentage of volume in solid rock equals its percentage of area in plane sections.

Delesse derived this from a beautiful geometrical insight [156, p. 380]. Consider the solid region occupied by the mineral of interest. The volume of a solid, of arbitrary shape, can be determined from the areas of its horizontal plane sections. If the rock is homogeneous, then all plane sections are "statistically equivalent". Using basic properties of expectation, it follows that the

* Achille Ernest Delesse (1817–1881), Ingénieur en chef des mines, Professor at the Ecole Normale (Paris), elected Académicien 1879.

average fraction of volume occupied by the mineral of interest in rock is equal to its *average* fraction of area on plane sections.

This led to the following practical method for determining the mineral composition of a rock. Viewing a polished plane section of the rock, the investigator should identify the different mineral components in the section, and trace their contours onto heavy wax paper or foil. The shapes are then cut out and weighed. By Delesse's principle (2.1), the *fraction of total weight* of wax paper, attributed to a particular mineral, is an estimate of the fraction of total rock volume occupied by this mineral.

Figure 2.1 *Polished plane section of rock containing iron pyrite (lighter, contoured regions) and silicates (darker). Optical microscope, reflected light, $KMnO_4$ oxidation. True field width about 3 mm. Reproduced by kind permission of Dr C. Mathison, University of Western Australia. Copyright © C. Mathison.*

For example, Figure 2.1 shows a polished plane section of rock from the Keillor 1 open cut mine in Western Australia. It is of interest to determine the volume fraction of pyrite, represented by the lighter-coloured phase in Figure 2.1. Using digital image analysis, we computed that the pyrite occupies a fraction

$$\frac{A(\text{pyrite in image})}{A(\text{image})} = 0.68$$

of the image area, so our estimate of the volume fraction of pyrite V_V is also 68%.

Needless to say, the fraction of area observed in a particular plane section image such as Figure 2.1 is merely an *estimate* of the true volume fraction V_V. Its value will vary from section to section. The accuracy of this estimate is a key question — and it is largely a statistical question. As we explain below (page 15), Delesse's principle implies that the observed area fraction is an *unbiased estimator* of the true volume fraction V_V.

Delesse's method represented an enormous saving in effort over the established mechano-chemical assay methods [559, p. 194]. It was gradually adopted in geology, with some practical improvements, after experiments showed it gave results in broad agreement with these techniques.

The style of notation used here has been widely adopted in stereology. Each geometrical quantity is represented by a standard letter: A for area, V for volume, and so on. A combination of two letters always indicates an average density: V_V is the "volume of interest per unit volume of reference" – the average volume of the mineral of interest, per unit volume of the reference space. See the list of notation for further examples.

2.1.2 *Lineal analysis*

Half a century later, Austrian geologist A. Rosiwal† developed a practical simplification of Delesse's technique [486]. A grid of equally-spaced parallel lines is superimposed on the plane section, as shown in Figure 2.2. The lines could be drawn on a transparent sheet which is then superimposed on a photographic print; they could be etched onto the microscope eyepiece; they could be scan lines traced by a moving spot; and so on.

Figure 2.2 *Rosiwal line transect method applied to Figure 2.1.*

The fraction of total *length* of these grid lines which lie above the phase of interest, is then an approximation to the area fraction. Assuming the rock is spatially homogeneous, Rosiwal proved

$$A_A = L_L \tag{2.2}$$

where the *length fraction* L_L is the average fraction of grid line length that lies over the mineral of interest. In fact this result is valid for any arbitrary arrangement of test lines (or curves) in the plane section. A regularly spaced grid of parallel lines is chosen mainly for practical convenience.

† August Rosiwal (1860–1923), professor of mineralogy and geology, Technische Hochschule, Vienna.

It follows that an estimate of L_L also serves as an estimate of A_A and therefore of V_V. In Figure 2.2 we measured directly the length of each transect of the pyrite phase, and calculated

$$\frac{L(\text{lines over pyrite})}{L(\text{lines})} = 0.57$$

where $L(\text{lines over pyrite})$ is the *total* length of these transects, and $L(\text{lines})$ is the *total* length of grid lines. Our estimate of V_V by Rosiwal's method is 0.57.

This method had actually been mentioned by Delesse [156, p. 388] but he rejected it as inaccurate. Nonetheless, Rosiwal's method was much more practical than the original Delesse technique, and was rapidly adopted in geology [341, 343]. It was also adopted in metallography where careful testing verified its reliability [466]. Closely related techniques measuring absolute and relative areas were adopted in cartography [350, p. 437] and discussed by the mathematician H. Steinhaus [534, 536].

2.1.3 *Test points*

A further simplification by the geologists Glagolev [200] and Thomson [559] was to superimpose on the plane section a grid of equally spaced points, as shown in Figure 2.3.

Figure 2.3 *Glagolev-Thomson point-counting method applied to Figure 2.1.*

Then simply by counting the *number* of these test points which lie above the phase of interest, we could estimate the area fraction. In shorthand,

$$A_A = L_L = P_P \tag{2.3}$$

where P_P is the (average) proportion of test points that are covered by the mineral of interest. Here P is the conventional notation in stereology for a count of the number of points.

In Figure 2.3, nine of the twelve test points lie over the pyrite phase, so our estimate of the volume fraction V_V of pyrite using the Glagolev-Thomson technique is $9/12 = 0.75$.

Thus the volume fraction parameter V_V can be determined by (empirical estimates of) the area fraction A_A, the length fraction L_L or the point count fraction P_P. While this progressive simplification may seem trivial to us, it is interesting that each successive step took half a century and encountered stiff resistance. One of the obstacles may have been that a deterministic view of measurement does not allow us to replace an 'accurate' geometrical quantity by an 'inaccurate' subsample.

Of the three estimators of volume fraction, the point-counting estimator usually requires the least effort. It is particularly useful in applications where the microscopic structure is difficult to delineate in the section — for example, where expert knowledge is needed to recognise a biological cell of a particular type. In such cases, point counting requires only a few decisions, judging whether each of the test points is inside or outside the phase of interest. By comparison, computation of the area fraction in Figure 2.1 required numerous image processing steps in order to obtain a satisfactory binary image.

Surprisingly, the point-counting estimator may be quite accurate, even for a coarsely spaced grid of test points. Very surprisingly, the point-counting estimator may actually be **more accurate** than the other two estimators, because of the effects of positive correlation. See page 17 and Chapter 13.

Point-counting techniques for estimating areas and relative areas were also developed elsewhere, notably in cartography [350, p. 394]. The pioneer of geometrical probability, M.W. Crofton, had already pointed out in 1885 that the area of a planar region Y can be estimated using a grid of test points, randomly superimposed on the plane [105].

2.1.4 Statistical aspects

We now examine the validity and accuracy of the three methods for estimating volume fraction.

To discuss such statistical issues, we must modify the classical notation and terminology which we have followed until now. Traditionally the same symbol V_V was used to denote both the volume fraction parameter and a sample estimate ('determination') of this parameter. This usage made it difficult to discuss the accuracy of stereological estimators! It ultimately led to some confusion over the mathematical content of the stereological equations themselves.

In the next few pages, we discuss statistical properties of stereological methods, using new notation.

Precise statement of Delesse's principle

The classical literature contains some confused and contradictory claims about Delesse's method: about the precise meaning of the equality sign in (2.1), the conditions under which (2.1) is valid, and the 'reproducibility' of estimates of volume fraction [78]. We hasten to clarify these issues.

It is important to understand that Delesse's relation (2.1) is an *exact* equation between the **average** fraction of rock volume and the **average** fraction of plane section area. Sampling variability is not an issue in (2.1) itself, since A_A and V_V are parameters describing the average composition of the rock.

Delesse [155, 156] effectively regards the contents of the rock as being random, and assumes the rock is 'spatially homogeneous' in the sense that the contents of two congruent spatial regions of rock have the same statistical properties. In a perfectly homogeneous rock, congruent regions would have exactly equal contents. The volume fraction V_V is then expressible as the *average* fraction of volume occupied by the mineral of interest inside any given 'reference' region of rock:

$$V_V = \frac{\mathbf{E}[V(\text{mineral in solid region})]}{V(\text{solid region})} \tag{2.4}$$

where V denotes volume. Here $\mathbf{E}$ denotes expectation over different possible outcomes of the random contents of the solid region. Since the rock is assumed to be homogeneous, this quantity does not depend on the choice of reference region, so that V_V is well defined. Thus V_V is the *average* density of the mineral in the solid rock.

Similarly the area fraction A_A is defined as the *average* fraction of area occupied by the mineral in any two-dimensional region within any plane section:

$$A_A = \frac{\mathbf{E}[A(\text{mineral in plane region})]}{A(\text{plane region})} \tag{2.5}$$

where A denotes area. In statistical terms, V_V and A_A are *parameters* describing the average contents of the rock.

Delesse then proved that $V_V = A_A$. There is a simple modern proof of (2.1) using indicator random variables. For any location u in three-dimensional space, let $I(u)$ be the indicator which equals 1 if the point u lies inside the mineral of interest, and 0 if not. Assume that these indicator variables have constant mean value, $\mathbf{E}[I(u)] = p$ say. Thus the probability that a given point u in space will lie inside the mineral of interest, is constant and equal to p. It follows fairly easily that $V_V = p$, since the volume occupied by the mineral of interest inside any reference region is equal to the integral of the function I over the reference region. Similarly we have $A_A = p$ since the area occupied by the mineral of interest in the plane section is equal to the integral of I over the section. Together these imply $V_V = A_A$. For full details, see Section 6.5.1.

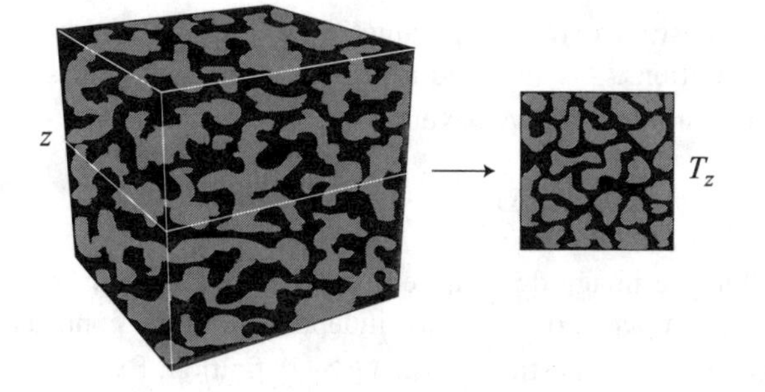

Figure 2.4 *Geometry for Delesse's proof.*

Delesse's original proof [155, 156] contains a valuable geometric insight. Consider a unit cube of rock X, and let Y be the solid region occupied by the mineral of interest inside X, sketched in Figure 2.4. The volume $V(Y)$ of the mineral region is equal to the integral of the areas $A(Y \cap T)$ of its horizontal plane sections:

$$\int A(Y \cap T_z)\,\mathrm{d}z = V(Y), \tag{2.6}$$

where T_z denotes the horizontal plane at height z. Taking the expectation on both sides of (2.6), exchanging the order of integration and expectation, and appealing to spatial homogeneity, it follows that $A_A = V_V$.

Delesse's geometrical insight was that three-dimensional reconstruction is not necessary in order to measure volume. The volume of a three-dimensional solid — of arbitrary shape — is the integral of the areas of its plane sections (by 'Fubini's theorem'). Two solid objects, which have equal cross-sections on every horizontal plane, have equal volume ('Cavalieri's principle'). This means that relative volumes of solid objects can be estimated from the relative areas of random plane sections, under the right sampling conditions, without reassembling the sections into a three-dimensional solid.

Unbiased estimation

Delesse's relation (2.1) states that the two *parameters* V_V and A_A are equal. It is implicit that, in practice, the parameter A_A will be *estimated* by the observed area fraction inside some microscope image, which we shall denote by

$$\widehat{A_A} = \frac{A(\text{mineral in image})}{A(\text{image})}. \tag{2.7}$$

Thus V_V will also be estimated by $\widehat{A_A}$. The key question is to understand the statistical properties of this estimator.

Recall that an estimator T of a parameter θ is called **'unbiased'** if $\mathbf{E}[T] = \theta$ under all conditions. An unbiased estimator is one which has no systematic error. For the estimator $\widehat{A_A}$ we have

$$\mathbf{E}\left[\widehat{A_A}\right] = \mathbf{E}\left[\frac{A(\text{mineral in image})}{A(\text{image})}\right] = A_A,$$

assuming that the image domain defining the area fraction is a fixed region in a fixed section plane, or is at least independent of the contents of the rock. Thus $\widehat{A_A}$ is an unbiased estimator of A_A by definition. By Delesse's principle $A_A = V_V$, it follows that also $\widehat{A_A}$ **is an unbiased estimator of** V_V.

Precision

Statisticians will be familiar with the distinction between bias and variance of an estimator. Notice that the theory outlined above is only a proof that the estimator $\widehat{A_A}$ is unbiased, and does not say anything about its variance.

Variance is a more difficult issue, to which we devote the whole of Chapter 13. The variance of stereological estimators is not simply a function of sample size; it depends on second-order properties of the material. Using the indicator random variables $I(u)$ again, the variance of the observed area fraction in an image field F is

$$\mathsf{var}\left[\frac{A(Y \cap F)}{A(F)}\right] = \frac{1}{A(F)^2} \int_F \int_F C(u,u')\, \mathrm{d}u \, \mathrm{d}u' \tag{2.8}$$

where $C(u,u') = \mathsf{cov}(I(u),\, I(u'))$ denotes the 'spatial covariance' of the mineral of interest at two points u and u' in the two-dimensional region F. See [514, chap. IX], [404].

Intuitively this says that the variance of the estimate depends on the size of the image field F relative to the scale of variability or 'texture' in the material. Comparing Figures 2.1 and 2.5 we may say that the second figure has a finer scale of texture, relative to the size of the image, than the first figure. The right hand side of (2.8) is the average, over all pairs of points u and u' in the image field F, of the covariance between $I(u)$ and $I(u')$. Thus we would expect the area fraction in Figure 2.5 to have a lower variance (higher precision) than the area fraction in Figure 2.1.

Under additional assumptions we can estimate the covariance $C(u,u')$ from the microscope image [444, chap. 5], [514, chap. IX] and plug this estimate into (2.8) to obtain an estimate of the variance. (Note also that the standard error would simply be the square root of this variance.) The methods of geostatistics [16, 82, 360, 362] provide general methodology for modelling and estimating these variances. See also Chapter 13 and the bibliographic notes to this chapter.

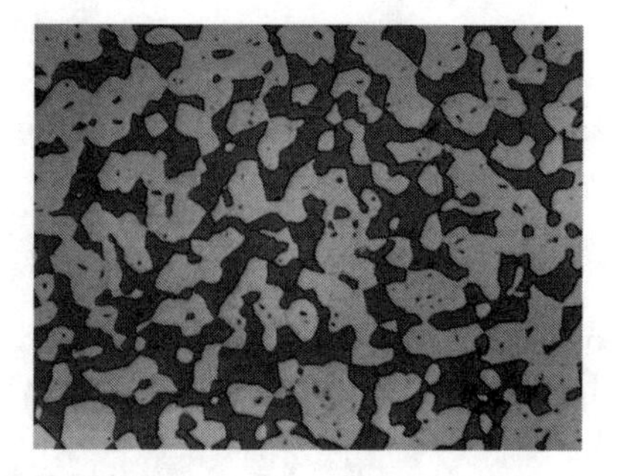

Figure 2.5 *Microstructure of annealed stainless steel. Ferrite phase (light) and austenitic phase (dark). True field width 0.1 mm. Reproduced by kind permission of Profs. J. Ohser and F. Mücklich. Copyright © J. Ohser.*

Efficiency

Of the three estimators of volume fraction described above, the point-counting estimator requires the least effort. It is virtually always the most efficient. ‡

Strikingly, the sub-sub-sample estimator P_P may have smaller variance than the sub-sample estimator L_L, which in turn may have smaller variance than the estimator A_A based on full information from the section plane. Formulae analogous to (2.8) for the variances of $\widehat{L}_L$ and of $\widehat{P}_P$ may be derived using indicator variables. There are positive covariances at small spatial scales, so that a point-counting estimator like P_P, with a coarse spacing between points, may indeed be more precise [32, 442, 506], [282, sect. 6]. This is still not widely understood in stereology. Naively we might think that an unbiased estimator of V_V that uses the full information from the section plane will always be more accurate than an estimator based on a subsample. A statistician might appeal to the Rao-Blackwell theorem, but this may not apply [32]. Paradoxes of this type are apparently well-known in the theory of wide-sense stationary random fields [435]. See Chapter 13 for details about variance comparisons.

2.2 Surface area density and length density

2.2.1 Sections of surfaces and curves

Apart from giving us information about the relative abundance of different minerals, plane sections of a rock also show its microscopic structure. The

‡ Throughout this book, 'efficiency' is meant in the statistical sense. An unbiased estimator T is more efficient than another unbiased estimator S if $n_T \mathrm{var}(T) \leq n_S \mathrm{var}(S)$ where n_T, n_S are measures of the sample size or effort required to obtain T and S, respectively.

curvilinear boundaries between the two minerals in Figure 2.1 or the boundaries between the two phases in 2.5 are plane sections of the curved two-dimensional boundaries between the minerals or phases in the solid material. In Figure 1.1, the points where three boundaries meet are plane sections of the triple junction lines, where three grain boundary surfaces meet at a curve in the solid rock.

Taking a plane section of a curved two-dimensional surface in three-dimensional space will generically yield a one-dimensional profile curve; taking a plane section of a one-dimensional curve in space will yield some isolated intersection points — a 'zero-dimensional profile'. A plane section of a k-dimensional subset of three dimensional space is generically a $k-1$ dimensional subset of the plane, for $k = 1, 2, 3$. See Figure 2.6.

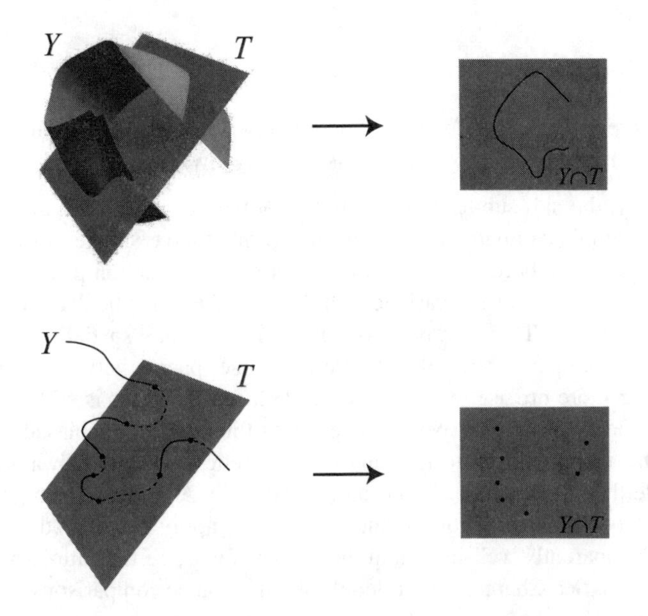

Figure 2.6 *Dimensions of spatial objects and dimensions of their plane sections. The left panels show the three-dimensional situation and the right panels are the plane sections. Top left is a 2D surface in 3D space, bottom left is a 1D curve in 3D space.*

The history of microscopy contains numerous examples of the incorrect geometrical interpretation of sections. For example, an important misunderstanding of the structure of quenched steel arose from the misinterpretation of plane sections. The needle-like streaks of Martensite seen in sections were interpreted as needle-like inclusions, whereas they must be plate-like in order to produce thin streaks on a typical section [495, Introd.]. Because of a similar error, the structure of mammalian liver was described incorrectly in medical research for a hundred years [171]. The pioneers of stereology played key roles in exposing these errors. We highly recommend perusing the real examples in [495, Introd.], [171] and the fictional example of *Flatland* [1] for further illustration.

2.2.2 *Stereology for surfaces and curves*

From Figure 2.6 it is a short step to guess that the *'amount'* of a structure that we see in a plane section is related to the *'amount'* of the structure that is present in three dimensions. Stereological methods for two-dimensional surfaces and one-dimensional curves in three-dimensional space were developed subsequently by the great Russian metallographer S.A. Saltykov [490, 491, 492, 493], American cancer researcher H. Chalkley [76, 77, 97], Tomkeieff and others [520, 560, 72]. The scientific community also became aware that many of the theoretical foundations for stereology already existed, in the field of *geometrical probability* [309], initiated in the late 18th century, and the closely related field of *integral geometry* [500]. The mathematician H. Steinhaus [534, 535] had already proposed applying these results to microscopy in the 1930's.

These new stereological methods were based on geometrical identities similar to (2.6), which state that

- the area of a surface in $\mathbf{R}^3$ can be determined from the lengths of its intersection curves with section planes;
- the area of a surface in $\mathbf{R}^3$ can be determined from the number of times it meets a straight line probe;
- the length of a curve in $\mathbf{R}^3$ can be determined from the number of intersection points it makes with section planes.

Full mathematical statements are postponed to Chapter 4. An important difference between these identities and (2.6) is that we must integrate over the spatial orientation of the section plane or line probe, as well as over its position.

In conventional stereological notation, S denotes the surface area of a curved surface in three dimensions, and S_V is the *surface density* defined as the expected surface area of the surface of interest per unit reference volume. A fundamental stereological formula is

$$S_V = \frac{4}{\pi} B_A, \tag{2.9}$$

where B (for 'boundary') denotes the length of profile curves observed on a plane section, and B_A is the average profile curve length per unit area of section. In turn, curve lengths in the plane section can be estimated by superimposing a grid of test lines and counting the number I of intersection points between test lines and curves. This yields

$$S_V = 2I_L, \quad (2.10)$$

where I_L is the average number of intersection points per unit length of test line.

For a curve or curves in three dimensions, L denotes length, and L_V is the *length density* defined as the average length of curve per unit reference volume. This is related to the frequency Q_A of intersection points (intersections per unit area) made by the curve with a section plane:

$$L_V = 2Q_A. \quad (2.11)$$

These formulae are valid under the assumption that the material is spatially homogeneous and '*isotropic*' (its statistical properties are unaffected by rotation as well as translation).

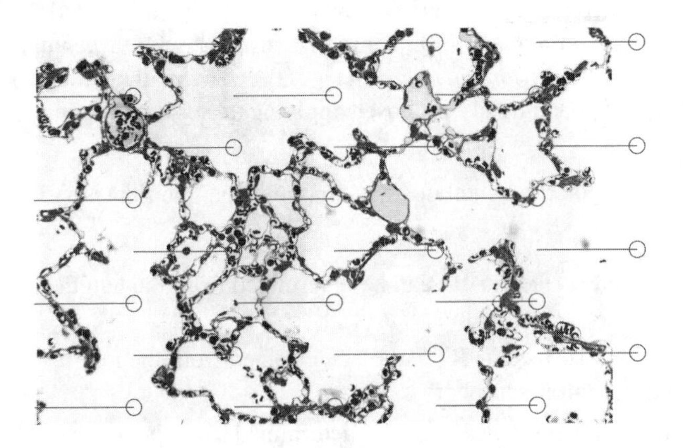

Figure 2.7 *Estimation of surface area per unit volume S_V using the formula $S_V = 2I_L$. Rat lung; white space is airway; dark blobs are red blood cells. Microtome thin section, optical microscope, true image width 150 microns. Standard test system on transparency, randomly translated over photographic print. Reproduced by kind permission of Prof. L.M. Cruz-Orive (University of Cantabria).*

Applications of these methods are typified by Figures 2.7 and 2.8. Figure 2.7 shows an optical microscope image from the lung of a rat. This is a *histological section* obtained by chemically preparing the tissue, embedding it in wax or resin, and cutting slices of thickness in the range 1–100 microns (0.001–0.1 mm), comparable to the diameters of cells. The section is translucent and is viewed in transmitted light, with the microscope's focal plane positioned somewhere inside the section. To a first approximation, this image can be regarded as an ideal plane section of three dimensional space. The edges seen in the image are traces of the gas exchange surface. The exchange surface area per unit volume of lung, S_V, could be estimated from this image using (2.9). Instead, we have superimposed a grid of test lines over the image, and estimate S_V using (2.10).

A stereological test grid has been superimposed on the image, consisting of 24 test points (circled endpoints) and 24 line segments. The original micrograph was printed at magnification 1500× and was 23 cm wide. On this scale each test line segment was 4 cm long. Since 6 out of 24 test points hit the tissue (rather than the empty airway) we estimate the volume fraction of tissue as $\hat{V}_V = \hat{P}_P = 6/24 = 25\%$. There are 49 positions where a line segment crosses the tissue-airway boundary, so the surface area of lung/air interface per unit volume of lung is estimated at $\hat{S}_V = 2\hat{I}_L = 2 \times 49/(24 \times 4/1500) = 1530\,\mathrm{cm}^{-1}$. By this estimate, a cubic centimeter of rat lung contains about 1500 square centimeters of lung/air interface.

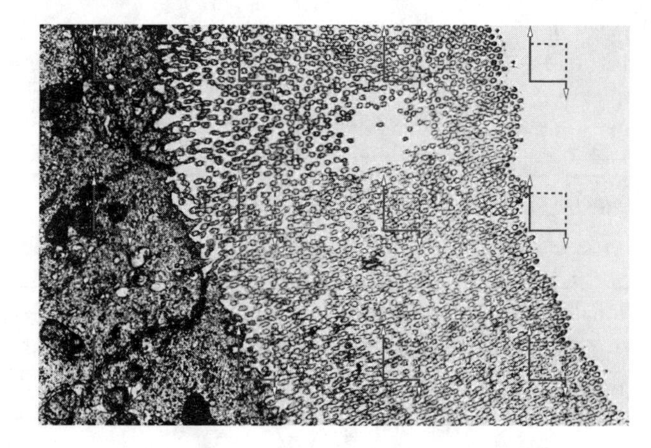

Figure 2.8 *Estimation of length density per unit volume L_V. Electron micrograph of thin section of the proximal tubule brush border in rat kidney. Round outlines are profiles of microvilli. True field width 11.6 microns. A systematic array of counting frames has been superimposed. Reproduced by kind permission of Prof. H.J.G. Gundersen, Dr. J.R. Nyengaard and the International Society for Stereology.*

Figure 2.8 shows an electron-micrograph of a thin section of rat kidney tissue. The fine dots are the profiles on the section plane of *microvilli*, minute cylindrical extensions of the cell surface. We estimate the length density L_V of microvilli per unit volume of kidney using (2.11). Since the profiles are very numerous, we subsample the image using a systematic array of rectangular counting frames. The total profile count for this image is 87. Since the microvilli are roughly parallel in any small region it is necessary to analyse several images, taken at different orientations of the section plane, in order to apply (2.11). The result is an estimate of $Q_A = 18.4$ profiles per $\mu\mathrm{m}^2$ yielding an estimate of $L_V = 36.8\,\mu\text{m}^{-2}$. A cubic micron of brush border contains about 40 microns of microvilli on average.

One-dimensional curves are an idealisation, of course. In applications of (2.11) the feature of interest is usually a thin tube or cylinder (capillaries, microvilli, microfilaments). These are represented on the section plane by small, pointlike profiles.

2.2.3 *Fundamental Formulae of Stereology*

Table 2.1 shows the conventional symbols for the most important geometrical quantities in stereology.

3-dimensional space	V volume S surface area M integral of mean curvature of surface L length of space curve N number of units
2-dimensional sections	A area B length of boundary or curve Q, N number of profiles or units
1-dimensional probes	L length I number of (intersection) points
0-dimensional test points	P number of test points

Table 2.1 *Conventional notation for geometrical quantities in stereology.*

The formulae given so far are traditionally called the 'Fundamental Formulae of Stereology'. There is a nice pattern to the scalar dimensions of these formulae which is clear when they are presented in the following form:

$$\begin{aligned} V_V &= A_A &&= L_L &&= P_P \\ S_V &= \tfrac{4}{\pi} B_A &&= 2 I_L \\ L_V &= 2 Q_A \end{aligned} \tag{2.12}$$

On each row of (2.12) the quantities have equal dimension: V_V is dimensionless (units $\text{length}^3/\text{length}^3 = \text{length}^0$); S_V has units $\text{length}^2/\text{length}^3 = \text{length}^{-1}$; L_V has units $\text{length}/\text{length}^3 = \text{length}^{-2}$. The left column of (2.12) contains quantities defined in three dimensions; the subsequent columns contain quantities observable from a section plane, test line and test point, respectively. Of course this pattern just reflects the fact that the intersection between a k-dimensional object in $\mathbf{R}^3$ and an m-dimensional plane is generically of dimension $k+m-3$.

The dimensional pattern also draws attention to one important omission from the table. The *number* per unit volume N_V of individual objects (cells, grains, nuclei, 'particles') in three dimensions does not appear in (2.12). It has scalar dimensions length^{-3} whereas none of the quantities conventionally observable in a plane section has this dimension.

The Fundamental Formulae of Stereology are the exception rather than the rule. There are no simple and general formulae for estimating most other three-dimensional parameters from plane sections.

2.2.4 *Derived formulae, test systems and ratio estimation*

Further stereological methods were derived from the Fundamental Formulae by taking ratios. Suppose for example that a biological membrane can be divided into 'active' and 'inactive' regions by some criterion, and we wish to estimate the fraction of membrane surface area that is active. This fraction, which deserves to be written S_S, can be defined as the ratio

$$S_S = \frac{\mathbf{E}\left[S(\text{active})\right]}{\mathbf{E}\left[S(\text{membrane})\right]} \tag{2.13}$$

of the expected surface area of active membrane in a reference region to the expected surface area of membrane in the same reference region. Dividing both numerator and denominator by the volume of the reference region, we get

$$S_S = \frac{S_V(\text{active})}{S_V(\text{membrane})}, \tag{2.14}$$

the ratio of the surface area density of the active membrane to that of the membrane overall. We can estimate $S_V(\text{active})$ and $S_V(\text{membrane})$ stereologically using a grid of test lines:

$$\begin{aligned} S_V(\text{active}) &= 2I_L(\text{active}) \\ S_V(\text{membrane}) &= 2I_L(\text{membrane}) \end{aligned}$$

Hence

$$S_S = \frac{I_L(\text{active})}{I_L(\text{membrane})}. \tag{2.15}$$

If we use the same test line grid to define both quantities I_L, then the lengths simply cancel and we have

$$S_S = I_I = \frac{\mathbf{E}\left[I(\text{active})\right]}{\mathbf{E}\left[I(\text{membrane})\right]} \tag{2.16}$$

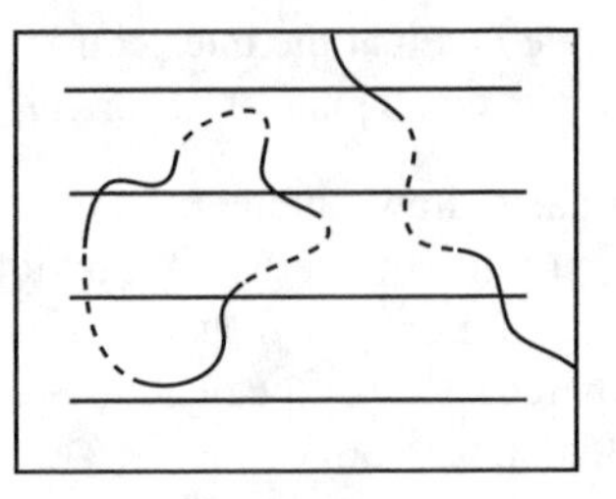

Figure 2.9 *Estimation of surface fraction S_S from intercept count fraction I_I.*

the ratio of average intersection counts, for the active membrane and for all membrane, obtained from the same test line. See Figure 2.9.

Alternatively, we may use different test line grids to count different structures. If the active membrane constitutes only a small fraction of total membrane surface, we may choose to use a denser line grid to count the active membrane. Then the simple equation (2.16) must be adjusted for the relative proportions of the two line grids. Suppose we count the active membrane and the total membrane using line grids 1 and 2, respectively. Let ℓ_1/ℓ_2 be the proportion of total grid line length for grid 1 relative to grid 2. Then we have

$$S_S = \frac{\ell_2}{\ell_1} I_I.$$

The foregoing has been presented in the classical style, ignoring the sampling variation. It is implicit in (say) the last equation that we shall estimate the surface area fraction S_S by

$$\frac{\ell_2}{\ell_1} \frac{I(\text{active; grid 1})}{I(\text{membrane; grid 2})}$$

where $I(\text{active; grid 1})$ denotes the observed total number of intersections between the active membrane and the test lines of grid 1, and $I(\text{membrane; grid 2})$ is the observed number of intersections between the membrane and grid 2. However, this raises many statistical issues, which are considered under the heading of Ratio Estimation in Chapter 9.

Tomkeieff [560] and Chalkley et al [77] developed stereological methods for estimating the ratio S_V/V_V of the surface density of one structure to the volume density of another structure. A regular grid of test points is used to estimate V_V by point counting, and a regular grid of lines to estimate S_V by counting intersections. The corresponding stereological identity is

$$\frac{S_V}{V_V} = \frac{2p}{\ell} I_P$$

where ℓ/p is the length of test line per test point, and I_P is the average number of intersections of the first structure with the test line grid for every test point that hits the second structure.

2.3 Particles

2.3.1 Particles and their profiles

'Particles' are any discrete physical objects in a three-dimensional material, such as cells in a biological tissue, crystalline grains in a metal, mineral grains in rock, bubbles in foam, cell nuclei or nucleoli, sand grains in concrete, inclusions in metals, reinforcements in polymers and composites, wear particles in lubricant, or machining chips. It is often important to know the number of particles in a material, their average 'size' (volume or diameter), or their size distribution. In metallurgy, the size of crystalline grains is an important structural quantity. In neuroscience, changes in the number of neurons in an anatomical region of the human brain are important indicators of development and disease.

For example, Figure 2.10 shows a histological section from the ovarian follicle of a mouse. The shapes visible in the Figure are plane sections of individual granulosa cells. It is of interest to know the number of granulosa cells in a follicle, and their average size.

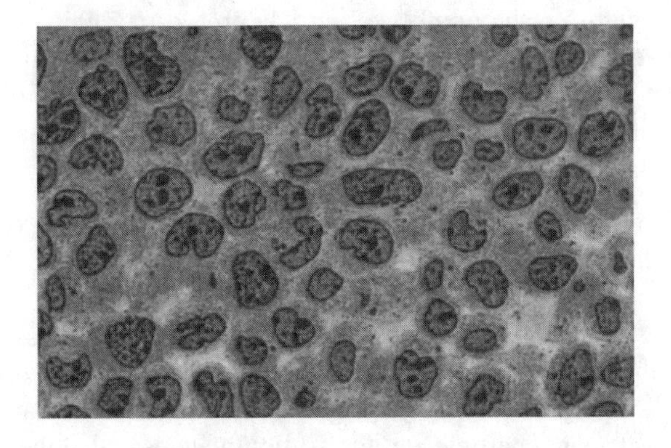

Figure 2.10 *Thin section of granulosa cells in murine ovarian follicle. Optical microscope, transmitted light. Image width 90 microns (0.09 mm). By courtesy of Dr J.R. Nyengaard, University of Aarhus, Denmark.*

Stereologists emphasise that, to avoid serious methodological errors, it is important to distinguish the three-dimensional particles from the flat shapes visible in plane sections, which are called *profiles*. In Figure 2.10, the individual shapes are *cell profiles*, not cells.

Our problem, then, is to draw inferences about a population of three-dimensional particles from a two-dimensional plane section of the population. Quantities of interest include the numerical density N_V (number of particles per unit volume), the fraction N_N of particles which belong to a given sub-population, the average volume $\bar{v}$ of individual particles, the population distribution of particle volume, etc. These are different from aggregate geometrical quantities like V_V. Sampling inference for a population of particles is also quite different from the foregoing stereological methods.

2.3.2 Particle number

Still following the classical approach, we assume that the particle population is 'spatially homogeneous'. The *numerical density* of particles, N_V, is defined as the (expected) number of particles per unit volume of the material,

$$N_V = \frac{\mathbf{E}[N(\text{particles in region})]}{V(\text{region})}$$

for any region of space. To resolve ambiguities at the edge of the region, we count a particle as being 'in' a reference region if its centroid (centre of mass) lies inside the region. The main problem is how to estimate N_V from plane sections of the material.

A common methodological error is to confuse N_V, the number of particles per unit volume, with N_A, the number of particle profiles per unit area in a plane section. It is easy to estimate N_A from plane section images, but N_A does not have a direct three-dimensional interpretation. This is clear from the dimensional units: it would be nonsensical to report that a biological tissue contains 1.2×10^5 cells per unit *area*.

Since it is so important to keep in mind the distinction between a solid particle and its two-dimensional sections, stereologists have adopted the discipline of writing Q for the number of profiles counted on a section, and reserving the symbol N for the number of objects in three dimensions only. In some applications where the particles are very irregularly shaped, there is the extra complication that a single particle may yield more than one profile. In such cases the intersection between a particle and the section plane is called the *trace* of the particle, and each particle trace consists of one or more separate *profiles*.

We shall use Q_A to denote the number of particle traces or particle profiles per unit area on a plane section. This density certainly depends on the number of particles in the material, but also on their *sizes*. The key relationship is that

$$Q_A = \mathbf{E}[H]\, N_V \tag{2.17}$$

(under the assumption of "spatial homogeneity") where $\mathbf{E}[H]$ is the population average 'height' of the particles. Here the 'height' of a particle is defined as the length of its projection onto a line normal to the section plane; see Figure 2.11. We write $\mathbf{E}[H]$ instead of the traditional notation $\overline{H}$ because of the potential for confusion with the sample mean.

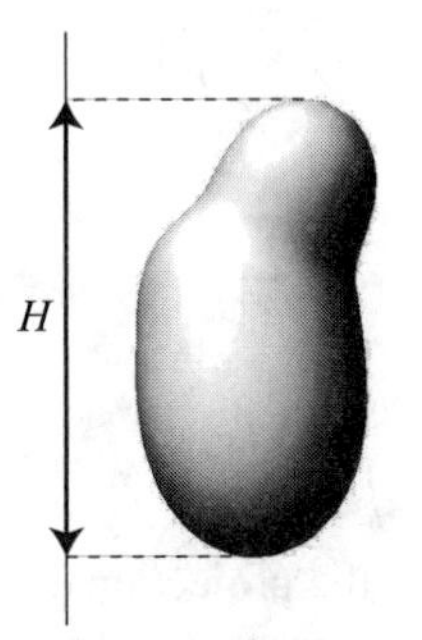

Figure 2.11 *The height H of a particle is the length of its projection onto a line normal to the section plane.*

The relation (2.17) is usually attributed to DeHoff and Rhines [153] who proved it for particles of convex shape. It has been rediscovered numerous times.

To prove relation (2.17) in the classical style, imagine a reference cube X of volume $V(X)$ containing N particles $Y_1, \ldots, Y_N$. See Figure 2.12.

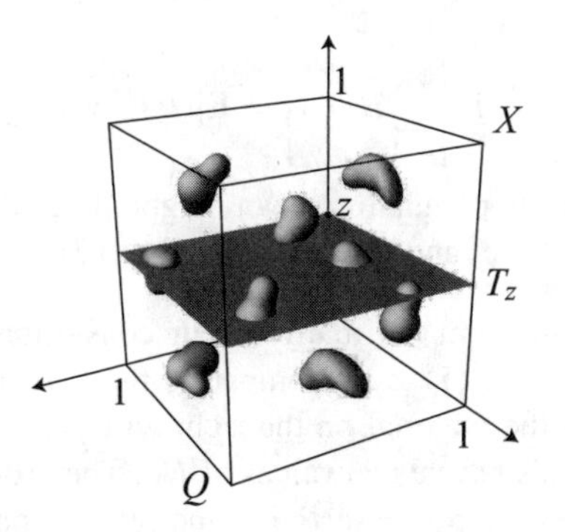

Figure 2.12 *Proof of Rhines-DeHoff relation.*

Write Y for the union of all the particles Y_i. For any horizontal plane T_z cutting this cube, the number of particles intersected by the plane is

$$Q(Y \cap T_z) = \sum_{i=1}^{N} \mathbf{1}\{Y_i \cap T_z \neq \emptyset\},$$

a sum of indicators for each particle. Integrating over different positions of the section plane, we have

$$\begin{aligned}\int Q(Y \cap T_z)\,\mathrm{d}z &= \int \sum_{i=1}^{N} \mathbf{1}\{Y_i \cap T_z \neq \emptyset\}\,\mathrm{d}z \\ &= \sum_{i=1}^{N} \int \mathbf{1}\{Y_i \cap T_z \neq \emptyset\}\,\mathrm{d}z \\ &= \sum_{i=1}^{N} H(Y_i)\end{aligned}$$

where

$$H(Y_i) = \int \mathbf{1}\{Y_i \cap T_z \neq \emptyset\}\,\mathrm{d}z \tag{2.18}$$

can be interpreted as the length of the projection of Y_i on the z-axis, as in Figure 2.11. This ignores cases where a particle lies partially outside the reference region X; they can be dealt with using a longer argument.

Assuming 'spatial homogeneity', $\mathbf{E}[Q(Y \cap T_z)]$ does not depend on z. Taking expectations and dividing by $V(X)$ we obtain

$$\frac{\mathbf{E}[Q(Y \cap T_0)]}{A(T_0)} = \frac{\mathbf{E}\left[\sum_{i=1}^{N} H(Y_i)\right]}{V(X)} \tag{2.19}$$

By a further appeal to 'homogeneity', the particle heights are identically distributed, so

$$\mathbf{E}\left[\sum_{i=1}^{N} H(Y_i)\right] = \mathbf{E}[H]\,\mathbf{E}[N]$$

where $\mathbf{E}[H]$ is again the population mean height. Thus the right hand side of (2.19) is equal to $\mathbf{E}[H]\,N_V$, and we have proved (2.17).

Since integrals are anathema to many of our consulting clients, a more effective way to explain (2.17) is by comparing the two scenarios sketched in Figure 2.13. Clearly the material on the right will yield a large value of Q_A, although both materials have equal values of N_V. The probability or frequency with which a particle is hit by the section plane depends on its size, since larger particles present larger targets. In fact the particles on the right were created by 'inflating' the particles on the left, and for a given section plane this inflation causes more particles to meet the section plane.

Alternatively, one may use the analogy with a tomato salad. The number of slices in a tomato salad depends not only on how many tomatoes we cut up, but also on how many slices we obtain from each tomato — that is, on the sizes of the tomatoes.

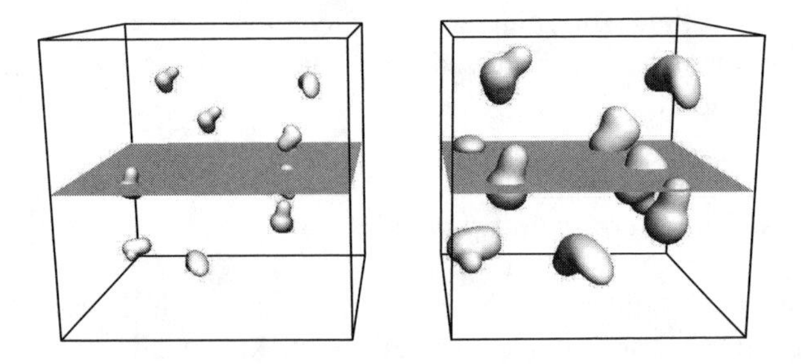

Figure 2.13 *One way to explain that Q_A must depend on the mean particle height* $\mathbf{E}[H]$. *The two boxes contain equal numbers of particles. The particles on the left are smaller than those on the right. A typical section plane cuts fewer particles in the left box than it does on the right.*

An important consequence of (2.17) is that an increase in the observed number of section profiles does not necessarily indicate an increase in the number of particles. Abercrombie [3] cites numerous examples in cell biology where observed changes in Q_A have been misinterpreted as changes in N_V. For example, in a study of changes in the pituitary gland after gonadectomy, basophils were observed to increase in size and in 'number' (i.e. section profile counts) by the same proportion. A more likely interpretation is that basophils increased in size without a change in the population number. See also [171, 495].

How, then, can we estimate N_V from plane sections? In fact this is *not possible, in complete generality,* from a plane section alone. The equation (2.17) or equivalently

$$N_V = \frac{Q_A}{\mathbf{E}[H]}, \tag{2.20}$$

involves the particle mean height $\mathbf{E}[H]$. Since particle heights are not directly observable in the plane section, $\mathbf{E}[H]$ cannot generally be determined, so N_V cannot generally be determined.

This problem can be circumvented only when we have extra information: where there is *a priori* knowledge about particle shape, or where particle heights can be measured, or where some form of three-dimensional observation is possible.

The classical literature on stereology contains many proposals for estimating N_V when only a single plane section is available. For a survey, see [580, sec. 2.6], [581, chap. 5]. Important classical references include [149, 153, 154, 587]. These techniques use shape assumptions to estimate $\mathbf{E}[H]$ from information observable in the section. They are closely connected to the problem of particle size discussed in Section 2.4.

If serial sections through the material are available, and if it is possible to identify profiles of the same particle on different sections, then there are robust stereological methods for estimating N_V. Stereological techniques of this kind for counting particles are discussed further in Sections 2.5 and 3.3.5, and in Chapters 10 and 11.

2.3.3 Relative numbers and sampling bias

Another quantity of interest is the fraction of particles which belong to a given sub-population — for example, the fraction of skin cells that are undergoing mitosis at a given time — or more generally, the ratio of population numbers for two types of particles — for example, the relative abundance of two types of constituent particles in concrete. An important example of the first kind is the Ito cell index $N(\text{Ito cells})/N(\text{hepatocytes})$ used in hepatology [36, 352].

A very pernicious methodological error is to estimate the relative proportion of particles by counting the relative frequency of *profiles* seen on a plane section. Since the number of profiles depends on particle size as well as particle number, the relative frequencies of two types of profiles depend on the relative size as well as relative frequency of particles.

Assuming spatial homogeneity, define the relative proportion of particles by

$$N_N = \frac{\mathbf{E}\left[N(\text{particles of type 1 in region})\right]}{\mathbf{E}\left[N(\text{particles of type 2 in region})\right]}$$

for an arbitrary reference region in space, and the relative frequency of section traces

$$Q_Q = \frac{\mathbf{E}\left[Q(\text{traces of type 1 in rectangle})\right]}{\mathbf{E}\left[Q(\text{traces of type 2 in rectangle})\right]}$$

for any reference rectangle in a plane section. Here types 1 and 2 may refer to separate populations of particles, or type 1 may be a sub-population of type 2. In the latter case, N_N is the fraction of particles belonging to this sub-population, and Q_Q is the observed fraction of section traces belonging to the sub-population.

We emphasise that N_N must not be confused with Q_Q. In fact (2.17) implies

$$Q_Q = \frac{\mathbf{E}[H_1]}{\mathbf{E}[H_2]} N_N \tag{2.21}$$

where $\mathbf{E}[H_1]$ and $\mathbf{E}[H_2]$ are the mean heights of the particles of type 1 and type 2, respectively. To prove (2.21), simply write $N_N = N_V^{(1)}/N_V^{(2)}$, where $N_V^{(1)}$ and $N_V^{(2)}$ are the numerical densities of the particles of type 1 and 2, respectively. Similarly $Q_Q = Q_A^{(1)}/Q_A^{(2)}$. Then apply (2.17) to obtain the result.

Thus, the relative frequencies of different cell types seen in a plane section of a biological tissue do not equal the relative abundance of those cell types in the tissue. One may again appeal to the 'tomato salad' analogy. Imagine a salad made from 3 large red tomatoes and 3 small green tomatoes. The red tomatoes, being larger, yield more slices than the green tomatoes. Hence, the proportion of red *slices* is much greater than one half, although the proportion of red *tomatoes* is exactly one half. Abercrombie [3] pointed out numerous errors in the early literature on cell biology arising in this way.

A very important consequence of (2.21) is that a plane section of particles is a biased sample of the particle population. When type 1 is a sub-population of type 2, equation (2.21) states that the observed fraction of section traces that belong to this sub-population, Q_Q, is biased relative to the true fraction of particles belonging to the sub-population, N_N, by a factor equal to the ratio of mean particle heights. Equivalently, when we take a plane section of a particle population, the particles are sampled with probability proportional to their height. There is a **sampling bias** in favour of larger particles. This is a fundamental methodological problem for microscopy which often goes unrecognised.

2.3.4 Sampling effects

The essential difficulty is that a plane section of particles cannot be regarded as a sample from the particle population in the usual statistical sense. The section plane does not contain the sampled particles themselves: it only contains plane sections of these particles. There are two sampling effects in operation:

selection: first, some particles are randomly selected (namely those particles which are cut by the section plane);

section: second, each sampled particle is cut by the section plane to yield a two-dimensional trace.

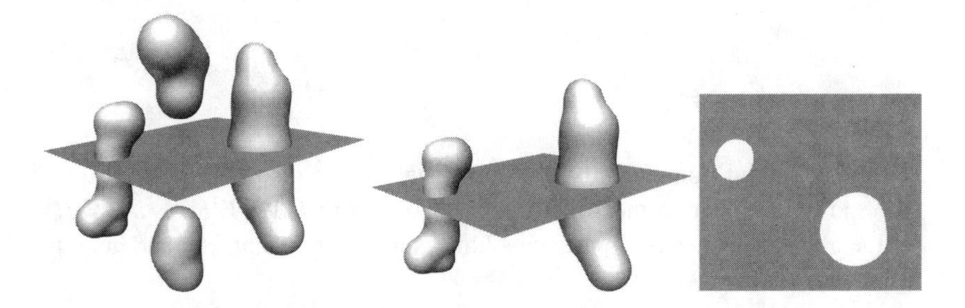

Figure 2.14 *A plane section of a particle population involves two sampling effects. First, from the original population (left), some particles are randomly selected by the section plane (middle panel) with sampling probability proportional to particle height. Second, each sampled particle (middle panel) is cut to yield a two-dimensional profile (right panel).*

These steps are sketched in Figure 2.14. As we have seen above, the first sampling step introduces a *sampling bias* proportional to particle height. It turns out that in the second step, given that a particle has been hit by the section plane, the position of the section plane is uniformly distributed (over the range of displacements of the plane in which it hits the particle). See Figure 2.15.

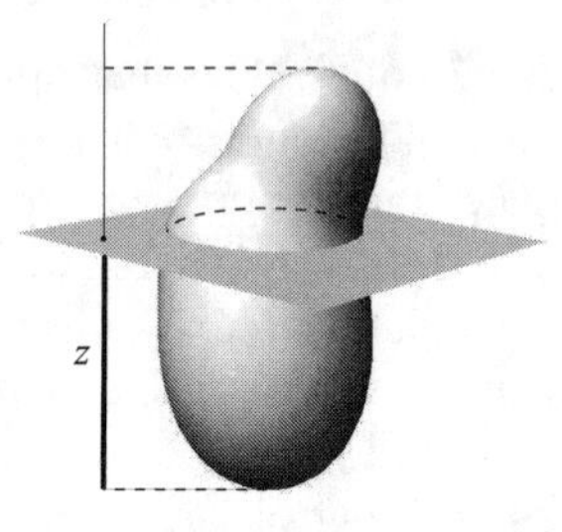

Figure 2.15 *Given that a particle has been cut by the section plane, the vertical displacement z of the section plane is uniformly distributed over all positions that lead to an intersection.*

Visualise again a tomato salad, and pick out one of the slices. This slice is the outcome of two sampling effects: first, larger tomatoes contribute more slices to the salad; and second, each tomato contributes a range of slices obtained by cutting the tomato uniformly from end to end.

2.4 Particle size

Next we consider inference about the 'size' of particles. Measures of particle size include the volumes, surface areas, diameters, and heights of individual particles.

2.4.1 Mean particle size

Assuming spatial homogeneity, we may speak of the distribution of particle size and the population mean particle size. We write $\mathbf{E}[V]$, $\mathbf{E}[S]$, $\mathbf{E}[D]$, $\mathbf{E}[H]$ for the population means of particle volume, surface area, diameter and height, respectively.

Since particles are non-overlapping, spatial homogeneity also implies that

$$V_V = \mathbf{E}[V]N_V \quad (2.22)$$

$$S_V = \mathbf{E}[S]N_V. \quad (2.23)$$

By the Fundamental Formulae of Stereology (Section 2.2.3) we have unbiased estimators of the following quantities:

$$P_P = L_L = A_A = \mathbf{E}[V]\,N_V \tag{2.24}$$

$$2I_L = \frac{4}{\pi}B_A = \mathbf{E}[S]\,N_V \tag{2.25}$$

$$Q_A = \mathbf{E}[H]\,N_V. \tag{2.26}$$

By taking ratios of these estimators, we have estimators of the ratios $\mathbf{E}[V]/\mathbf{E}[S]$, $\mathbf{E}[V]/\mathbf{E}[H]$ and $\mathbf{E}[S]/\mathbf{E}[H]$ without any assumptions about particle shape.

For example, Tomkeieff [560] noted that

$$\frac{\mathbf{E}[S]}{\mathbf{E}[V]} = \frac{4}{\bar{\ell}_1}$$

where $\bar{\ell}_1$ is the average length of intercepts between test lines and particle profiles. The factor 4 arises because after combining the formulae $S_V = 2I_L$ and $V_V = P_P$, we must account for the fact that each linear intercept has two endpoints. A related technique for estimating $\mathbf{E}[S]/\mathbf{E}[V]$ was developed by Chalkley et al [77].

However, such techniques do not lead to a method of estimating $\mathbf{E}[V], \mathbf{E}[S]$ and $\mathbf{E}[H]$ that can be applied for completely general shapes. Again the numerical density and the particle size are confounded.

Figure 2.16 *Sections of particles which can be approximated by ideal geometrical shapes.* Left: *Metal powder (Cu35Sn, light) which has been embedded in resin (dark) and sectioned in order to estimate the particle size distribution by stereological methods. Particles can be idealised as spheres. Field height 0.9 mm. From [444, Fig. 6.1, p. 164].* Right: *Ceramic-metal composite material consisting of particles of tungsten carbide (light) in a cobalt matrix (dark). Particles may be idealised as cubes. Field width 0.14 mm. From [444, Fig. 7.1, p. 200]. Copyright © 2000 John Wiley and Sons. Reproduced by kind permission of Dr. U. Täffiner, Profs. J. Ohser and F. Mücklich, and John Wiley and Sons.*

If all particles are known to have the same shape (i.e. every particle is a rescaled and translated copy of the same solid) or are known to belong to a one- or two-parameter family of solids, then the quantities $\mathbf{E}[V]/\mathbf{E}[S]$, $\mathbf{E}[V]/\mathbf{E}[H]$ and $\mathbf{E}[S]/\mathbf{E}[H]$ may be enough to determine $\mathbf{E}[V], \mathbf{E}[S]$ and $\mathbf{E}[H]$, and therefore N_V.

For example, if all particles are spheres, then $\mathbf{E}[H]$ is the population mean diameter, and $\mathbf{E}[S]$, $\mathbf{E}[V]$ are proportional to the second and third moments of diameter. If the spheres all have equal size, then any one of the ratios $\mathbf{E}[V]/\mathbf{E}[S]$, $\mathbf{E}[V]/\mathbf{E}[H]$ and $\mathbf{E}[S]/\mathbf{E}[H]$ determines $\mathbf{E}[H]$ and hence $\mathbf{E}[S]$ and $\mathbf{E}[V]$. See [580, sec. 2.6], [581, chap. 5], [149, 153, 154, 587]. Figure 2.16 illustrates some particles for which these methods would be appropriate.

2.4.2 Particle size distribution

Next we consider inference about the population distribution of particle size.

Spherical particles

Swedish mathematical statistician S.D. Wicksell [602, 603, 604] first treated the problem of estimating the size distribution of spherical particles from plane sections[§]. In a study of cancer of the spleen, it was desired to estimate the distribution of sizes of small focal tumours from observations in plane sections of tissue. Equivalent problems were mentioned for corpuscles in the thymus, Islets of Langerhans in the pancreas, and for structures in astronomy and geology.

Wicksell initially assumed [603] that the tumours are spherical, with centres homogeneously scattered in space, and with diameters having a common distribution with cumulative distribution function F. The problem is to estimate F from a sample of observed diameters of section profiles.

Wicksell elucidated [603, p. 86] the two sampling effects described in Section 2.3.4. Firstly, spherical tumours are hit by the section plane with probability proportional to their diameters. Thus the diameters of tumours selected by the section plane follow the size-weighted distribution

$$F_1(t) = \frac{1}{\mu}\int_0^t s\,\mathrm{d}F(s) \tag{2.27}$$

where $\mu = \int_0^\infty t\,\mathrm{d}F(t)$ is the mean tumour diameter in the original population. [Integrals involving $\mathrm{d}F$ are Stieltjes integrals [178, p. 131]: if the distribution has a probability density $f(t) = F'(t)$ then $\int_0^t s\,\mathrm{d}F(s) = \int_0^t s\,f(s)\,\mathrm{d}s$.]

§ Wicksell's main results were rediscovered independently by Scheil [504, 505], Fullman [185], Reid [475], Santaló [499] and Krumbein and Pettijohn [325].

Secondly, given that a spherical tumour of diameter t is cut by the section plane, it yields a circular section profile of diameter

$$s = \sqrt{t^2 - 4d^2}$$

where d is the distance from the centre of the sphere to the section plane. It can be shown that d is (conditionally) uniformly distributed from 0 to $t/2$. Thus the conditional distribution function of circular profile diameter s given spherical particle diameter t is

$$H(s \mid t) = 1 - \sqrt{1 - \frac{s^2}{t^2}}, \quad 0 \le s \le t$$

with conditional probability density

$$h(s \mid t) = \frac{s}{t\sqrt{t^2 - s^2}}, \quad 0 \le s \le t. \tag{2.28}$$

Combining these two sampling effects, the observed distribution of circular profile diameter has probability density

$$\begin{aligned} g(s) &= \int h(s \mid t)\,\mathrm{d}F_1(t) \\ &= \frac{s}{\mu} \int_s^\infty \frac{1}{\sqrt{t^2 - s^2}}\,\mathrm{d}F(t). \end{aligned} \tag{2.29}$$

Note that the size distribution of circular profiles always has a probability density g, even if the original size distribution of particles F is discrete.

Wicksell inverted the integral equation (2.29) to give F explicitly in terms of g and μ:

$$F(t) = 1 - \frac{2\mu}{\pi} \int_t^\infty \frac{g(r)}{\sqrt{r^2 - t^2}}\,\mathrm{d}r. \tag{2.30}$$

Assuming $F(0) = 0$ so that there are no particles of zero diameter, it follows from the case $t = 0$ in (2.30) that the particle mean diameter μ is related to the *harmonic mean* of profile diameter:

$$\mu = \frac{\pi}{2} \left[\int_0^\infty \frac{g(s)}{s}\,\mathrm{d}s \right]^{-1}. \tag{2.31}$$

Hence we could, in principle, recover the true distribution of tumour diameter F from the distribution of profile diameters G, by first estimating particle mean diameter μ using (2.31), then applying the inversion (2.30). Numerical instability makes this difficult in practice: the inversion is an ill-posed problem. The sample harmonic mean has infinite variance [479, p. 210]. A great deal of effort has been expended subsequently to develop numerically stable and statistically efficient procedures for estimating F given a sample from g. Important contributions were made by Saltykov and Spektor [494, 529] and Bach [21, 22]. For reviews see [113], [444, Chap. 6] and [543, sect. 11.4, pp.

353–369], or the classic surveys in [562, chap. 5] and [580, chap. 5]. See also [514, pp. 350 ff.].

Just to unsettle our intuition, it is possible for the mean circle diameter to be *greater* than the mean sphere diameter; see Example 2.4.

Wicksell's problem has attracted considerable recent attention among mathematical statisticians. Techniques proposed for estimating F include numerical analysis [12, 13] and Gauss-Chebyshev quadrature methods for solving the Abel integral equation [354], kernel density estimation [241, 553], smoothing splines [440], singular value decomposition [450, 516] and wavelets [14]. Others have studied the asymptotic convergence rate [207] and the estimation of the weighted distribution F_1 [268].

However the practical relevance of these theoretical efforts is severely restricted by their reliance on very fragile geometrical assumptions. In most applications, particles are not spherical. Even for spherical particles, there are observation effects, such as the failure to observe small profiles, which may invalidate the model [113].

The numerical density N_V of spherical particles can be estimated from sections, using the DeHoff-Rhines relation $Q_A = \mathbf{E}[H]N_V$, since in this case the mean caliper diameter $\mathbf{E}[H]$ is simply the mean sphere diameter, which we may estimate using the methods above [444, pp. 166, 173], [185].

Sphere size distribution can also be estimated from the distributions of line transects [71, 347].

Weighted moments of sphere diameter

A few key points about Wicksell's result should be noted. Given that a sphere of radius t is cut by the section plane, the random diameter S of its section profile can be represented as

$$S = Rt$$

where the multiplier R is a random variable with probability density

$$f_R(r) = h(r \mid 1) = \frac{r}{\sqrt{1-r^2}}, \quad 0 \le r \le 1.$$

where $h(s \mid t)$ was defined in (2.28). This may be clear from geometrical intuition, or can be deduced from the scaling property $h(s \mid t) = \frac{1}{t}\, h(\frac{s}{t} \mid 1)$. Thus the diameter S of a typical circular profile on the plane section can be represented as

$$S = RT \tag{2.32}$$

where R has the same distribution as before, T is a realisation from the size-weighted distribution of sphere radii F_1, and R and T are independent. In words, each profile diameter S is obtained by selecting a sphere diameter T according to the diameter-weighted distribution, then multiplying T by a random fraction R.

Hence the moments of S and T are related by

$$\mathbf{E}[S^k] = c_k \, \mathbf{E}[T^k] \tag{2.33}$$

where c_k are the known constants

$$c_k = \mathbf{E}[R^k] = \frac{\sqrt{\pi}\Gamma(\frac{k}{2}+1)}{2\Gamma(\frac{k}{2}+\frac{3}{2})}.$$

For example $c_{-1} = \frac{\pi}{2}$, $c_0 = 1$, $c_1 = \frac{\pi}{4}$ and $c_2 = \frac{2}{3}$. Setting $k = -1$ yields (2.31).

Note that $\mathbf{E}[T^k]$ is the kth moment of diameter in the *weighted* distribution of sphere size:

$$\begin{aligned}\mathbf{E}[T^k] &= \int_0^\infty t^k \, \mathrm{d}F_1(t) \\ &= \frac{1}{\mu}\int_0^\infty t^{k+1} \, \mathrm{d}F(t).\end{aligned}$$

Thus, *weighted* moments of particle size can easily be estimated from the section profiles. The sample mean of observed profile diameters, multiplied by $4/\pi$, is an unbiased estimator of the *diameter-weighted* mean particle diameter

$$\mathbf{E}[T] = \int_0^\infty t \, \mathrm{d}F_1(t) = \frac{1}{\mu}\int_0^\infty t^2 \, \mathrm{d}F(t).$$

(This is also a consequence of the generic formulae $(4/\pi)B_A = \mathbf{E}[S]N_V$ and $Q_A = \mathbf{E}[H]N_V$.) Similarly for other weighted moments.

Ellipsoids

For spherical particles, equations (2.29) and (2.30) establish a 1–1 correspondence between the distribution of sphere diameters F and the distribution of circular profile diameters G. For non-spherical particles there is an additional difficulty that the joint distribution of particle size and shape may be *unidentifiable* from the distribution of observed particle profile size and shape. That is, the former may not be uniquely determined by the latter. If the parameters are not identifiable, consistent estimators do not exist [564, p. 62].

Wicksell [604] generalised some of the foregoing results to ellipsoidal particles of variable size but fixed shape. The particles were assumed to be prolate spheroids (i.e. ellipsoids with major axes a_1, a_2, a_3 satisfying $a_1 = a_2 < a_3$) or oblate spheroids ($a_1 = a_2 > a_3$). The key integral equations (2.29) and (2.30) remain true for ellipsoids if the diameter of an elliptical profile is taken as the geometric mean of its major and minor axes, and the diameter of the ellipsoid is the largest diameter (as just defined) of any plane section parallel to the same set of planes.

However, Cruz-Orive [107, 108] later showed that a general size-shape distribution of ellipsoids is unidentifiable from plane sections. That is, the joint distribution of the ellipsoid major axes (a_1, a_2, a_3) cannot be recovered uniquely from the joint distribution of elliptical profile major axes (b_1, b_2). Similar comments apply to parallelepipeds [444, p. 165].

2.5 Thick sections

In microscopy, the 'section' is the physical slice of material that we place under the microscope. Slices of biological tissue are usually 10–100 microns (0.01–0.1 mm) thick, but may be cut as fine as 50 nanometres (5×10^{-5} mm). Thin foils of metallic or crystalline materials are slices typically 1 nanometre (10^{-6} mm) thick.

Sections are called 'thin' if their thickness can be neglected in comparison to the microscopic structures of interest, so that the slice may be treated as an ideal two-dimensional section plane. Otherwise, the slice is called a 'thick section' or 'slice' and must be regarded as three-dimensional rather than two-dimensional.

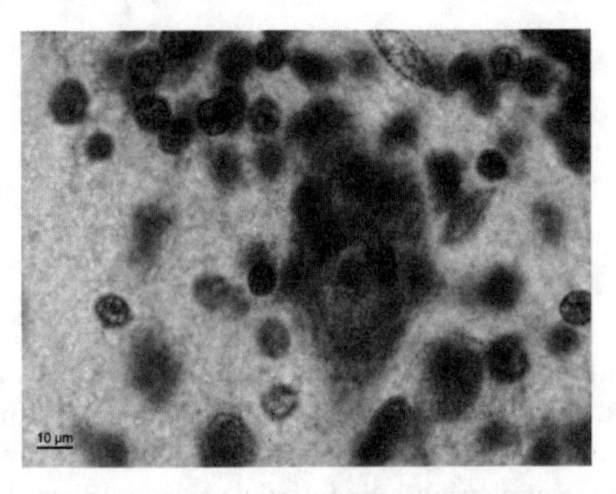

Figure 2.17 *Thick section of human cerebellum. Section thickness 70 µm (0.07 mm). Plastic embedding, Giemsa stain, light microscopy. Reproduced by kind permission of Dr J.R. Nyengaard, University of Aarhus.*

Figure 2.17 shows a light microscope image of a thick section of human brain tissue. Only some of the cells are in sharp focus. The observer can adjust the microscope controls so that the focal plane moves up and down in the thick section. In this way the observer can inspect the entire contents of the thick section, except for a zone near the upper and lower faces of the section.

Stereological methods require modification for thick sections. Idealise the thick section as the region between two parallel planes separated by a known distance t, the section thickness. A thick section of a particle population may contain entire particles and pieces of particles. See Figure 2.18.

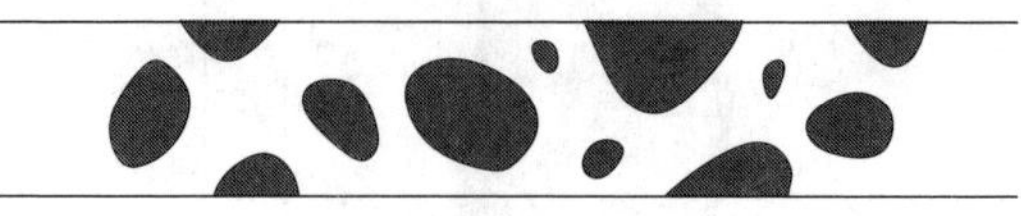

Figure 2.18 *Sketch of the contents of an idealised thick section of a population of particles. The section is shown end-on, with the upper and lower faces depicted as straight lines.*

Many pioneering studies in cell biology counted the number of particles that were visible (wholly or partly) inside the thick section, and divided by the *volume* of the thick section to obtain a figure for the numerical density of particles per unit volume, N_V. However, this is incorrect; this estimate has a large positive bias.

Assume the particle population is 'spatially homogeneous'. The expected number of particles which are visible (wholly or partly) inside the thick section, per unit volume of the section, is $N_V^+ = Q_A^+/t$, where Q_A^+ is the expected number of particles visible (wholly or partly) per unit *area* of the upper face of the slice. The superscript $+$ will refer to densities for the three-dimensional thick section that are referred to the area of one face. It was effectively shown by Abercrombie [3] and others [51, 193, 344] that

$$Q_A^+ = (t + \mathbf{E}[H])N_V \tag{2.34}$$

where t is the section thickness and $\mathbf{E}[H]$ is the mean particle height in the direction normal to the faces of the section. This can be proved in the same way as (2.17). The range of vertical displacements of the thick section, in which it intersects a given particle of height h, is equal to $t + h$, as sketched in Figure 2.19.

It follows that the putative numerical density

$$N_V^+ = \frac{Q_A^+}{t} = \frac{t + \mathbf{E}[H]}{t} N_V \tag{2.35}$$

is actually larger than the true numerical density N_V. Typically the section thickness t is of the same order as the particle height $\mathbf{E}[H]$, so that the relative bias may be as much as 100%.

Note that the thick section also introduces a sampling bias, in the sense that the relative proportion Q_Q of two types of particles as observed in the thick section is not equal to their relative population proportion N_N. A particle of height H is represented in the thick section with a probability proportional to $t + H$.

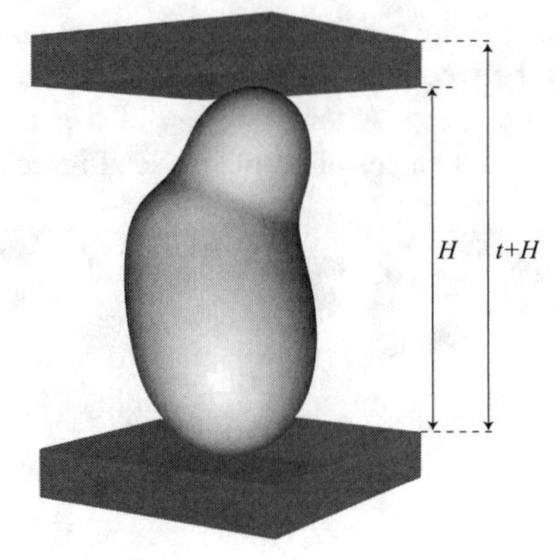

Figure 2.19 *The range of vertical displacement of a thick section, of thickness t, in which it intersects a given particle of height H, is equal to* $t+H$.

Abercrombie [3] and some earlier writers [51, 193, 344] proposed that N_V be estimated by rewriting (2.34) as

$$N_V = \frac{1}{t + \mathbf{E}[H]} Q_A^+ \tag{2.36}$$

and substituting sample estimates of the quantities on the right-hand side. Techniques of this type were the methods of choice for counting cells in histological sections from the 1940's until the late 1980's.

In Abercrombie's [3] Method 1, $\mathbf{E}[H]$ is estimated by cutting sections perpendicular to the original direction, treating the particles as spheres of equal size, and estimating $\mathbf{E}[H]$ by $4/\pi$ times the mean diameter of profiles. Abercrombie's Method 2 requires sections of two different thicknesses $t_1 < t_2$. The corresponding densities $Q_A^{+(1)}, Q_A^{+(2)}$ satisfy

$$Q_A^{+(2)} - Q_A^{+(1)} = (t_2 - t_1) N_V$$

by (2.34). Hence

$$N_V = \frac{Q_A^{+(2)} - Q_A^{+(1)}}{t_2 - t_1} \tag{2.37}$$

may be estimated by substituting sample estimates of the right-hand side.

Two major weaknesses of Abercrombie's technique arise in the estimation of mean particle height $\mathbf{E}[H]$ and the control or measurement of section thickness t. Indeed the procedures for measuring these quantities are somewhat ill defined. It is not easy in practice to measure heights normal to the image plane, without changing the experimental conditions. Tissue shrinkage and compression will be different, usually greater, in the direction normal to the section plane. It is inappropriate to assume that cells are spherical. See Section 2.8.2 for further statistical critique.

More complicated versions of this approach have been proposed by Floderus [182], Konigsmark [320] and others. These are based on modifications of (2.34) to allow for other observation effects such as the failure to observe small end pieces of particles ("lost caps").

In biological applications these techniques seem to perform poorly. There are wide discrepancies, by a factor of up to 5, in published estimates of the total number of neurons in the human brain [11, 453, 606, 607], primary sensory neurons in rat dorsal root ganglia [87, Table 1, p. 191], [521, Table 1, p. 740], [517], and neurons in the hippocampus [601, Table 6, p. 494], obtained by different techniques. Coggeshall *et al.* [87] compared six classical methods for number estimation by applying them to the same tissue material, which was also cut exhaustively into serial sections to determine the exact number of cells. Most of the methods yielded substantial overestimates. Mendis-Handagama and Ewing [391] applied four classical methods to the same tissue in a controlled experiment, and found that they produced divergent and biologically implausible results. Coggeshall [85] compiles further evidence that the biases may be severe.

Alternatives to the classical techniques are discussed in Section 3.3.5 and in Chapters 10 and 11.

2.6 Counting profiles in two dimensions

A restricted field of visibility in the two-dimensional plane section also introduces sampling effects. Suppose we wish to measure or count the individual profiles in the rectangular reference frame in Figure 2.20. Some profiles extend partially outside the reference frame and it is unclear whether these profiles should be included in the sample. Including all such profiles would introduce a sampling bias in favour of larger profiles. Excluding them, that is, restricting consideration to those profiles which lie completely inside the frame, would also introduce a sampling bias, this time in favour of smaller profiles.

The need to eliminate such 'edge effects' in planar sampling has long been recognised, in many disciplines. One of the most common proposals for eliminating bias is to count all of the profiles that lie inside the frame, and only some of the profiles crossing the frame's perimeter. For example, a current industry standard in metallography [20] recommends counting exactly one-half of the profiles crossing the perimeter of a circular reference frame. This eliminates the bias approximately — to first order, when the profiles are small relative to the reference frame — but is acknowledged to have inherent bias in general

Figure 2.20 *Edge effects in planar sampling. Should the profiles that extend outside the frame be included or excluded from consideration?*

[437, 572]. Other popular rules for rectangular frames include adding or subtracting a count of the profiles that contain one of the four corners of the frame. These also have some inherent bias.

Successful strategies for eliminating edge effect bias are presented in Section 3.3.5 and in Chapters 10 and 11.

2.7 Projections

Figure 2.21 illustrates a different kind of microscopy in which translucent solid objects are seen in projection on a flat image plane. Stereological methods are also available to deal with this situation.

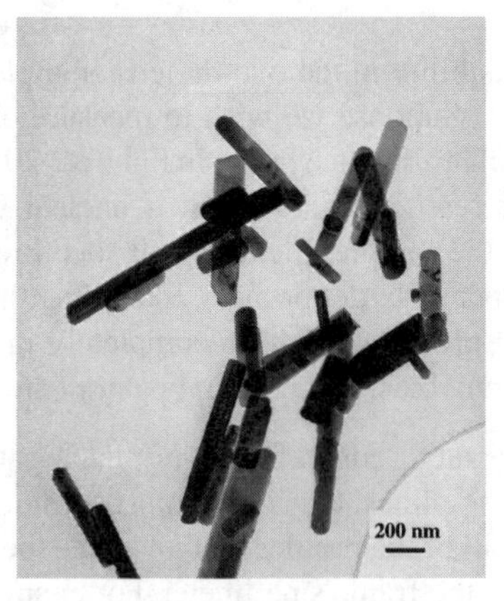

Figure 2.21 *Projected image of three-dimensional material. Rod-like particles of zirconium oxide, synthesized by mechanochemical processing. Transmission electron microscopy. Reproduced by kind permission of Prof. P. McCormick and Dr. J. Robinson (Advanced Nano Technologies and University of Western Australia).*

We assume the projection is orthogonal, that is, the line joining a point in space to its projected image is perpendicular to the projection plane. Other types of projection can also be treated [192].

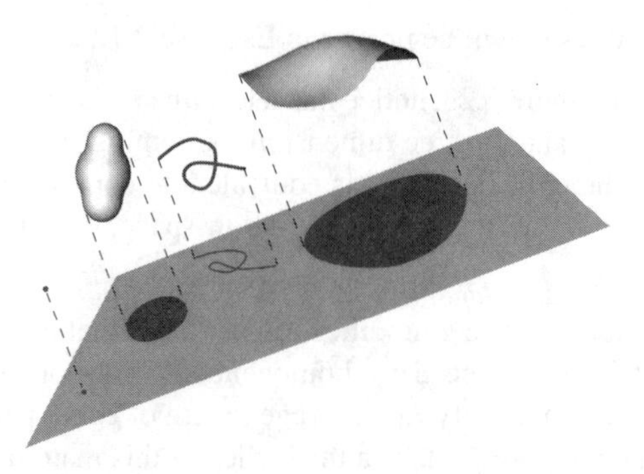

Figure 2.22 *Effect of projection on the dimension of a spatial feature.*

The effect of projection on the dimension of a spatial feature is shown in Figure 2.22. It is different from the effect of plane sections (Figure 2.6). Under projection, zero-dimensional features (pointlike objects) are projected onto zero-dimensional features; one-dimensional features (space curves) are projected onto one-dimensional features (plane curves) except at certain orientations; two-dimensional features (surfaces) are projected onto two-dimensional features (flat regions) except at certain orientations; and three-dimensional features (solid regions) are projected onto two-dimensional features (flat regions) as well.

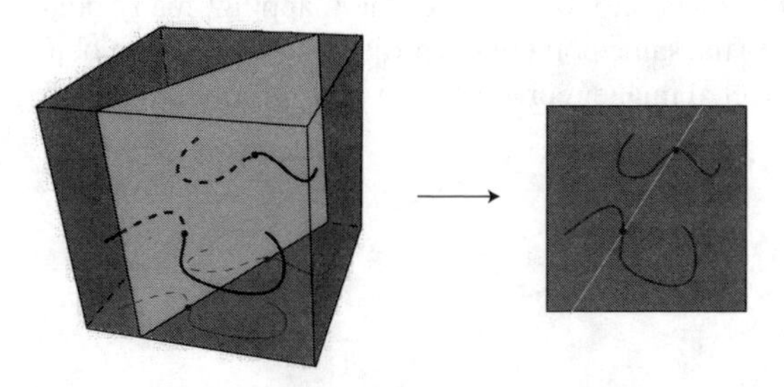

Figure 2.23 *Stereology for projections of curves in space. A test line in the projection plane (right) corresponds to a test plane in three dimensions (left).*

Stereology for projections of curves is fairly straightforward. First consider a bounded curve or curves in space with finite total length L. See Figure 2.23. Suppose the spatial orientation of the plane of projection is isotropically random (the direction normal to the projection plane is randomised, with uniform distribution over all possible directions in space). If $\mathbf{E}[L']$ denotes the expected length of the projected curve over all orientations of the projection plane, then

$$\mathbf{E}[L'] = \frac{\pi}{4} L \tag{2.38}$$

This fundamental result will be proved in Exercise 7.12.

Referring again to Figure 2.23, notice that a test line in the projected image corresponds to a test plane in three dimensions. Counting intersections between the test line and the projected curve is equivalent to counting intersections between the test plane and the original curve in space. This is another way to understand (2.38).

In most applications we use a modification of (2.38) that applies to thick sections seen in projection. Consider a homogeneous, isotropic material containing curves with length density L_V (average length of curve per unit volume of reference space). Suppose we take a thick slice of this material, of thickness t, as in Section 2.5, by cutting two parallel section planes a distance t apart and taking all the material between these planes. Next we project the thick slice orthogonally onto one of the two section planes. Then the projected curves have length density (average length of projected curves per unit area of projection plane)

$$L'_A = \frac{\pi}{4} t L_V. \tag{2.39}$$

For example this result is widely used to study filamental structures in thin metal foils.

Stereology for projections of surfaces and solid bodies is more complex. Figure 2.21 shows that the images of different surfaces or bodies may overlap nontrivially. Projection is a many-to-one mapping; many points in space are projected onto the same point in the projection plane. If the objects are opaque, then the projected image represents a loss of information.

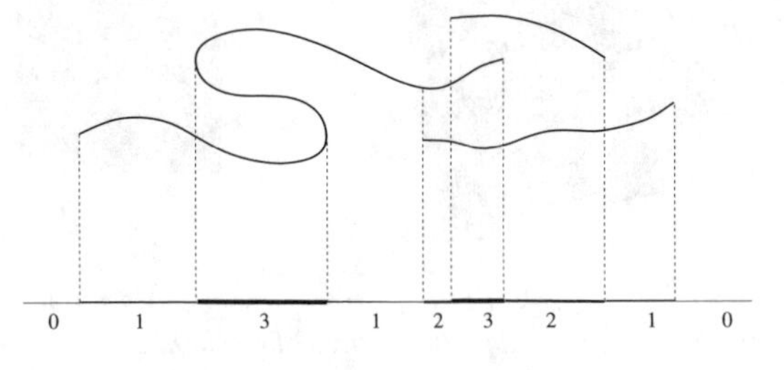

Figure 2.24 *Concept of total projection.*

A mathematical solution is to use the *total projection* in which each part of the projected image is assigned a multiplicity, equal to the number of spatial features which are projected onto it. Figure 2.24 sketches the concept. Stereological formulae hold for total projections under assumptions similar to those required for stereology of plane sections. These formulae are applicable in practice if the objects are translucent (as in Figure 2.21). If the objects are completely opaque, the projected image is the ordinary mathematical projection, called the *opaque projection*. Stereological formulae for opaque projections hold only under extra model assumptions (as in any missing-data problem).

For a bounded surface or surfaces of total area S, the expected area of the *total* projection onto an isotropic projection plane is

$$\mathbf{E}[A'] = \frac{1}{2} S. \tag{2.40}$$

See [367]. One way to derive this result is to notice that test points in the projected image correspond to test lines in space (Figure 2.25) so that test point counts in the projected image — in total projection — correspond to intersection counts with test lines in space, yielding (2.40).

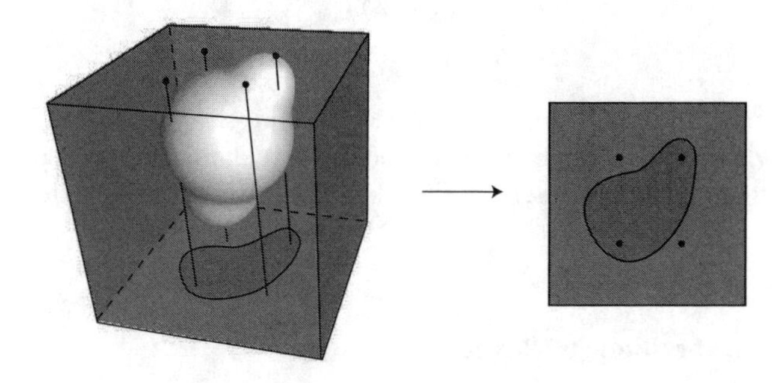

Figure 2.25 *Stereology for projected area. Test points in the projection plane (right) correspond to test lines in three dimensions (left).*

It follows from (2.40) that in a homogeneous and isotropic material, taking a slice of thickness t as described above, the density of *total* projected area per unit area of section A'_A is

$$A'_A = \frac{1}{2} t S_V. \tag{2.41}$$

Equation (2.40) is a generalisation of a classical result of Cauchy [74, p. 167] for convex bodies. Cauchy showed that the expected area of the *opaque* projection of a convex body is $A'' = \frac{1}{4}S$ where S is the true surface area of the body. The *total* projection area is $2A'' = \frac{1}{2} S$, because every point in the interior of the opaque projection has multiplicity 2, being the image of the 'front' and 'back' surfaces of the convex body.

Such properties of convex sets can be utilised in analysing opaque projections. Assume that the feature of interest is a union of convex sets, whose centroids are the points of a uniform Poisson point process (cf. [316]) with intensity λ

centres per unit area, and whose shapes and sizes are independent and identically distributed. This is the "Boolean model" [366, 415, 514, 543]. Now take a slice of thickness t from this material as described above, and project the slice onto one of its boundary section planes. Then the opaque projected set has area fraction

$$A''_A = 1 - \exp\left\{-\lambda\left(\mathbf{E}[V] + \frac{1}{4}t\mathbf{E}[S]\right)\right\} \tag{2.42}$$

where $\mathbf{E}[V], \mathbf{E}[S]$ are the expected volume and surface area of one of the constituent convex sets. The boundary of the opaque projection has length density

$$L''_A = \pi\lambda\left\{\frac{1}{4}\mathbf{E}[S] + t\mathbf{E}[H]\right\}(1 - A''_A) \tag{2.43}$$

where $\mathbf{E}[H]$ is the expected height of one of the convex sets (see page 27). For the same model, the true volume fraction V_V and surface density S_V are

$$V_V = 1 - \exp\{-\lambda\mathbf{E}[V]\} \tag{2.44}$$

$$S_V = \lambda\mathbf{E}[S]\,(1 - \exp\{-\lambda\mathbf{E}[V]\}). \tag{2.45}$$

These results, due to Miles [400], generalise findings by DeHoff [151], Giger and Hadwiger [198] and others. Using these and related results it is possible to estimate the model parameters λ, $\mathbf{E}[V], \mathbf{E}[S]$ and $\mathbf{E}[H]$ from opaque projected images.

2.8 Statistical critique of classical stereology

2.8.1 *Variability and homogeneity*

The classical theory of stereology had no statistical content, in the sense that it did not deal explicitly with variability. The fundamental formulae, such as Delesse's equation

$$V_V = A_A$$

are equations holding between two *parameters* of the material (defined in equations (2.4) and (2.5)). Implicitly, the parameter A_A can be *estimated* from plane sections by the corresponding sample fraction (2.7).

Delesse's [156] original definition of homogeneity was a very narrow one, in which the fluctuations of the sample area fraction (2.7) are bounded, that is, the maximum absolute error is small. This was gradually extended by others to a distributional assumption of 'spatial homogeneity' which was not precisely defined in classical proofs. Hence, there was an incomplete understanding of the conditions under which stereological methods are valid. A precise probabilistic formulation of 'spatial homogeneity' and 'isotropy', and a rigorous proof of the Fundamental Formulae of Stereology, were developed in the 1980's by

J. Mecke, D. Stoyan and others using the theory of stationary random measures. This is set out in [543, Chap. 10], see also [388, 510].

Classical approaches did not yield general theoretical results about the behaviour of stereological estimators. There were many empirical investigations of variance (see Section 13.9).

The variance of classical stereological estimators can also be analysed using the theory of random measures. However, as we indicated on page 16, these variances depend on second-order moment properties of the random material, which are not generally known, although they can also be estimated from sample data.

The assumption of 'spatial homogeneity' is appropriate in many applications in geology, metallurgy and materials science. However, it is hard to check, and it is certainly inappropriate for other applications such as anatomy and soil science. If the material is not homogeneous, classical stereological methods do not apply.

2.8.2 Particles

The basic theory for stereology of a particle population again ignores sources of variability. The key equations (2.17) and (2.34) are relations between parameters, and do not deal directly with the estimation of these parameters.

The assumptions required for validity of these key equations were often unclear. Wicksell's results and the equations (2.17) and (2.34) can be established under very weak stationarity conditions [543, p. 347 ff.].

Because of the confounding or non-identifiability of the parameters N_V and $\mathbf{E}[H]$, parameter estimation is more complicated than in Delesse's method.

A key point to understand about Abercrombie's method for estimating N_V and its relatives (Section 2.5) is that, while Abercrombie's *relation* (2.36) is an exact equation with very general validity, Abercrombie's *method* involves the *estimation* of the numerator and denominator in (2.36) in a manner that is not precisely defined. Abercrombie [3] offered two such techniques (Section 2.5) which are intentionally rather general. Thus, the statistical connection, if any, between the estimates of numerator and denominator in either (2.36) or (2.37) is undefined, so the statistical properties of the method are unknown.

On the other hand, typical implementations of Abercrombie's method estimate the numerator and denominator of (2.36) by observations on different samples. In such cases the ratio (2.36) is estimated by a ratio of two random variables which are likely to be approximately uncorrelated. The performance of such ratio estimators is known to be very poor: see Chapter 9.

Geometrical assumptions about particles — for example that the particles are all spheres — are overly simplistic for most applications.¶ Some of the methods presented above are highly sensitive to departures from the shape assumptions. For example, estimation of the mean particle diameter using the harmonic mean of profile diameter (2.31) relies heavily on the assumption of spherical shape, and assumes perfect observation of the sections. Biological tissues are compressible, and spherical cells may easily be compressed into ellipsoids by the forces exerted during microtome slicing. Abercrombie warned about relative bias of the order of 5–10% in his technique. Methods which assume fixed size and shape are extremely sensitive to variability. Thick sections are often prepared by embedding the tissue in paraffin wax, which expands with heat, so that section thickness is temperature-dependent!

Geometrical idealisations are implicit in the description of the plane section and the sectioning process. These idealisations may be untenable in applications. A microtome cutting through biological tissue should be visualised as 'ploughing' rather than 'slicing' [606]: it may push cells aside, or be pushed aside, rather than cutting cells at a glancing angle. This reduces the number of small section profiles. Small pieces of cells ("lost caps") may also fall out of the thick section. Section thickness t varies from place to place; the two cut faces of the section are not parallel; different sections may have different thicknesses; and actual section thickness may be very different from the nominal thickness chosen by the microtome operator.

Numerous strategies for 'correcting' these effects have been proposed; see the survey [113].

Alternative approaches in modern stereology, which avoid many of these problems, are described in Chapter 3.

2.9 Advice to consultants

Following is a checklist of methodological issues which a statistical consultant may wish to consider. We continue these notes in Chapter 3 and in subsequent chapters.

2.9.1 Qualitative interpretation of sections

The correct interpretation of two-dimensional plane sections is an important methodological issue for all scientists who use microscopy. Errors in the qualitative geometrical interpretation of plane sections are surprisingly common.

¶ They are appropriate in some applications, where shape is rigidly determined by phenomena such as surface tension and crystallisation.

The two-dimensional shapes seen in a plane section may erroneously be treated as if they were actually the three-dimensional objects of interest. This error is usually betrayed either by language ("the number of cells in a section") or by the use of inappropriate units, for example, the average volume of mineral grains may be reported in μm^2, and the density (number per unit volume) of brain cells may have been reported in μm^{-2} (number per unit area).

Fallacious geometrical interpretations of sections can lead (and have led) to invalid scientific conclusions. Misunderstandings of the microstructure of quenched steel and of the spatial organisation of mammalian liver were mentioned in Section 2.2.1. We recommend reading the discussion of these errors in [495, Introd.], [171].

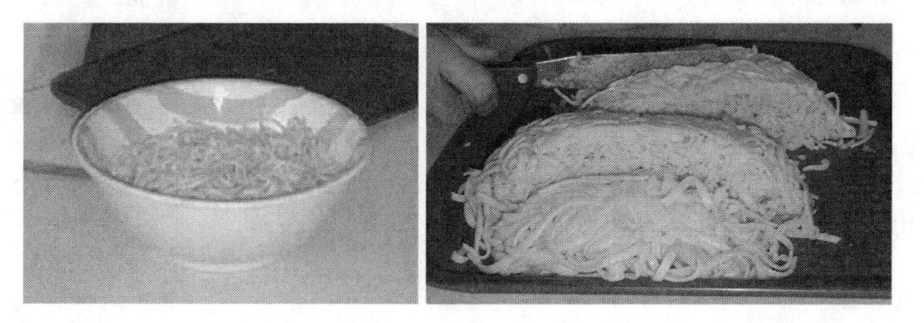

Figure 2.26 *Slicing a pot of noodles, to illustrate equation (2.11). Photograph by Ian Harper, Stereology Workshop, Monash Micro Imaging, Monash University.*

It is a common error to count profiles of tubes or capillaries on a section and attempt to interpret them as giving the 'number of capillaries'. Correct stereology relates the profile count to the *total length* of capillaries via (2.11). One effective explanation is to cook a bowl full of spaghetti, let it harden, then slice it. Most spectators will then agree that the number of spaghetti profiles has little relation to the number of pieces of spaghetti. See Figure 2.26.

In general the most important advice is to *think in three dimensions*.

2.9.2 Parameters

The parameters which are to be estimated by stereology must be chosen and defined carefully, in the light of the scientific goals and the exigencies of the experiment. The parameters should be three-dimensional quantities ("particle number per unit area" is not suitable). They should be sufficient to answer the scientific question (N_V is not sufficient to answer questions about changes in N). If the desired parameters cannot easily be estimated from our experiment, then alternatives may be sought: for example, from plane sections, it is hard to estimate the relative proportions of two cell types N_N, but easy to estimate their relative volumes V_V.

2.9.3 Bias

Sampling bias is the most important problem in stereology. Virtually all microscopy involves sampling the material, and when considered in this light, many seemingly innocuous techniques have an inherent sampling bias.

We have mentioned sampling bias for particle populations, due to sectioning or due to restriction of the field-of-view. Bias can also be introduced by selection, such as the preferential selection of good/interesting material, selective positioning of the microscope field-of-view (e.g. so that the image is fully inside the boundary of an organ), preferential orientation of the section plane (e.g. transverse sections of a tube), and selective deletion of observations (e.g. ignoring micrographs which contain little of the feature of interest).

Bias is also introduced by various observational effects such as overprojection (the "Holmes effect"), the physical loss of material from the section, and the unobservability of some parts of a biological structure due to the lack of optical contrast.

2.10 Exercises

Practical exercises

Exercise 2.1 Take a single slice of Swiss cheese (Emmenthaler) and estimate (a) the volume fraction of holes, (b) the surface density of holes.

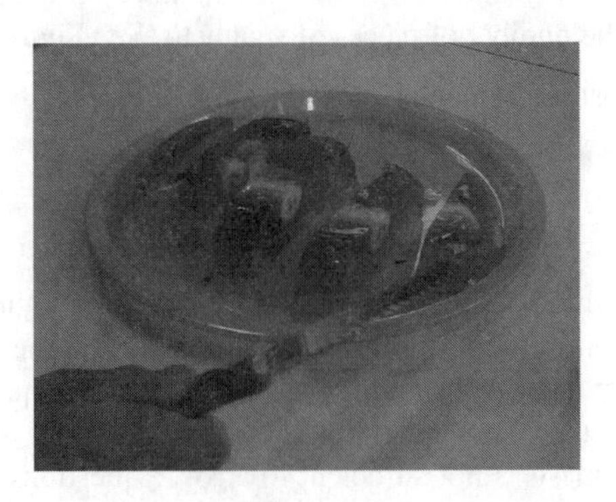

Figure 2.27 *Slices of an artificial population of particles (grapes, bananas and peas in a gelatine matrix) for Exercise 2.2. Photograph by Ian Harper at Stereology Workshop, Monash Micro Imaging, Monash University.*

Exercise 2.2 Make an artificial population of particles by peeling two bananas and twenty grapes, say, and placing them in a bowl of gelatine. When the gelatine is set, remove it and take a slice with a sharp knife. Count the number of banana profiles and the number of grape profiles. Compare this with the (known) number of bananas and grapes in the original population. See Figure 2.27.

Theoretical exercises

Exercise 2.3 A team of investigators has counted the number of hepatocytes visible per unit area of thin sections of human liver, for biopsies from patients in different stages of a degenerative illness. There is a clear increase in the number of 'hepatocytes per unit area' in the later stages of illness. Although the investigators are aware that it is not stereologically valid to count section profiles, they argue that comparisons between different disease stages are valid, because the same counting method was used throughout. The investigators publish an article entitled *'Hepatocyte Proliferation In Late Stages of X's Disease'*. Comment on this finding.

Exercise 2.4 In a plane section of a population of spheres (Wicksell's problem, page 34) show that it is possible for the mean diameter of the circular section profiles to be **larger** than the mean diameter of the spheres. For example, consider the case where 10% of the spheres have diameter 1 micron and 90% have diameter 10 microns. Explain this paradox.

Exercise 2.5 Find the coefficient of variation ($\text{CV} = \text{SD}/\text{mean}$) of the diameters of the circular section profiles of a population of spheres of equal diameter D. [*Hint:* use equation (2.33).]

Exercise 2.6 A particular mineral exists in two different crystalline types I and II, which can be distinguished on a plane section. It is desired to estimate the fraction V_V of volume *of this mineral* that is of type I. The researcher measured, on several plane sections, the area of each grain profile of the mineral, noting whether it was of type I or II. Then the researcher calculated $\bar{A}_I/\bar{A}$, the average grain profile area of type I grains divided by the average profile area of grains of either type. Is this a valid estimate of the volume fraction V_V?

Exercise 2.7 According to a stereology website,

> "Delesse showed that the profile area of a random section through a population of objects is proportional to the expected value of the objects' volume."

On the contrary, show that the mean area of plane sections of a particle population is *not* proportional to the mean volume of the particles, by considering the special case of spherical particles and applying equation (2.33).

2.11 Bibliographic notes

History

A definitive history of stereology has not been written and indeed early work is still coming to light. Excellent surveys include [580] and [118]. Early work in stereology includes the historic papers of Delesse [155, 156], Rosiwal [486], Thomson [559] and Glagolev [200]. The technique of determining area by weighing paper or foil cutouts used by Delesse dates from the sixteenth century [350, p. 64]. An early account of geological applications is given by Thomson [559]. For the early history of stereology in metallography see [495, pp. 57–62] and [562, Introd.]. The mathematical work of Steinhaus [534, 535, 536] anticipated some stereological techniques.

Classic literature

An excellent early introduction to stereology is that of Weibel and Elias [584, 583]. Weibel et al [591] mounted a compelling argument for the adoption of stereological methods in biological science. Weibel et al [588] first showed practical calculations of the effort and efficiency involved in stereological sampling technique. Weibel [578] articulates the value of stereology for understanding biological structure and function, while Amstutz and Giger [8] describe its benefits in mineralogy, petrology, reserve estimation and ceramics.

Classic books on stereology are those by Saltykov [493], Weibel and Elias [585], DeHoff and Rhines [154], Underwood [562] and Weibel [580, 581].

Much original research on stereology was published in the Proceedings of the International Congresses on Stereology [249, 170, 590, 561, 5]. Other important proceedings volumes are [79, 412, 438].

Early findings

Apparently Weibel [576] was the first to point out the formula for the dimension of a section, in a stereological context.

H. Steinhaus [534, 535] first proposed estimating the length of a curve by counting its intersections with test lines. The relation $S_V = \frac{4}{\pi} B_A$ is usually attributed to Tomkeieff [580, p. 31], [581, p. 6] although Saltykov [495] claims priority, see [490, 491]. It was rediscovered by Smith and Guttman [519], Duffin et al [166] and Horikawa [269]. See also [98].

The relation $L_V = 2Q_A$ is due to Saltykov [491] and was rediscovered by Smith & Guttman [520]. According to Weibel & Elias [583] the relation $L = 2dI$, for estimation of 3-D length from parallel section planes, is due to Smith & Guttman [520]. Hennig [257, 256] first used the formula $L_V = 2Q_A$ for cylindrical tubes.

Early work on stereology for projections includes that of P.A.P. Moran [418, 417], DeHoff [151] and Giger and Hadwiger [198]. See [562, chap. 6], [563].

The calculation of moments of A_A and V_V using the indicator function I originates with Kolmogorov [318, p. 41] but is often attributed to Robbins [480, 481, 482]. See also [64, 404, 524, 525].

Particle size

Particle size distribution and mean particle size were studied notably by Wicksell [602, 603, 604], Spektor [529], Saltykov [494], DeHoff [148, 149], Bach [21, 22] and Cruz-Orive [107, 108, 113]. See also Coleman [91, 92, 93, 94, 95]. For reviews see [113], [444, Chap. 6] and [543, sect. 11.4, pp. 353–369], or the classic surveys in [562, chap. 5] and [580, chap. 5]. See also [514, pp. 350 ff.].

Statistical theory of Wicksell's problem is explored in [12, 13, 14, 566, 207, 241, 268, 278, 354, 434, 439, 440, 450, 516, 553, 573, 574].

For Wicksell's problem for thick sections see [90, 92, 364, 367].

Particle number

Particle counting and sampling rules have been discussed for well over a century [44, 251, 254]. This chapter has covered only the classical theory of estimating numerical density N_V. For a survey of classical methods see [580, sect. 2.6], [581, chap. 5]. Influential classical work includes [3, 149, 153, 154, 182, 185, 587]. See also [444, pp. 166, 173]. Some of the forgotten early history of particle counting methods in biology is reported in [254, 44]. The tomato salad metaphor is due to G. Bach [444, Preface]. The terminology 'trace' and 'profile' was proposed by L.M. Cruz-Orive.

Mathematical basis

In geometrical probability the founding work is that of Buffon [68], Crofton [105, 104, 103], Barbier [37], Czuber [142] and Deltheil [157]. Accessible textbooks were written by Kendall and Moran [309] and Solomon [526]. Literature reviews on geometrical probability include [421, 423, 345, 24].

Rigorous mathematical statements of the Fundamental Formulae of Stereology (2.12) were given by Giger [196, 197] and Miles [408]. On stereological methods for projections, see [198, 367, 400]. Literature reviews on mathematical stereology include [28, 281, 542, 592].

Key texts in integral geometry are those of Blaschke [57], Hadwiger [233], Santaló [500] and Schneider & Weil [509].

Image analysis and stereology

The quantitative analysis of digital images has a close affinity with stereological principles. The array of pixels which constitute an image forms a sample of two-dimensional space — and also of three-dimensional space, if the image is a plane section. The Fundamental Formulae of Stereology (2.12) and other stereological principles have direct counterparts for digital images in two and three dimensions [513], [514, pp. 191–196]. Counts of pixels give estimates of area, and counts of certain configurations of pixels give estimates of curve length and other quantities. Configuration counts can also be used to estimate the Euler Poincaré characteristic and orientation distribution of a surface, cf. [311, 445, 444, 514, 515]. The development of mathematical morphology [514, 515] and random set theory [307, 366, 543, 299, 510] was strongly motivated by applications in microscopy. It greatly influenced and was influenced by the field of stereology [514, pp. 21–22, 26, Chaps. VIII, XIII].

Current stereological research

Much new methodological research in stereology is published in the *Journal of Microscopy* and *Image Analysis and Stereology*. These are official journals of the International Society for Stereology (`www.stereologysociety.org`) which also supports forums, congresses, courses and research activity in stereology.

Recent books on the theory and practice of stereology include those of Bertram and Wreford [49], Elias and Hyde [173], Hilliard and Lawson [262] Howard and Reed [272], Kurzydłowski and Ralph [328], Mouton [427], Russ and DeHoff [489], and Saxl [503]. Supplementary to [328], the paper [471] gives an illuminating discussion of general issues in the quantitation of materials.

The work of Pierre Gy [230, 465] provides a comprehensive theory (of sampling correctness, sources of error, and variance estimation) for sampling particulate materials in mining. It has features in common with stereology although the two fields have developed quite separately.

Stereology is still being taken up into new areas, for example into botany (e.g. [167, 612, 565]) and dentistry (e.g. [275, 611]).

CHAPTER 3

Overview of modern stereology

This chapter gives an overview of modern stereology. It may serve as an introduction and guide to the more detailed accounts in the following chapters.

A transition from 'classical' to 'modern' stereology occurred in the late 1970's. The modern theory differs by having rigorous statistical foundations (Section 3.2), involving a variety of random sampling designs (Section 3.3), offering a wider range of stereological identities than the Fundamental Formulae of Stereology (Section 3.4), and using a wider range of stochastic models (Section 3.5). We begin by explaining the motivation for these developments in Section 3.1.

3.1 Motivation

After its establishment as a science in the early 1960's, stereology enjoyed an exciting period of expansion, finding very broad application in science and technology [154, 584, 583, 585]. The pioneers emphasised that stereology rests on 'generic' principles, of a geometrical and statistical nature, which do not depend on specific knowledge about the structures and materials under study, and which can therefore be applied very widely.

However, there were two major weaknesses: the assumption that materials were 'homogeneous', and the reliance on idealised geometrical models of particle shape.

The classical methods described in Chapter 2 assume the material under study is 'homogeneous'. This assumption is clearly inappropriate for many biological tissues and organs, which are highly organised. The classical methods are simply not applicable in such cases (cf. [172], [575, p. 10], [168, 577, 579]).

Since 'homogeneity' was not precisely defined in the classical literature, it became unclear whether stereological methods were valid for more-or-less homogeneous materials in geology and materials science [78]. It became important to clarify the nature of stereological estimation, to identify the precise conditions under which stereological methods are valid, and to assess their accuracy.

A rigorous statistical foundation for stereology was eventually established, and we sketch this in Section 3.2.

At the same time, an alternative approach evolved. It was mooted (e.g. [575, p. 15]) that the classical stereological methods might still be applied to an inhomogeneous biological tissue, if 'representative samples' of the tissue were taken. Instead of relying on the assumed homogeneity of the tissue to guarantee that the section plane is 'representative' of the tissue, we would instead take random plane sections using a sampling design that guarantees they are 'representative'. This alternative 'random sampling' approach to stereological inference is also discussed in Section 3.2.

Random sampling designs were then developed for a wide variety of biological tissues and materials. Section 3.3 sketches the main types of sampling design.

New ground was broken in the basic principles of stereology. The Fundamental Formulae of Stereology were once believed to be the only stereological relations holding for very general spatial structures. However, other formulae of an equally general nature were discovered or rediscovered. These new geometrical identities are sketched in Section 3.4.

The 'random sampling' approach opened up a world of new stereological techniques, based on new random sampling designs and on alternative interpretations of the existing stereological formulae. The probability theory underlying these new techniques is sketched in Section 3.5.

3.2 Statistical foundations

3.2.1 Connection with sampling theory

In the late 1970's, Australian statisticians R.E. Miles and P.J. Davy pointed out the very strong analogy between a stereological experiment and a sample survey [146, 147, 396, 401, 404, 408].

A sample survey is a well-established technique for investigating a '*population*' of individuals (such as the population of Scotland) by gathering information from only a *sample* of individuals from this population (such as a telephone survey of Scottish householders). See Appendix A or [83, 548, 555].

Consider the simplified stereological experiment sketched in Figure 3.1. The '*reference space*' Z (for example, a rock) contains a '*feature of interest*' Y (for example, a particular mineral). We generate a *probe* T (for example, a section plane) intersecting Z and we are able to observe the intersections $Z \cap T$, $Y \cap T$ of the probe with Z and Y.

Miles and Davy pointed out that this experiment is analogous to a sample survey. The reference space Z corresponds to the *population* studied in the survey;

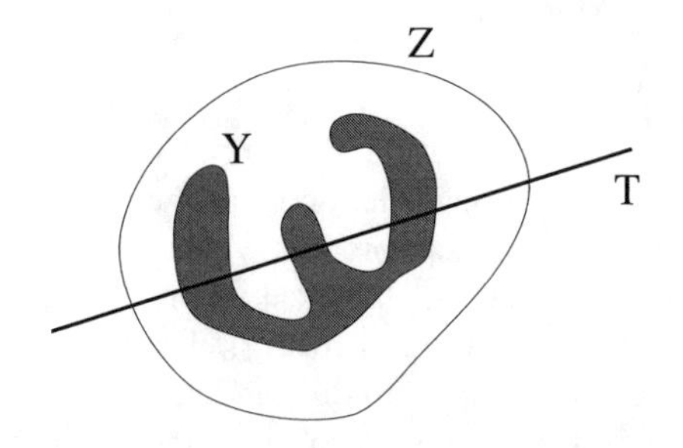

Figure 3.1 *A simplified stereological experiment: the reference space Z, containing a feature of interest Y, is intersected by a probe T.*

the probe T yields a *sample* $Z \cap T$ from this population; and Y is a subpopulation of interest to us. Our goal is to draw inferences about the population from the sample.

While sampling theory is normally applied to a population of discrete units such as individual people, it also applies to populations or 'sampling universes' of a more general kind, including the continuous reference space in a stereological experiment. This direct connection between stereology and sampling theory allows us to enlist well-tried techniques and ways of thinking from statistics into stereology.

3.2.2 Kinds of inference

Statistical inference from a sample can be formulated in two different ways, depending on whether the sampling variation arises from the random selection of the sample (*design-based* or 'randomisation' inference) or is assumed to arise from intrinsic randomness in the population (*model-based* or 'superpopulation' inference). These two approaches have very different practical implications.

Accordingly, there are two broad approaches to inference in stereology. In *model-based stereology*, the contents of the material under study are treated as random. Inference is based on the assumption that the material is homogeneous. In *design-based stereology*, the material is treated as fixed and arbitrary, while the sampling probe T is selected randomly by the experimenter. Inference is based on the fact that the probe was randomised.

'Unbiased stereology' is a synonym for design-based stereology, used by some writers in biological science [272, 427]. From the statistical viewpoint this is misleading: see Section 3.6.5.

3.2.3 Model-based stereology

Classical stereology was model-based. Following Delesse, stereological formulae were developed by assuming the material was 'statistically homogeneous'. This is effectively a stochastic model of the material. The material is treated as a realisation of a random process within the fixed reference space. We assume the random process is 'statistically homogeneous' (in some sense that was not spelt out in the classical theory). The stereological parameters V_V, A_A are expectations with respect to this random process. The Fundamental Formulae of Stereology are statements connecting different parameters of the random process.

A rigorous and elegant probabilistic foundation for model-based stereology was established about 1980 by the former East German stochastics school of J. Mecke, D. Stoyan and collaborators [539, 540, 541, 544] following foundational work by Matheron [364, 366] and others. The feature of interest Y is described as a realisation of a random set or random measure, using the ideas of *stochastic geometry*. Fundamentals come from the theory of random sets [366] and the theory of moments of stochastic processes, especially point processes. The assumption of 'homogeneity' is clarified by requiring that the process be stationary and isotropic. The proofs of key stereological results are simplified and unified. For a detailed presentation see [543, Chap. 10], [510, Chap. 5].

One of the main practical benefits of this theory is that it gives precise conditions under which model-based stereological techniques are valid. The Fundamental Formulae of Stereology were shown to hold for all *first-order* stationary and isotropic random sets Y. This requirement is considerably weaker than the 'statistical homogeneity' assumed in the classical literature, which corresponds to *strict* (almost sure) stationarity and isotropy.

For the estimation of particle size distribution (Section 2.4.2), it has often been assumed that the particles have independent identically distributed radii and that their centres form a Poisson point process. This turns out to be mathematically unnecessary, and indeed unrealistic (since in such a process the particles would sometimes overlap). Stoyan [540, 541, 544] showed that the particle process is only required to be statistically stationary; the particles may be stochastically dependent on each other. The Poisson assumption is in no sense a cornerstone of stereology as has been claimed by some writers (cf. [46]).

If desired, explicit models of the random contents of materials can also be constructed using stochastic geometry, random functions, and mathematical morphology. See [299, 300, 301].

Another benefit of stochastic geometry is that it provides a foundation for analysing variances and higher moments of stereological estimators, and spatial covariance properties of homogeneous materials.

Note that, in model-based stereology, the target parameters are 'superpopulation' quantities. For example V_V is the *expected* volume fraction of the feature of interest, averaged over hypothetical realisations of the random contents of the material,

$$V_V = \frac{\mathbf{E}[V(Y)]}{V(Z)}.$$

Estimators are unbiased in the 'superpopulation' sense. The empirical area fraction $\widehat{A}_A = A(Y \cap T)/A(Z \cap T)$ is an unbiased estimator of the volume fraction V_V in the sense that

$$\mathbf{E}\left[\widehat{A}_A\right] = V_V \tag{3.1}$$

where $\mathbf{E}$ denotes expectation with respect to the random process, that is, over hypothetical realisations of the contents of the material.

Unless we are careful, this viewpoint conflates all sources of variability. If Delesse's principle is applied to 10 microscope images of stainless steel and the results are averaged to obtain a final estimate of V_V, any variation between samples that is attributable to differences in the temperature, annealing schedule, initial composition, etc, is conflated with large-scale and small-scale spatial variability and with variation in the orientation of the section plane. The expectation in (3.1) is an average over all these sources of variability.

We might hope that the estimators are also *consistent*, meaning that the empirical estimates $\widehat{A}_A$ obtained from larger and larger samples would converge to the true value $A_A = V_V$. Some classical writers [78] interpreted Delesse's principle in this way. However, consistency requires much stronger probabilistic assumptions, which imply that there are no external random effects.

Model-based stereology is well suited to most applications in materials science, geological science, food science and other fields which focus attention on the 'typical' contents of a 'homogeneous' material, having virtually infinite extent at the scale of observation. Figure 2.5 is an example.

3.2.4 Design-based stereology

The assumption of spatial homogeneity is often inappropriate, especially in bioscience. Many biological structures are highly organised, and are anything but 'homogeneous' or 'random' in their spatial arrangement [579]. Likewise, many synthetic materials exhibit a gradient or inhomogeneity at large scales. Classical stereological methods do not apply to them.

The parameter of interest may be an absolute measure of size, such as the total area of glomerular filtration surface in the kidney. This is analogous to a population total in sampling theory. Classical stereology gives only indirect access to such parameters.

It is often important to draw sampling inferences about a *fixed* spatial structure of finite size, such as the kidney of a specific human patient. Model-based stereology is not well suited to this problem. The clinician treating a kidney patient wants to know the total glomerular filtration surface area S for this patient, not an average over the hypothetical superpopulation of kidney tissue from similar patients.

This impasse was broken in the pioneering work of Miles and Davy [146, 147, 396, 408]. They pointed out that, instead of assuming a spatially homogeneous material, we may alternatively conduct the experiment so that our sampling of the material is spatially homogeneous. In 'design-based' stereology, the feature Y and reference space Z of Figure 3.1 are arbitrary, fixed sets, and the probe T is randomised.

Example 3.1 (Design-based estimation of volume) *Suppose Y and Z are arbitrary, fixed, solid objects with $Y \subset Z$. The volume of Y can be expressed as the integral of the areas of its horizontal plane sections:*

$$V(Y) = \int_0^s A(Y \cap T_u)\,\mathrm{d}u, \tag{3.2}$$

where T_u denotes the horizontal plane at height u. Assume Z lies between the planes T_0 and T_s. Now suppose we choose a section plane T_u randomly *by letting the height u be a uniformly distributed random variable in the range $[0,s]$. The probability density of U is*

$$f_U(u) = \begin{cases} \frac{1}{s} & \text{if } 0 \le u \le s \\ 0 & \text{otherwise} \end{cases} \tag{3.3}$$

The intersection between the fixed solid Y and this random plane $T = T_U$ has expected value

$$\begin{aligned} \mathbf{E}[A(Y \cap T)] &= \int_0^s A(Y \cap T_u)\, f_U(u)\,\mathrm{d}u \\ &= \frac{1}{s} V(Y) \end{aligned}$$

by (3.2). It follows that

$$\mathbf{E}[sA(Y \cap T)] = V(Y). \tag{3.4}$$

In other words, $\widehat{V}(Y) = sA(Y \cap T)$ is an **unbiased estimator** *of $V(Y)$, for* any *set Y, provided T has the required probability distribution.*

The key point is that *design-based inference does not require assumptions about the material*. Recall that, in design-based survey sampling, estimators of population parameters are unbiased by virtue of the randomisation of the sample, without any need for assumptions about the population structure (cf. [555, p. 3]). Similarly, in design-based stereology, estimators of three-dimensional quantities are unbiased under appropriate randomisation of the section plane, *regardless* of the spatial arrangement of the material. This vastly increases the scope and power of stereological methods, since we no longer need to assume the material is 'spatially homogeneous' in any sense at all.

Design-based stereology is suited to most applications in biological science, mainly because biological tissues are usually inhomogeneous.

On the other hand, the validity of design-based stereology depends crucially on adherence to the random sampling protocol. The plane sections must have been selected according to the random sampling design that is specified in the theory; otherwise the relevant stereological identity will be false. For example (3.4) holds only when the section plane is uniformly distributed according to (3.3).

This may exclude some forms of sampling in biology and medicine, such as needle biopsies, that are selected by an informal or arbitrary procedure. It definitely forbids preferential selection of section images based on what they contain or do not contain, or on their image clarity, photographic beauty, or other attributes that might be correlated with the stereological observations.

Hybrids of model-based and design-based inference are also used. For example, the human lung is 'inhomogeneous' at large scales but might be treated as 'homogeneous' at small scales and away from the bronchial tree. Thus design-based estimation of the total lung surface area might involve some model-based reasoning applied to the lowest levels of the sampling. Many rocks and materials may be considered spatially homogeneous but not isotropic. The orientation of the section plane must be randomised in order to estimate S_V and L_V (by design-based reasoning) but the location of the section plane is arbitrary (by model-based reasoning).

We return to these fundamental issues in Chapter 6.

3.2.5 Parameters

This reform of the statistical foundations of stereology also focused attention on the meaning of stereological parameters such as V_V. These parameters need to be defined differently in the design-based context. It is natural to reinterpret them as ratios, for example V_V is simply the ratio

$$V_V = \frac{V(Y)}{V(Z)}$$

of the volumes of the two fixed sets Y and Z.

It is important to distinguish between absolute quantities (like the volume $V(Y)$ of a solid object) and relative quantities (like the volume fraction V_V or

$V(Y)/V(Z)$ of the same object with respect to a reference space Z). These are analogous to population totals and population means, respectively, in survey sampling theory.

Interpretations of absolute and relative parameters are quite different. A change in numerical density N_V of particles does not necessarily have the same interpretation as a change in total number N of particles. Consider the numerical density of cells in an organ,

$$N_V = \frac{N(\text{cells})}{V(\text{organ})}.$$

If N_V increases, we cannot conclude that the number of cells has increased. For example, it may be simply that the organ volume $V(\text{organ})$ has decreased, while the number of cells is unchanged.

An important example of this reference space artefact or **reference trap** concerns the finding that the total number of neurons in the human brain declines steadily with age. There is an indisputable decline in brain weight, size, and function with normal ageing. However the finding (dating from [63]) that the absolute *number* N of neurons declines with age, was typically based on estimates of numerical density N_V from samples of brain tissue. Until the 1990's, most studies did not report the total volume of the entire brain, either for individual brains or for different age groups. Therefore, this conclusion was not a valid inference from the published data, as stereologist and neurologist H. Haug [250, 251] courageously pointed out. Typically the reference space for N_V estimates was the whole brain *after* processing with chemical fixative. Haug established that younger brain tissue tends to shrink under fixation more than older tissue, so in effect the volume of the reference space was increasing with age. When this effect is corrected, the data showed no evidence of a decline in the total number of neurons. The observed decrease in N_V appeared to be attributable to the dependence of tissue shrinkage on age.

The question of age-related neuron loss is now under renewed investigation. Recent results suggest that there is a loss of approximately 10% of neurons during life — considerably less than thought previously [10, 454, 462]. Age-related changes also appear to be different in different regions of the brain.

The problems of reporting changes in N_V rather than in absolute number are canvassed by many writers. In the placenta, Mayhew [381] cites numerous examples. In pathology, Casley-Smith [73] shows that oedema (swelling of tissue) would tend to lower N_V, even if no other changes to the tissue have occurred. In neuroscience this theme is also stressed by Oorschot [447] and Lange *et al.* [334]. Atrophy of tissue in the central nervous system may take the form of a shrinkage of glial (supporting) cells with or without a loss of cell numbers. Shrinkage may result in an increased numerical density of glial cells, leading to reports of gliosis (proliferation of glial cells). Shrinkage may also result in increased numerical density of neurons even when there is a loss of neurons. This has great potential to confound the study of pathological processes [334, p. 402]. Another example concerns the comparison of tissues from different species. Neuron density N_V in the brain does not change greatly across species, but total N is very different between species [334].

In the biological sciences it is almost always the absolute quantity (cell number N, membrane surface area S) that is of primary scientific interest [334, p. 405], [73, 225, 447]. In such cases, the relative quantity (numerical density N_V, surface density S_V, etc) should not be used as a surrogate for the absolute quantity, because of these problems of interpretation.

It is advisable to measure the volume of the reference space V(organ) so that the absolute quantities can be recovered by ratio estimation,

$$\begin{aligned} \widehat{N}(\text{cells}) &= \widehat{N_V} \times \widehat{V}(\text{organ}) \\ \widehat{S}(\text{cells}) &= \widehat{S_V} \times \widehat{V}(\text{organ}). \end{aligned}$$

3.3 Random sampling designs

The development of design-based stereology opened up a world of new techniques, all based on random sampling of the material under study. A range of different 'sampling designs' became available to suit different experimental conditions. Indeed, different situations *required* different sampling designs, as we now explain.

3.3.1 Estimating a ratio

Suppose we wish to estimate volume fraction in an inhomogeneous biological organ from a random sample of the organ. The tissue compartment of interest Y is an arbitrary solid region inside the organ Z. The target parameter is

$$V_V = \frac{V(Y)}{V(Z)} \tag{3.5}$$

which we shall estimate using

$$A_A = \frac{A(Y \cap T)}{A(Z \cap T)} \tag{3.6}$$

where T is a section plane, or perhaps a microscope image field in the section plane. The 'random sampling' analogue of Delesse's principle would then be

$$\mathbf{E}\left[\frac{A(Y \cap T)}{A(Z \cap T)}\right] = \frac{V(Y)}{V(Z)} \tag{3.7}$$

where **E** denotes expectation with respect to the randomised position of the section plane or probe T in an appropriate sampling design.

However, there is a complication. The random sampling design under which we have unbiased estimators of absolute volume (3.4) is *different* from the random sampling design for unbiased estimation of volume fraction (3.7). This happens because $A(Y \cap T)/A(Z \cap T)$ is now a ratio of two random quantities. The expectation of this ratio is usually not equal to the ratio of the expectations $\mathbf{E}[A(Y \cap T)]/\mathbf{E}[A(Z \cap T)]$. This may be interpreted as a problem of variable sample size, since the denominator $A(Z \cap T)$ is effectively the 'sample size' of the plane section $Z \cap T$.

There are two common remedies for the problem of variable sample size in sampling theory: sampling with probability proportional to size (pps), and systematic sampling. A pps sample gives us unbiased estimators of population means, while a systematic sample gives unbiased estimators of population totals. See Appendix A.

3.3.2 Weighted sampling

Miles and Davy pursued the analogue of pps sampling. They showed, for example, that the design-based Delesse principle (3.7) holds when the section plane is selected with probability proportional to the section area $A(Z \cap T)$. This might be called an 'area-weighted' random section.

Example 3.2 (area-weighted section with fixed orientation) *In the context of Example 3.1, suppose the random section plane $T = T_U$ has a non-uniform distribution, with probability density*

$$f_U(u) = \frac{A(Z \cap T_u)}{V(Z)}$$

proportional to the area of the intersection with the reference space. The denominator $V(Z)$ is the normalising constant which ensures that the density f_U integrates to 1. Then [146]

$$\begin{aligned}
\mathbf{E}\left[\frac{A(Y \cap T)}{A(Z \cap T)}\right] &= \int \frac{A(Y \cap T_u)}{A(Z \cap T_u)} f_U(u)\,\mathrm{d}u \\
&= \int \frac{A(Y \cap T_u)}{A(Z \cap T_u)} \frac{A(Z \cap T_u)}{V(Z)}\,\mathrm{d}u \\
&= \frac{1}{V(Z)} \int A(Y \cap T_u)\,\mathrm{d}u \\
&= \frac{V(Y)}{V(Z)}
\end{aligned}$$

by (3.2), so that the Delesse principle (3.7) holds.

Miles and Davy developed a complete theory of stereology for randomised sampling designs using "size-weighted" sampling. They obtained design-based analogues of all the Fundamental Formulae of Stereology (2.12) for these designs. They also suggested how to implement weighted sampling in practice [402, 410].

Swiss anatomist and pioneering stereologist E.R. Weibel championed the design-based approach. His authoritative treatise on stereology [580, 581] expounded the design-based theory and laid out practical random sampling protocols for many of the standard stereological methods, using weighted sampling designs.

3.3.3 Systematic sampling

The importance of systematic sampling in stereology was realised somewhat later, in the mid-1980's*. A simple example of systematic sampling is *serial sectioning* — cutting the material of interest with a series of equally-spaced parallel plane sections. See Figure 3.2. One may think of an egg-slicer, or a microtome.

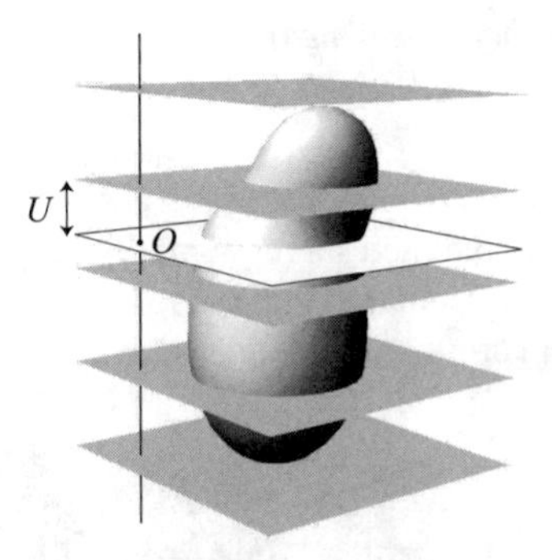

Figure 3.2 *The serial section plane design. The distance between successive planes (grey) is s. The parallel white plane is a reference plane, passing through the origin O.*

Example 3.3 ('Cavalieri' estimator of volume) *A stack of serial section planes at constant spacing $s > 0$ may be written as*

$$\mathcal{T} = \{T_{U+ks} : k \in \mathbf{Z}\}$$

where as usual T_u denotes the horizontal plane at height u. The k-th plane in the stack is $T_{(k)} = T_{U+ks}$ where the index k is a signed integer. There are infinitely many planes in this stack, but in practice only a finite number of planes will actually be required. The position of the entire stack is determined by its

* Systematic sampling was proposed as early as 1963 by Weibel [575, p. 15].

'starting position' U, which we assume is random and uniformly distributed in the range $[0,s)$.

The volume $V(Y)$ of an arbitrary solid Y can be estimated by

$$\widehat{V}(Y) = s \sum_k A(Y \cap T_{(k)}). \tag{3.8}$$

This will be familiar as a simple finite-sum approximation to the integral (3.2) and therefore the volume. However, it is also an unbiased *estimator of $V(Y)$ under appropriate randomisation:*

$$\begin{aligned}
\mathbf{E}[\widehat{V}(Y)] &= s\,\mathbf{E}\left[\sum_k A(Y \cap T_{(k)})\right] \\
&= s \int_0^s \sum_k A(Y \cap T_{u+ks}) \frac{1}{s}\,\mathrm{d}u \\
&= \int_{-\infty}^{\infty} A(Y \cap T_v)\,\mathrm{d}v \\
&= V(Y).
\end{aligned}$$

The only conditions are that Y be a measurable set with finite volume, and that U be uniformly distributed on $[0,s]$.

This is a direct analogue of the estimation of a population total from a systematic sample in finite population survey sampling (Appendix A). Thompson [557, p. 21] describes it as 'similar to that used in the estimation of the volume of the hull of a ship from cross-sectional plans to scale.'

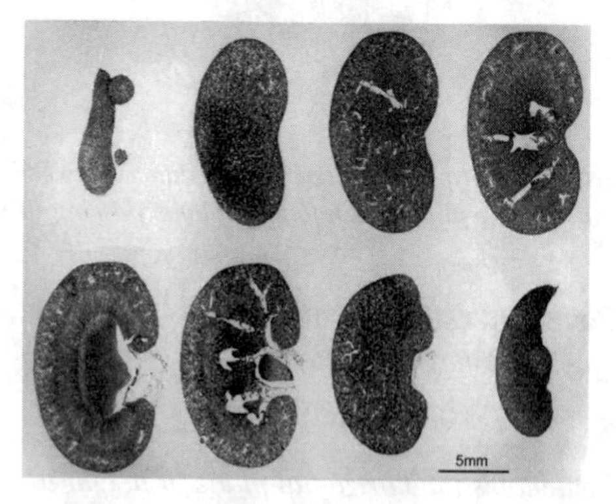

Figure 3.3 *Full set of serial sections of a rat kidney. The areas of all sections are summed to provide an estimate of the volume of the kidney, and of kidney cortex (outermost, coarser region). Reproduced by kind permission of Dr. J.R. Nyengaard, University of Århus, Denmark.*

There are important applications of the Cavalieri estimator throughout biological science. A typical application would be to estimate the volume of an organ (cf. Figure 3.3) or a tumour [35].

3.3.4 Discrete designs

While the design-based methods sketched above do not require any assumption of 'homogeneity' for the material or tissue under study, they do make some geometrical idealisations: the sections are perfectly flat planes, they are equally spaced, and so on. The reality may be quite different. As noted in Section 2.8.2, the thickness of a histological section varies between sections, and varies from place to place within the section, in a manner which may depend on the contents of the section. Such effects may contribute substantial bias.

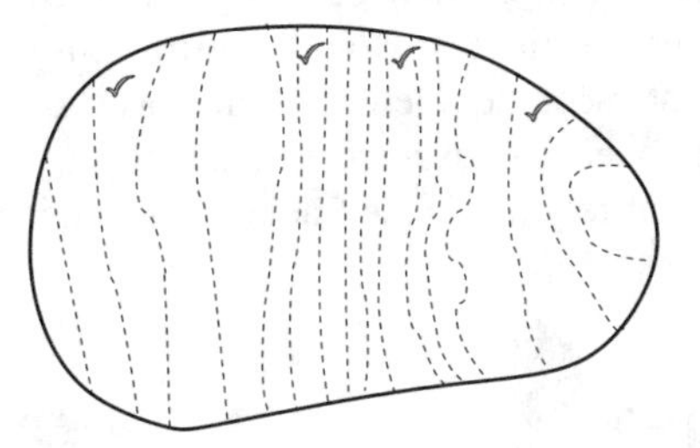

Figure 3.4 *A discrete sampling design. After the material is cut into pieces in an* arbitrary *fashion (dotted lines), the pieces are treated as sampling units, and a random sample of pieces is taken. The ticks show a systematic random sample of the pieces with period 4.*

This can be avoided by using *discrete* random sampling conditional on the cutting process, as illustrated in Figure 3.4. Let the sections be cut in any fashion at all — including irregular shape, very variable thickness, very unequal thickness, and section placement depending on the contents of the material — provided that no material is actually destroyed. After the material is cut, we treat the pieces as individual sampling units, and take a random sample of this finite population.

Figure 3.4 illustrates the particular case of a systematic random sample of the pieces, obtained by selecting one of the first 4 pieces from the left, with equal probability, then taking every fourth piece in the sequence. The selected pieces constitute our sample.

Consider any location in the material (before or after cutting); there is a probability of $1/4$ that this point will be part of the random sample, since the cut piece of material containing this point will be selected with probability $1/4$.

Thus, all parts of the material have equal chance of being represented in the sample.

This simple principle makes it possible to eliminate many sources of bias in stereological technique, especially for studying particle populations. We discuss discrete sampling techniques further in Chapter 10.

3.3.5 Counting rules

Counting profiles in 2D

Figure 3.5 sketches a two-dimensional region, perhaps a microscope section image, containing some profiles and a rectangular reference frame. We saw in Section 2.6 that there is a problem in deciding how to count or sample profiles that cross the edge of the frame. It would be erroneous to count or include all the profiles which intersect the reference frame, as shown in the left panel. It would be equally erroneous to exclude all the boundary cases and count only those profiles lying entirely inside the frame, as shown in the right panel.

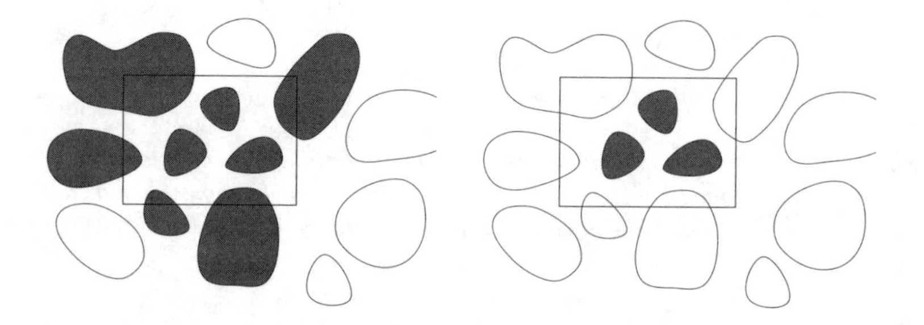

Figure 3.5 *Two biased rules for counting or sampling profiles in two dimensions. In* plus-sampling *(Left) we count or sample all profiles which intersect the reference frame. In* minus-sampling *(Right) we count or sample only those profiles which lie completely inside the reference frame.*

Miles [399] analysed this problem. The two sampling operations sketched above were dubbed *plus-sampling* and *minus-sampling* respectively. He showed that plus-sampling over-counts the number of profiles per unit area, and introduces a sampling bias: larger profiles are more likely to be hit by the reference frame. Minus-sampling, on the other hand, under-counts the number of profiles per unit area, and introduces a sampling bias in favour of smaller profiles.

In general, there are two (correct) remedies for sampling bias: we may either change the sampling rule to eliminate the bias, or we may correct the bias by reweighting the sampled items.

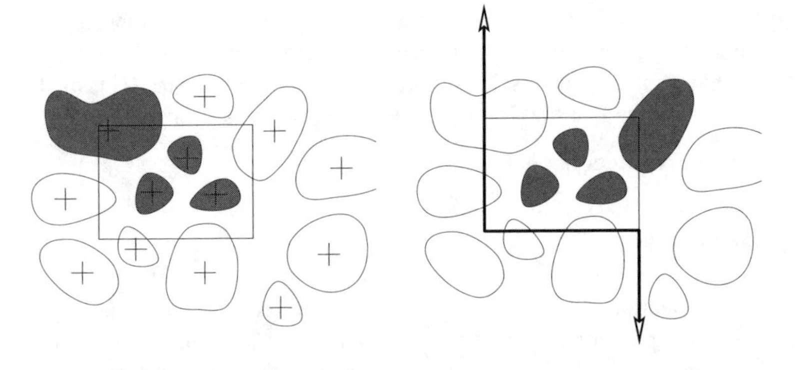

Figure 3.6 *Two unbiased sampling rules.* Left: *associated point rule.* Right: *tiling rule.*

Sampling rules which eliminate edge effect bias, for plane profiles of completely general shape, were proposed by Miles [399] and Gundersen [215, 216]. They are sketched in Figure 3.6.

The left panel shows Miles' [399] *associated point rule*. To each profile P we associate a unique point $a(P)$, for example, the centroid. We count or sample those profiles whose associated point falls in the reference frame. This is an unbiased sample of profiles. An easy way to see this is to tessellate the entire plane by copies of the reference frame; each profile will be counted in exactly one of these frames, namely the frame in which its associated point falls. Thus there is no over- or under-counting.

The right panel in Figure 3.6 shows Gundersen's *tiling rule* or 'unbiased counting frame'. A profile is counted or sampled if it intersects the rectangular frame, and does not intersect the solid boundary lines (which extend to infinity in both directions as indicated by the arrows). This is an unbiased sample of profiles.

To see that the tiling rule is an unbiased sampling rule, again imagine the entire plane tessellated by copies of the reference rectangle. Let the frames be ordered lexicographically by the x and y coordinates (in that order) of their bottom left corner. Then a profile is counted or sampled by the first frame, in this ordering, which intersects it. Since there is a unique 'first' frame, each profile is counted or sampled by exactly one frame.

The associated point rule and tiling rule require some information about profiles extending outside the reference frame. They are both suitable for applications in which a 'guard region' is visible outside the reference frame.

An alternative strategy must be employed when we can only observe what is inside the reference frame — for example when the data are a digital image of the microscope field of view. In this case, we are forced to perform minus-sampling (i.e. to take only those profiles lying inside the reference frame). Miles [399] and Lantuéjoul [335, 336] showed that the sampling bias inherent in minus sampling can be corrected by numerically re-weighting each individual sampled profile by a factor inversely proportional to its sampling bias. This method is analogous to the Horvitz-Thompson device in survey sampling [29, 270].

Further details are set out in Chapter 10.

3D counting

We saw in Section 2.3.3 that sampling bias is introduced by taking a plane section of a population of particles. Again, two ways to eliminate this bias are by changing the sampling operation, or to reweight the sampled particles.

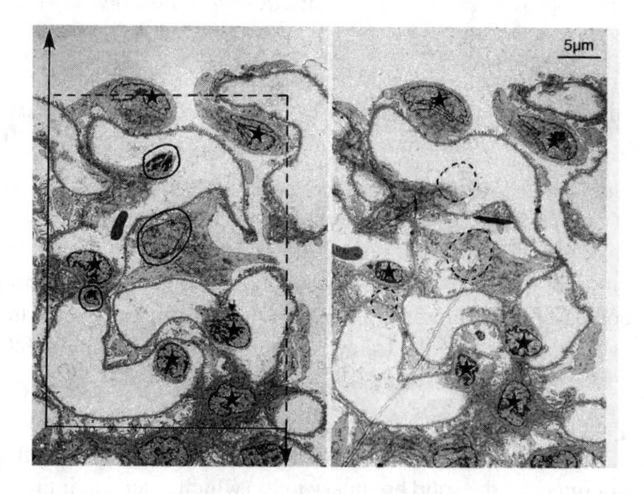

Figure 3.7 *Physical disector technique for estimating cell number. Rat kidney glomerulus, electron microscopy. Left: counting section, with reference frame for tiling rule. Right: lookup section. Six glomerular cell nuclei (marked by stars) are present in both images and not sampled. Three of the nuclei (encircled) have disappeared on the lookup section, and are therefore sampled by the disector. Reproduced from [538] by kind permission of Blackwell Publishing and D.C. Sterio. Copyright © 1984 Blackwell Publishing.*

Figure 3.7 illustrates a technique of the first kind, called the *disector* [538]. Two successive thin sections a known distance apart are examined. The number of particles that appear on one section (the 'counting section') but not on the other (the 'lookup section') is observed. If we think of the sections as horizontal planes, with the lookup section above the counting section, then this rule is equivalent to counting the number of particles whose 'top' or uppermost point lies in the intervening space between the two sections. This counts particle 'tops' without introducing a bias. Thus, to count how many tomatoes were used to make a tomato salad, we may simply count the number of tomato tops in the salad, since each tomato has a single top. This elegant and simple principle has been discovered many times in the biological and materials literature since 1895, but did not become widely known and accepted until the 1990's. See the bibliographic notes to this Chapter.

A technique of the second kind is to sample all the particles cut by a particular section, and reweight them. This, too, has a long history, dating from 1896 (see the bibliographic notes). Consider a stack of serial sections (not necessarily parallel or evenly-spaced) through a particle population. Assume we are able to recognise which profiles on two successive sections belong to the same particle. Choose one of the section planes at random with equal probability. For each particle represented on this section, find all the profiles of the same particle on the other sections. A particle which is represented on n sections should then be weighted by $1/n$. This is equivalent to the Horvitz-Thompson [270] estimator for adaptive cluster sampling [555, 556]. It yields an unbiased estimator of particle number, provided no particle is so small that it may fail to be sectioned. This elegant and effective solution was proposed in detail in the 1930's by Harvard pathologist W.R. Thompson, in a far-sighted article in the statistical journal *Biometrika* [557, 558]. It, too, was overlooked by stereologists and biologists until the 1980's when (with some variations) it became known as the *empirical method* [86, 87, 467]. Further details are discussed in Chapter 10.

3.3.6 Further developments

The close connection between stereology and survey sampling theory was further developed by Cruz-Orive, Gundersen, Jensen, Baddeley and others [29, 112, 140, 282, 293]. In spite of the central role that sampling theory has played in the development of modern stereology, there has not been much interplay between stereologists and researchers in sampling theory. Indeed many of the well-known techniques of sampling theory have been reinvented by stereologists. For example, the Horvitz-Thompson [270] principle for correction of sampling bias (see Appendix A) has a fundamental place in stereology (see [29, 96, 293]) and adaptive cluster sampling [555] has important applications, as exemplified above.

3.4 Geometrical identities

Our goal is to estimate geometrical quantities such as volume and surface area for an object in three-dimensional space, formally, a subset Y of $\mathbf{R}^3$. In order to estimate a geometrical quantity $\beta(Y)$ stereologically using a particular type of probe T, we must be able to express $\beta(Y)$ as an integral of some other quantity $\alpha(Y \cap T)$ applied to the section $Y \cap T$. There must be an identity of the form

$$\int \alpha(Y \cap T)\,\mathrm{d}T = \beta(Y), \tag{3.9}$$

where the integration ranges over all possible positions of the probe T, and $\mathrm{d}T$ is the appropriate 'uniform integration' over positions of T.

For example, we are able to estimate the volume $V(Y)$ of a three-dimensional object Y from information on two-dimensional planes T, because the volume can be expressed in terms of the areas $A(Y \cap T)$ of plane sections:

$$\int A(Y \cap T_u)\,\mathrm{d}u = V(Y), \tag{3.10}$$

where T_u denotes the horizontal plane at height u. Similarly we can estimate the surface area $S(Y)$ of a surface Y in three-dimensional space from information on two-dimensional planes T because

$$\frac{2}{\pi^2}\int\int L(Y \cap T_{\omega,p})\,\mathrm{d}p\,\mathrm{d}\omega = S(Y) \tag{3.11}$$

where L denotes length, $T_{\omega,p}$ is the plane with normal unit vector ω at a distance p from the origin, and $\mathrm{d}\omega$ denotes uniform integration over the unit sphere. Notice that in this formula, unlike (3.10), we need to integrate uniformly over all directions (orientations of the section plane) as well as over parallel displacements of the plane. Thus, the integral in (3.9) has different geometrical meanings in different contexts.

The study of representations of the form (3.9) is called *integral geometry* [500, 509, 592]. Closely related fields are geometric measure theory [177] and geometric tomography [192].

Stereology depends crucially on integral geometry. Identities like (3.10) and (3.11) provide the essential link between three-dimensional parameters and two-dimensional observations. Stereology owes its broad applicability to the fact that these identities hold under minimal regularity conditions on Y (from a practical viewpoint) so that there are effectively no restrictions on the 'shape' of Y.

The Fundamental Formulae of Stereology (2.12) are applications of identities analogous to (3.10) and (3.11). These identities all have the form

$$c_{k,m}\int V_{k+m-3}(Y \cap T)\,\mathrm{d}T = V_k(Y) \tag{3.12}$$

known as *Crofton's formula*, where Y is a set of dimension k in three-dimensional space, $V_k(Y)$ is its k–dimensional 'volume' or 'content' (V_3 is volume, V_2 is surface area and V_1 is length), while T is a plane of dimension m in space (a section plane, test line, or test point according as $m = 2, 1, 0$), and $\mathrm{d}T$ denotes integration over all positions and orientations of T. Full details are postponed to Chapter 4.

Possible combinations of the dimensions k and m in (3.12) are

$$(k,m) = (3,2), (3,1), (3,0), (2,2), (2,1), (1,2)$$

with corresponding values of the constant

$$c_{k,m} = \frac{1}{2\pi}, \frac{1}{2\pi}, \frac{1}{2\pi}, \frac{2}{\pi^2}, \frac{1}{\pi}, \frac{1}{\pi}. \tag{3.13}$$

For example, the identity (3.11) is the case $(k,m) = (2,2)$ of (3.12).

Some quantities cannot be estimated

There is a limited set of quantities β for which an identity of the form (3.9) can be derived. Stereological estimation of other quantities β can, in general, only be performed under additional assumptions about Y. The most important quantity that is inaccessible from plane sections without extra information is the number N of particles in a population.

When we slice a three-dimensional structure we lose some information that simply cannot be recovered by averaging measurements from its plane sections. A theorem of Hadwiger [233] implies that Crofton's formula (3.12) cannot be extended to any other geometrical quantities without losing some desirable properties. This was interpreted (e.g. [196]) as implying that the Fundamental Formulae of Stereology are the only possible ones: under reasonable assumptions they are the only possible stereological formulae that allow us to estimate properties of completely general shapes by averaging uniformly over locations and orientations of plane sections.

Also, it is important to realize that some two-dimensional measurements may have no direct three-dimensional interpretation. The diameters of section profiles of cells, the relative numbers of cell profiles of two types, and other *ad hoc* measures, have no simple three-dimensional meaning.

A particularly persistent urban legend, sometimes called the "breadcrust theorem", claims that the surface area of an arbitrary curved surface in $\mathbf{R}^3$ is equal to the integral of the lengths of its intersection curves with a stack of *parallel* planes — that is, the analogue of (3.11) taking the integral over p only, fixing the unit normal direction ω. This statement is false (see Section 4.1.5). Only uniform integration over all orientations will yield the surface area.

New identities

As mentioned above, it was once thought that Crofton's formula (3.12) was the only identity of the form (3.9) with practical relevance for stereology. However, many other integral geometric identities do exist, and stereological applications were eventually found for some of them.

For example, a classical result of Crofton expresses the *squared* volume of a convex set Y in $\mathbf{R}^3$ as the integral, over all straight lines T in $\mathbf{R}^3$, of the fourth power of the chord length, $L(Y \cap T)^4$, up to a constant factor. This result can be generalised to sets of arbitrary shape, and to higher moments of volume. This makes it possible to estimate moments of particle volume from plane sections of a population of particles of arbitrary shape. Such formulae arise as the result of *alternative factorisations* of known results. For example, we may regard (3.10) as a consequence of representing volume in Cartesian coordinates; instead consider the representation of volume in polar coordinates,

$$V(Y) = \int_Y r^2 \, \mathrm{d}r \, \mathrm{d}\omega \qquad (3.14)$$

where (r, ω) are the polar coordinates of a point in three dimensions. If (for simplicity) Y is a convex set containing the origin, then we may integrate (3.14) with respect to r to obtain

$$V(Y) = \frac{1}{3} \int L(Y \cap R_\omega)^3 \, \mathrm{d}\omega \qquad (3.15)$$

where R_ω denotes the ray (half-infinite line) from the origin in the direction ω, which intersects Y in a line segment of length $L(Y \cap R_\omega)$. Since the total measure of all directions is $\int \mathrm{d}\omega = 4\pi$, it follows that an unbiased estimator of the volume of Y is

$$\widehat{V}(Y) = \frac{4\pi}{3} L(Y \cap R)^3, \qquad (3.16)$$

where R is the ray drawn from an origin inside Y in a randomly-chosen direction. This estimator is the volume of a sphere of radius $L(Y \cap R)$. In the general case, when Y has arbitrary shape, a result similar to (3.15) still holds, with $L(Y \cap R)^3$ replaced by a more complicated expression. See Section 8.3.

Notice that equation (3.15) has the same general form (3.9) as other integral geometric identities, but with a different interpretation of '$\mathrm{d}T$'. In (3.15) we integrate uniformly over all possible directions ω for the ray R_ω *through the fixed origin*, while in Crofton's formula (3.12) we integrate over all orientations *and locations* of the section plane T. Stereological applications of (3.15), developed in the 1970's and 1980's, involve sampling experiments in which a line or plane is constrained to pass through a fixed point. They are discussed in Section 3.5 and in Chapters 5 and 8.

In a similar way, the representation of volume in cylindrical coordinates leads to new stereological estimators, also discussed in Chapter 8.

3.5 Geometrical probability

In order to apply the geometrical identities of Section 3.4 to stereology, we must give them a stochastic interpretation. We postulate a probability model in which a fixed set Y is intersected by a randomly-positioned section plane T, or a fixed section plane T is intersected by a random set Y. The geometrical identities give the expected values $\mathbf{E}[\alpha(Y \cap T)]$ of quantities that are observable in the plane section. This is the domain of geometrical probability [309] and its modern successor, stochastic geometry [543].

In fact, each identity of the form (3.9) has many different stochastic interpretations. Just as we may estimate a finite population total in survey sampling by using simple random sampling, systematic sampling, pps (probability proportional to size) sampling and so on, each geometrical quantity can be estimated under a variety of stereological sampling designs. In the example of volume estimation, we gave three different interpretations of (3.10) in Section 3.3, Examples 3.1–3.3.

Classical geometrical probability [309] was concerned mainly with *uniformly distributed* random sampling probes. Example 3.1 is typical. Elements of this theory are presented in Chapter 5.

3.5.1 Particle sampling

One of the main achievements of recent decades in the stereological analysis of particle systems has been the development of stereological methods for estimating number and moments of particle size that do not depend on specific assumptions about the shape of the particles [293, 538]. These methods represent an important gain in flexibility, compared to earlier methods.

One strategy is weighted sampling. Consider a reference region Z containing disjoint particles $Y_1, \dots, Y_n$ and let Y be their union,

$$Y = Y_1 \cup Y_2 \cup \dots \cup Y_n$$

with volume $V(Y) = V(Y_1) + \dots + V(Y_n)$.

Example 3.4 (Volume-weighted sampling of particles) *Take a random section plane $T = T_U$ with the area-weighted distribution based on Y, i.e. U has probability density*

$$f(u) = \frac{A(Y \cap T_u)}{V(Y)}.$$

The intersection $Y \cap T$ consists of one or more particle profiles $S_i = Y_i \cap T$. Let us select one of these profiles at random, with probability proportional to area.

That is, given T, we sample the profile $S_i = Y_i \cap T$ probability

$$\mathbf{P}[S_i \textit{ sampled} \mid T] = \frac{A(S_i)}{\sum_j A(S_j)} = \frac{A(Y_i \cap T)}{A(Y \cap T)}.$$

Then the marginal probability that we select particle i is

$$\begin{aligned}\mathbf{P}[Y_i \textit{ sampled}] &= \int \frac{A(Y_i \cap T_u)}{A(Y \cap T_u)} f(u)\,\mathrm{d}u \\ &= \int \frac{A(Y_i \cap T_u)}{A(Y \cap T_u)} \frac{A(Y \cap T_u)}{V(Y)}\,\mathrm{d}u \\ &= \frac{V(Y_i)}{V(Y)}.\end{aligned}$$

That is, this mechanism selects three-dimensional particles with probability proportional to their volume. *This holds for particles of arbitrary shape and spatial arrangement.*

By comparison, if the random section plane T has a uniform distribution as in Example 3.1, the probability of selecting a given particle is proportional to the particle height.

3.5.2 Local stereology

In some experiments it is desired to constrain the position and/or the orientation of the section plane. For example, it may be required that the section plane pass through a fixed point (a section through the nucleus or nucleolus of a cell), or contain a fixed axis (an axial section of a tube), or lie perpendicular to a fixed plane (a vertical section of a layered structure). Classical stereological methods do not apply to such sampling designs.

Prompted by the needs of users, stereologists developed new integral geometric identities for these situations. *Local stereology* [293] uses sampling designs in which the section plane or line probe must pass through a reference point, and allows stereological inference about the neighbourhood of this reference point. The model example is a biological cell with the nucleus or the nucleolus as the reference point. For $\beta =$ volume and surface area, it is possible to derive an identity of the type (3.9) with T being a section plane through the reference point. There is a rich collection of local stereological techniques, based on generalisations of the Blaschke-Petkantschin formula derived in the early 1990's [297, 616]. Special cases are the representations of volume in polar and cylindrical coordinates [220, 295].

In addition to mean particle volume and mean particle surface area, higher order moments of particle size can also be estimated using local techniques, without specific shape assumptions [284]. These techniques are most easily implemented if optical sectioning is available.

It is also possible by means of the techniques from local stereology to estimate second order properties of particle systems, such as the average number of particles lying within a given distance from a typical particle. In a model-based formulation this is the K-function from point process theory, cf. [543]. The local estimator is based on information in a slice centred at the reference point of the typical particle. In the derivation of the estimator, a general formula for hitting probabilities of random slices is used [293, Chap. 6]. The statistical properties of various estimators of second order characteristics have been compared in [546].

3.5.3 Vertical sections

A 'vertical' section is a plane section whose orientation is constrained to be perpendicular to a fixed plane (the 'horizontal'), or equivalently is constrained to be parallel to a fixed axis (the 'vertical axis') without a constraint on its location.

The vertical and horizontal directions may be special directions in the material under study (e.g. longitudinal sections of muscle tissue, sections of skin perpendicular to its surface, sections of a metal fracture surface perpendicular to the macroscopic plane of fracture) or they may be directions that are convenient for the experimental technique (e.g. sections of a large object cut perpendicular to the laboratory bench). Vertical sections are sometimes necessary in order to see or recognise the structures of interest.

Figure 3.8 shows a vertical section of the femoral growth plate of a rabbit, the section taken approximately perpendicular to the growth plate. It shows the spatial organisation of the chondrocytes (white ovals) in vertical columns, and enables an anatomist to delineate the boundaries of different regions in the growth plate.

The classical stereological formulae for estimating surface area density and length density do not apply to vertical sections. Vertical sections introduce a sampling bias dependent on the orientation of features relative to the vertical direction.

However, it is possible to correct this sampling bias. The classical formula

$$S_V = 2I_L \tag{3.17}$$

remains true if the intersection count fraction I_L is estimated from a test system of curves as shown in Figure 8.7. The curve is the *cycloid* of classical geometry.

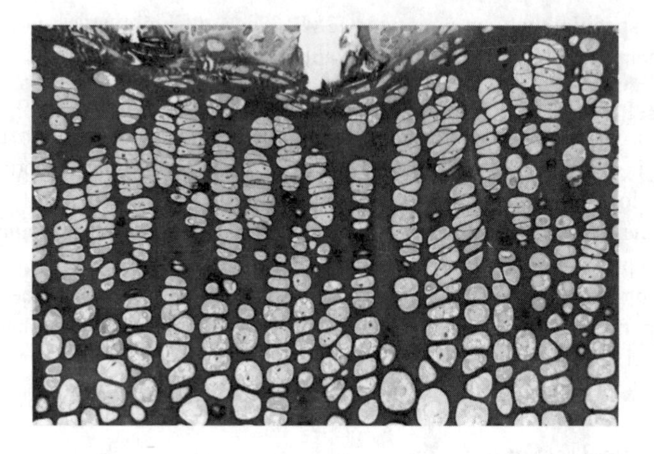

Figure 3.8 *Vertical section of rabbit femoral growth plate. Reproduced from [136] by kind permission of Blackwell Publishing and Prof. L.-M. Cruz Orive. Copyright © 1986 Blackwell Publishing.*

Methods of this kind were derived by Spektor [531] and Hilliard [260, 464] under the assumption that the material is homogeneous and is isotropic with respect to rotation around the vertical axis. It was realised much later [29, 33] that there is a design-based version, i.e. that the technique applies to an *arbitrary* surface when random vertical sections are taken according to an appropriate sampling design.

Recall the formula (3.11) on page 72 expresses surface area $S(Y)$ as the integral of the length $B(Y \cap T)$ of intersection with all section planes T. The key is to derive an alternative expression for $S(Y)$ as the integral over all *vertical* planes T of another quantity $W(Y \cap T)$. This quantity can in turn be estimated using a test system of cycloids. Further details are in Chapter 8.

3.6 Advice to consultants

This section continues our checklist of issues begun in Section 2.9.

3.6.1 Bias

Bias refers loosely to the two related concepts of sampling bias (non-constant or incorrect probability of sampling different units in the population) and estimation bias (systematic deviation of the estimated value from the true value of a quantity).

Cochran [83, chap 13] gives an overview of sources of error in sample surveys. They include noncoverage, nonresponse, measurement error, bias, and intrasample correlation. Cochran notes that "a constant [estimation] bias passes undetected" [83, pp. 380, 396] and is the most dangerous of all sources of error.

Bias is a pernicious problem in stereology for the same reasons. We saw in Section 2.9 that some quite plausible counting and sampling rules for microscopy of particles lead to an inherent sampling bias and estimation bias. It is therefore important to study experimental protocols carefully for inherent bias of this kind. Subtle fallacies about random sampling design may also lead to biased sampling protocols; see Section 5.8.

It is often said in defence of biased techniques that they may be employed to compare two experimental groups, since the difference between groups will eliminate the bias. Cochran [83, chap 13] counters this argument forcefully. It assumes that the numerical value of the bias is the same in both groups, which is very unlikely.

Students of statistics learn there is often a tradeoff between bias and variance of estimators. A biased estimate with low variability might be preferable to an unbiased estimate with high variability. However, performing this tradeoff requires that we know the size of the bias in the biased estimate. This is not currently realistic in stereology. The most dangerous aspect of bias in stereology (and in sample surveys) is that its magnitude is unknown. In current stereology we usually avoid using biased estimates, not because of a naive insistence on unbiased estimates to the detriment of their slightly biased competitors, but because (a) the size of the bias is unknown and cannot be inferred from observations, (b) bias may differ in experimental and control groups, thereby confounding the experiment, and (c) an unbiased estimate can always be made more accurate than a biased estimate, by increasing the sample size.

3.6.2 *Interpretation of ratios*

Errors are very common in the interpretation of ratio parameters such as $N_V = N(\text{cells})/V(\text{organ})$. An increase in a ratio does not imply that the numerator has increased; instead the denominator may have decreased, or both numerator and denominator may have changed. A degenerative disease process may cause cell density N_V to *increase*, simply by causing the reference volume V to decrease. It may also cause S_V to increase because of the different scaling properties of surface and volume. Examples were cited in Section 3.2.5. Similar problems occur for particle average size and particle size distributions.

3.6.3 *Variability*

Sources of variability in stereological experiments include sampling variability (at each level of the sampling design), spatial inhomogeneity of the material, and variability between replicates, such as the biological variability between individual animals.

Some physical materials in mineralogy and materials science may be called spatially homogeneous, and we may infer the average composition of the material from an arbitrary plane section. However, most structures in biology are highly organised and spatially inhomogeneous, so that sampling inferences cannot be drawn from a single arbitrary sample. Typically it is not possible to make sampling inferences about a complete organ from a biopsy.

Sampling variability is contributed by each level of the sampling design. A typical stereological experiment in biology involves extracting an organ from each animal, dividing the organ into blocks, selecting some of the blocks and cutting them into sections, selecting some of the sections and processing them for microscopy, and placing microscope fields-of-view at random locations on the sections. There must be adequate replication at each level.

Biological (inter-animal) variability implies that there is a difference between inference about an individual and about the species. Thus "the rat lung surface area" is imprecise. Current evidence suggests that inter-animal variability is quite large in comparison to sampling variability in many experiments.

Very inefficient allocation of sampling effort is common in experimental studies. Extreme examples include the intensive sampling and painstaking reconstruction of a single cell; the use of digital image analysis to measure the area of a plane section "exactly" on just one section; and the inspection of thousands of kidney glomeruli from one individual rat.

Classical stereological methods included a class of "corrections" for bias in various estimators. These corrections were derived by making geometrical assumptions about the material being sampled (for example that particles are spheres), and assuming spatial homogeneity, leading to analytic expressions for the bias, which could then be subtracted from the original biased estimator to obtain an unbiased estimator. This approach is now deprecated because it ignores all sources of variability, and rests on untenable geometrical assumptions.

3.6.4 *Accuracy*

It may appear that stereology is a low-budget, inaccurate substitute for exact three-dimensional reconstruction.

Stereology is traditionally a cheap technique, and that is its practical appeal. It exploits the fact that three-dimensional reconstruction is not necessary if the objective is to measure certain quantities such as volume.

"Exact three-dimensional reconstruction" in the context of microscopy usually refers to the reconstruction of a single tiny region (say, up to 1 mm across) within the material. This is still a sample of the material, so sampling issues are unavoidable. There is enormous variation between different parts of a biological material, and between different microscope images of the same material. The microscope's field of view covers only a tiny part of the experimental material. To draw conclusions about the entire material, statistical issues must be addressed, no matter which technique is used.

On the contrary, it is unclear whether three-dimensional imaging has reached the stage where it is accurate enough for quantitative microscopy. There are nonlinear geometric artefacts in magnetic resonance imaging at fine scales [187]. In optical confocal microscopy, measurement of displacement in the third dimension (normal to the image plane) is still problematic.

Stereological techniques can also be applied usefully in three-dimensional and other noninvasive imaging techniques to measure anatomical features *in vivo* [329, 382, 452, 483].

3.6.5 Polemics about unbiased stereology

There has been vigorous debate in some scientific forums over the validity of different methods for particle counting [46, 45, 23, 84, 194, 195, 508, 600], in particular in defence of Abercrombie's method [46]. The notions of 'unbiased estimation' and 'unbiased sampling' have become controversial in this context, so we offer the following advice for consultants.

First we recall that design-based stereological estimators are unbiased, under the right sampling conditions, *without any assumptions* about the geometry of the specimen, other than the minimal regularity conditions required for (3.9). This is a great strength of modern stereology. It is exactly analogous to the existence of unbiased estimators of population totals in finite population survey sampling. The unbiasedness of the estimator relies on the uniformity of the random sampling design, rather than on assumptions about the population.

'Unbiased stereology' is a synonym for design-based stereology introduced by some expositors [272, 427]. This is an unfortunate misnomer, since model-based methods also yield unbiased estimators under suitable conditions (including partial randomisation), not all techniques in design-based stereology do yield unbiased estimators, and not all sampling techniques considered in design-based stereology are free of sampling bias. It also confuses the concepts of unbiased estimation and unbiased sampling. The technical term 'unbiased' has been stretched well beyond its proper meaning here.

'Unbiased' is also interpreted to mean "free of assumptions" (cf. Exercise 3.4), an oversimplification of the statement that design-based inference does not depend on assumptions about the material or tissue being sampled, in contrast to the homogeneity assumption required in model-based stereology and to the contentious geometrical assumptions required for bias correction in classical stereology for particle populations.

Of course no method is entirely free of assumptions, and all methods are susceptible to additional bias from unexpected sources. The extraordinary suggestion [272, p. 6] that randomised design based methods do not require careful scrutiny and comparison with other experimental results, deserves healthy skepticism.

3.7 Exercises

3.7.1 Practical exercises

Exercise 3.1 Discuss whether the following materials might be called 'homogeneous' and 'isotropic' for stereological purposes: Emmenthaler cheese; ice cream; white bread; cooked spaghetti; mille feuilles pastry; white cabbage; broccoli.

Exercise 3.2 Cut serial sections through a banana at constant spacing 2 cm. (Note that the sections must be parallel to one another, not normal to the axis of the banana). Use Cavalieri's principle to estimate the total volume of the banana. Similarly estimate the volume of the inner flesh and the volume of the peel.

Exercise 3.3 Bake a cake containing glacé cherries in a rectangular cake tin. Cut the cake in slices of even thickness parallel to one of the rectangular faces.

(a) Use the disector principle to estimate the number of cherries in the cake.

(b) Can this technique be applied if the cake is round and we take radial slices in the traditional way?

Exercise 3.4 A stereology software package is sold with the warning that "All stereological functions are unbiased with respect to feature, shape, orientation and spatial distribution, provided sections of appropriate orientation are presented to the system." Comment on this statement.

3.7.2 Theoretical exercises

Exercise 3.5 Consider a plane section of a population of spherical particles as described in Section 2.4.2. Show that the area-weighted mean area of profiles is proportional to the volume-weighted mean volume of the spheres. [Use equation (2.33).] Compare this with Exercise 2.7.

Exercise 3.6 Use (3.11) to derive unbiased estimators of $S(Y)$ and $S(Y)/V(Z)$ in Examples 3.1 and 3.2, respectively.

3.8 Bibliographic notes

Books

Recent books on stereology include (with a focus on applications) [49, 173, 272, 328, 427, 444, 489] and (more theoretical) [262, 503].

For integral geometry, see [500, 509, 192] and for stochastic geometry [48, 388, 543]. Good general texts on sampling theory are [83, 555]. See also Appendix A.

Surveys

See the recent surveys of stereology by Cruz-Orive [124, 126, 127, 128, 582] and Gundersen et al [225, 224]. Biological scientists will also find the surveys [221, 374, 375, 376, 377, 378] succinct and useful. A good discussion of issues involved in the quantitation of materials is given by Ralph and Kurzydłowski [471].

Older surveys by statisticians and mathematicians include [28, 281, 592], [543, chap. 10]. Baddeley [30] gives a brief course in stochastic geometry.

History

Thompson [557, p. 21] used the Cavalieri method (not by name) to estimate the volume of islets of Langerhans in the pancreas. The earliest mention of Cavalieri's name in the stereological context appears to be [341].

Giger [196, 197] elucidated the geometrical basis of the 'fundamental formulae of stereology' and proved [196] that the Euler-Poincaré characteristic cannot be estimated from sections in general. Curved test systems (e.g. the Merz grid [392]) were treated by Miles [397].

Modern stereology really began with the work of P.J. Davy and R.E. Miles [146, 147, 399, 398, 401, 402, 403, 404, 408, 410]. See also [424, 364], [514, chap. VIII], [196, 197], [89, 90]. Design-based theory was further developed by L.M. Cruz-Orive, A.O. Myking and E.R. Weibel [109, 112, 119, 138, 139, 140, 141], by E.B.V. Jensen and H.J. Gundersen [282, 284, 287] and many others.

Particle number

The early history of particle counting (reviewed in [44, 251, 254]) includes some very early dates for the invention of the disector counting rule and the 'empirical method' (related to the Thompson/Cruz-Orive sample) [44, 254].

Modern stereology for particle counting began with R.E. Miles' study [399] of edge effects in counting planar profiles. There followed work by Lantuéjoul [335], Gundersen [215, 216, 217, 218, 219, 220, 538], Cruz-Orive [110, 111, 119] and others [290, 273]. A general theory of edge effects is presented by Baddeley [31].

Sampling bias due to edge effects in spatial sampling has a much older history [99, 100, 143]. See the work of Buckland *et al* [66, 65]

Controversies about unbiased methods

Editorial policy in some journals now requires the use of unbiased counting methods [88, 501]. This has not been without its critics [213]. The term 'unbiased' is seen as unfortunate because of its connotations outside the technical definition [502]. Mouton [427, p. 2] acknowledges the same problems and cautions that 'unbiased' should be used only in the technical sense. [Both [502, 427] appear to interpret bias only in the sense of estimation and not the sense of sampling bias.]

In stereological literature it is often asserted incorrectly that an unbiased estimator is consistent (e.g. [272, p. 4], [427, p. 25]).

Particle size

Modern stereology of particle size, for particles of general shape, was initiated by Miles [405, 406, 407] and developed by Jensen and Gundersen [220, 293, 283, 284, 286, 287, 295] and Cruz-Orive [114, 116, 304, 124]. See also [69]. Röthlisberger [487] has priority for the technique of point sampled intercepts [228].

Variances

Variances of stereological estimators have been studied since the 1940's. See Chapter 13 especially the Bibliographic Notes.

New sampling designs

See the Bibliographic Notes to Chapter 8 for references on vertical sections, vertical projections, vertical slices, local stereology, and the nucleator and rotator techniques.

Sampling theory foundations

Sampling theory in continuous space can be developed formally using the concepts of point processes. This was first suggested by Krickeberg [323, p. 208] and is described in detail by Reiss [476, chap. 5]. The Horvitz-Thompson principle in continuous space was described in [29, 96]. Its role in stereology was explained by Baddeley [29].

CHAPTER 4

Geometrical identities

Our ability to estimate geometrical parameters rests mainly on the *section formulae* of integral geometry [500, 510, 592]. These have the general form

$$\int \alpha(Y \cap T)\, \mathrm{d}T = \beta(Y), \tag{4.1}$$

where α and β are geometrical quantities, and Y is the spatial object of interest. The 'probe' T may be a plane, a line, a grid of parallel lines, etc. The integral is over all possible positions of T, and $\mathrm{d}T$ is the appropriate 'uniform integration' over positions of T.

Key features of these identities are, firstly, their **generality:** they hold under minimal regularity conditions on Y. Stereological estimation techniques based on these identities apply without regard for the 'shape' of the objects of interest. Secondly these identities involve **uniform integration** using the density $\mathrm{d}T$. This will later dictate the probability distribution of the random sampling probe T in stereological sampling techniques. Thirdly, these identities are **scarce:** there is a very limited set of quantities β for which such results exist. Stereological estimation of other target quantities $\beta(Y)$ (for example, diameter) is not possible in the same generality. A quantity $\alpha(Y \cap T)$ observable in the sample (for example, profile diameter) need not have a simple interpretation in terms of the geometry of Y.

This chapter presents the best known identities of the form (4.1). Two general classes of identities are the *Crofton formula*, holding when T is an m-dimensional plane, and the *kinematic formula*, holding when T is a bounded set. They are discussed in Sections 4.1 and 4.2, respectively. There is also a distinction between formulae holding for probes T of *fixed orientation* in space, and those which require *isotropic orientation*, that is, integration over all possible orientations of T.

4.1 Plane sections and line transects

Crofton's formula concerns a k-dimensional subset Y cut by m-dimensional planes T in d-dimensional space. We give a general statement of the formula in Subsection 4.1.1, then elaborate special cases in Subsections 4.1.2–4.1.5.

For line probes in two dimensions ($d = 2$, $m = 1$), Crofton's formula states that:

- the area of a plane region can be determined from the lengths of its intersections with straight lines ($k = 2$);
- the length of a curve in the plane can be determined from the number of intersection points with straight lines ($k = 1$);

see Subsection 4.1.2 for details. For plane probes in three dimensions ($d = 3$, $m = 2$),

- the volume of a solid can be determined from the areas of its plane sections ($k = 3$);
- the area of a surface in $\mathbf{R}^3$ can be determined from the lengths of its intersection curves with section planes ($k = 2$);
- the length of a curve in space can be determined from the number of intersection points it makes with section planes ($k = 1$);

see Subsection 4.1.3. For line probes in three dimensions ($d = 3$, $m = 1$),

- the volume of a solid can be determined from the lengths of its intersections with straight line probes ($k = 3$);
- the area of a surface in $\mathbf{R}^3$ can be determined from the number of times it meets a straight line probe ($k = 2$);

see Subsection 4.1.4.

The two statements about the area of a plane region and the volume of a solid have counterparts for planes of fixed orientation, which we present in Subsection 4.1.5.

All versions of the Crofton formula relate the k-dimensional 'volume' of a k-dimensional subset Y to the $m+k-d$ dimensional 'volume' of its $m+k-d$ dimensional sections $Y \cap T$. However, there are other section formulae in which the volumes are replaced by quantities of lower dimension. A case with practical interest occurs for plane sections of curved surfaces in space ($d = 3$, $k = m = 2$) where the total curvature of the section $Y \cap T$ is related to the integral of mean curvature of the surface Y. This identity is treated in Section 4.3.

4.1.1 General statement

Consider a k-dimensional subset Y in $\mathbf{R}^d$, with $0 < k \leq d$, intersected by m-dimensional section planes T, where $d - k \leq m < d$. The intersection $Y \cap T$ has dimension $m+k-d$ generically (i.e. for almost all T). Then under minimal regularity conditions on Y,

$$c_{k,m,d} \int_{m\text{–planes}} V_{m+k-d}(Y \cap T)\,\mathrm{d}T = V_k(Y). \tag{4.2}$$

where V_k denotes k-dimensional volume (Hausdorff measure) and $c_{k,m,d}$ is a known geometrical constant. The integral is over all m-dimensional planes T in $\mathbf{R}^d$, and $\mathrm{d}T$ denotes the appropriate sense of 'uniform integration' over all m-dimensional planes.

An understanding of the 'uniform density' $\mathrm{d}T$ is important. This is an analogue of Lebesgue measure for the set of all m-dimensional planes. Recall that Lebesgue measure ('volume') is the only measure on $\mathbf{R}^d$, up to a constant factor, which is unaffected by translation and rotation. Similarly the general theory of invariant measures [500, 510, 592] states that there is a unique measure $\mathrm{d}T$, up to a constant factor, on the set of all m-dimensional planes in $\mathbf{R}^d$, which is invariant under the effect of translations and rotations of $\mathbf{R}^d$. The particular form of $\mathrm{d}T$ is explained in Subsections 4.1.2–4.1.4 for each of the special cases of interest.

4.1.2 Lines in $\mathbf{R}^2$

Invariant density for lines in the plane

The position of a straight line T in the plane $\mathbf{R}^2$ may be determined by polar coordinates (θ, p). For $\theta \in [0, \pi)$ and $p \in \mathbf{R}$, define the straight line

$$T_{\theta,p} = \{(x,y) : x\cos\theta + y\sin\theta = p\}.$$

In geometrical terms, θ is the angle of incidence between the normal to the line and the x-axis, while p is the signed distance from the origin to the line. See Figure 4.1.

The appropriate uniform measure for lines in the plane is [500, p. 28]

$$\mathrm{d}T = \mathrm{d}p\ \mathrm{d}\theta. \tag{4.3}$$

This has the nice property that it is unaffected by rotations and translations of the plane. See Exercise 4.2.

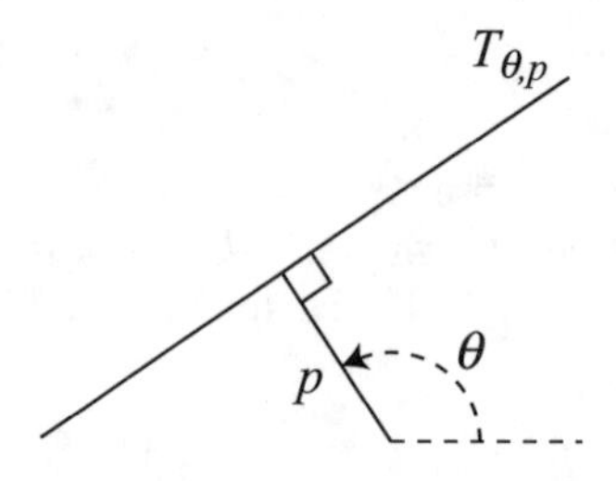

Figure 4.1 *Parametrization of lines in the plane.*

Crofton formula for lines in the plane

In this context there are two versions of the Crofton formula, namely

$$\int_0^\pi \int_{-\infty}^\infty L(Y \cap T_{\theta,p})\, \mathrm{d}p\, \mathrm{d}\theta = \pi A(Y), \tag{4.4}$$

holding for any measurable set $Y \subset \mathbf{R}^2$ with finite area, and

$$\int_0^\pi \int_{-\infty}^\infty N(C \cap T_{\theta,p})\, \mathrm{d}p\, \mathrm{d}\theta = 2L(C), \tag{4.5}$$

for any rectifiable curve $C \subset \mathbf{R}^2$. Here, A denotes area, L length and N number. The identities (4.4)–(4.5) allow us, for example, to estimate areas and lengths in a microscope image using a grid of test lines (cf. Section 2.1.2 and Exercises 7.5 and 7.10).

Proofs of these two statements are instructive. The identity for plane area (4.4) is a consequence of Fubini's theorem, since for any fixed $\theta \in [0, \pi)$,

$$\int_{-\infty}^\infty L(Y \cap T_{\theta,p})\, \mathrm{d}p = A(Y). \tag{4.6}$$

Integrating over θ yields (4.4).

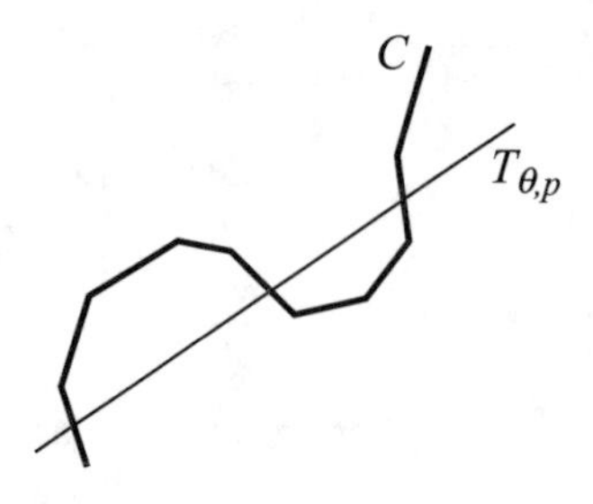

Figure 4.2 *A polygonal curve intersected by a line in the plane.*

The identity for curve length (4.5) has a nice proof using indicator variables.

Suppose C is a polygonal curve, consisting of line segments s_i with lengths ℓ_i for $i = 1, \ldots, n$. See Figure 4.2. We may express $N(C \cap T)$ as a sum of indicators:

$$N(C \cap T_{\theta,p}) = \sum_{i=1}^{n} I(s_i \cap T_{\theta,p}) \tag{4.7}$$

where $I(s_i \cap T_{\theta,p})$ is equal to 1 if $T_{\theta,p}$ intersects s_i, and 0 otherwise. The line containing s_i is an exception which we may exclude: thus (4.7) holds for almost all lines (namely for those lines which avoid the vertices of the polygonal curve).

Now for fixed θ, the range of values of p for which $T_{\theta,p}$ intersects s_i is equal to the length of the orthogonal projection of the segment s_i onto a line in the direction θ. See Figure 4.3.

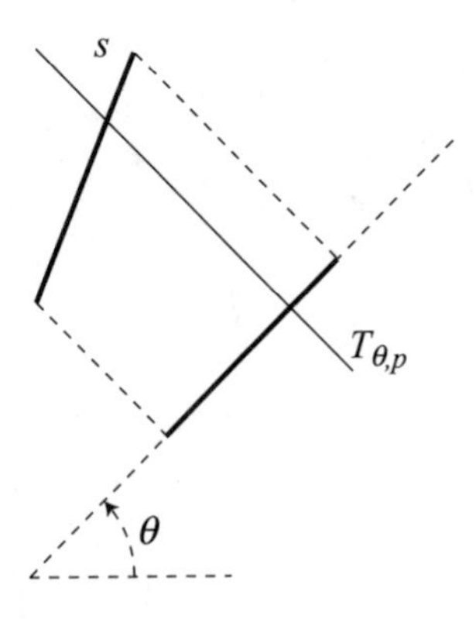

Figure 4.3 *A line segment s is intersected by a line T if and only if T intersects the projection of s onto the line perpendicular to T.*

This length is

$$\int_{-\infty}^{\infty} N(s_i \cap T_{\theta,p})\, \mathrm{d}p = \ell_i \,|\cos(\theta - \alpha)|$$

where s_i has length ℓ_i and inclination α to the x-axis. Integrating over θ,

$$\begin{aligned} \int_0^{\pi} \int_{-\infty}^{\infty} N(s_i \cap T_{\theta,p})\, \mathrm{d}p \; \mathrm{d}\theta &= \ell_i \int_0^{\pi} |\cos(\theta - \alpha)|\, \mathrm{d}\theta \\ &= 2\ell \end{aligned}$$

regardless of α. Summing over all line segments,

$$\begin{aligned} \int_0^{\pi} \int_{-\infty}^{\infty} N(C \cap T_{\theta,p})\, \mathrm{d}p\, \mathrm{d}\theta &= \sum_{i=1}^{n} 2\ell_i \\ &= 2L(C). \end{aligned}$$

Thus, (4.5) is true for polygonal curves C. The general result (4.5) holds by approximation for all rectifiable curves. Note that we must integrate uniformly over θ to obtain (4.5).

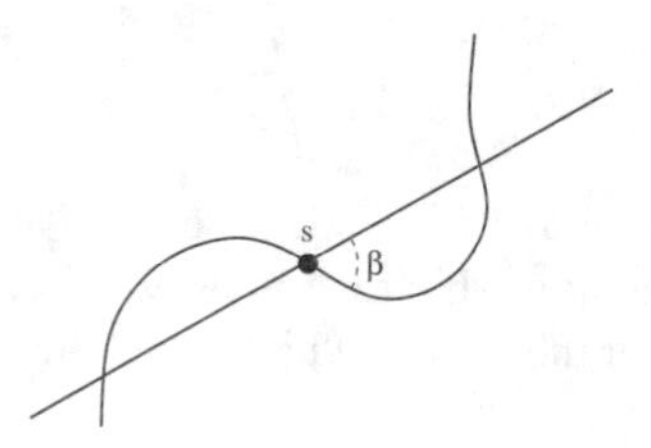

Figure 4.4 *A straight line drawn through a point s at an angle* β *to the smooth curve C.*

An alternative proof of (4.5) using calculus is also instructive. Assume that the curve C is differentiable. For any point s on the curve, and any $\beta \in [0, \pi)$, consider the straight line through s which makes an angle β with the tangent to the curve C. See Figure 4.4. This straight line is $T_{\theta,p}$ where $\theta = \tau - \beta$ and $p = x\cos\theta + y\sin\theta$, where (x, y) are cartesian coordinates of s, and $\tau = \tau(s)$ is the angle between the x-axis and the tangent to the curve at s. This is a differentiable many-to-one mapping from (s, β) coordinate pairs to (θ, p) coordinate pairs. The Jacobian of this mapping is $\sin\beta$. A change of variables gives

$$\int_C \int_0^\pi \sin\beta \, \mathrm{d}\beta \, \mathrm{d}s = \int_0^\pi \int_{-\infty}^\infty N(C \cap T_{\theta,p}) \, \mathrm{d}p \, \mathrm{d}\theta$$

and since the inner integral on the left side is equal to 2 we obtain (4.5).

4.1.3 *Planes in* $\mathbf{R}^3$

Invariant density for planes in space

The position of a two-dimensional plane T in $\mathbf{R}^3$ is determined by its unit normal vector ω and its distance s from the origin:

$$T_{\omega,s} = \{u \in \mathbf{R}^3 : u \cdot \omega = s\},$$

where $\cdot$ denotes inner product of vectors. See Figure 4.5.

It is convenient to take the unit normal ω to lie on the upper hemisphere

$$S_+^2 = \{(x, y, z) : x^2 + y^2 + z^2 = 1, z \geq 0\}$$

and to allow signed distances $s \in \mathbf{R}$. It can be shown [510] that the appropriate uniform measure for integrating over all planes is

$$\mathrm{d}T = \mathrm{d}s \, \mathrm{d}\omega \tag{4.8}$$

where $\mathrm{d}\omega$ is the usual uniform (area) measure on the hemisphere, with total value 2π. The measure $\mathrm{d}T$ is invariant under translations and rotations in $\mathbf{R}^3$.

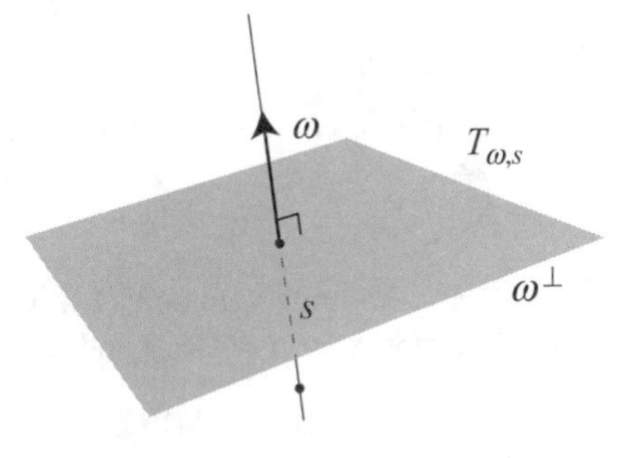

Figure 4.5 *Parametrization of planes in space.*

A direction ω in S^2_+ may be specified in terms of spherical coordinates (θ, φ),

$$\omega = (\sin\theta\cos\varphi, \sin\theta\sin\varphi, \cos\theta), \quad \theta \in [0, \frac{\pi}{2}), \quad \varphi \in [0, 2\pi), \tag{4.9}$$

see Figure 4.6. In cartographical terms, φ is the longitude (angle around the z-axis) and θ is the co-latitude (angle away from the z-axis). The area measure $d\omega$ can be expressed in spherical coordinates as

$$d\omega = \sin\theta \, d\varphi \, d\theta. \tag{4.10}$$

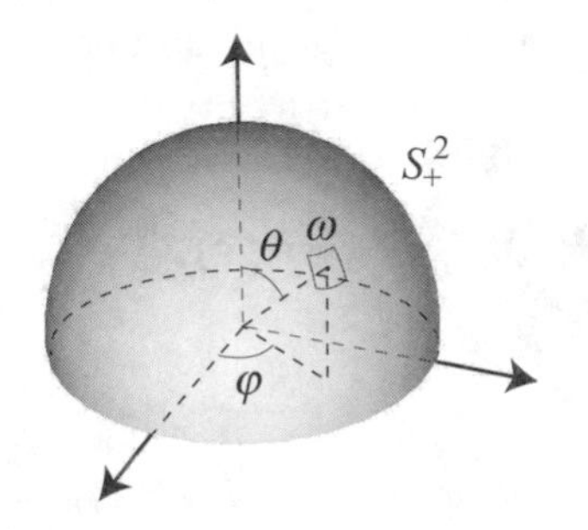

Figure 4.6 *Spherical coordinates* φ *and* θ. *The area of the infinitesimal quadrilateral is* $\sin\theta \; d\varphi d\theta$.

This representation is very important in some stereological applications. One must understand that the apparently 'nonuniform' term $\sin\theta$ in (4.10) ensures that $d\omega$ is the uniform area measure on the sphere. This may be explained as an artifact of the coordinate system or cartographical distortion. A rectangle in (longitude, latitude) coordinates corresponds to a deformed quadrilateral on the earth's surface, whose area decreases with increasing latitude (Figure 4.6).

Crofton formula for planes in space

For planes normal to a fixed direction $\omega \in S^2_+$, Fubini's theorem gives

$$\int_{-\infty}^{\infty} A(Y \cap T_{\omega,s})\,\mathrm{d}s = V(Y), \tag{4.11}$$

for any measurable set $Y \subset \mathbf{R}^3$ with finite volume $V(Y)$. Integrating uniformly over ω, we obtain Crofton's formula for $d = k = 3$ and $m = 2$,

$$\int_{\text{planes}} A(Y \cap T)\,\mathrm{d}T = 2\pi V(Y), \tag{4.12}$$

where the integral is now over all planes in $\mathbf{R}^3$. This states that the volume of a set Y (satisfying the minimal condition of measurability) can be determined by integrating over all planes T in $\mathbf{R}^3$ the area of intersection between Y and T. This obviously has direct application to stereology; see Chapters 7 and 8.

Other results are perhaps more surprising. Let Y be a two-dimensional surface in $\mathbf{R}^3$ of finite surface area $S(Y)$, satisfying certain regularity conditions. Then

$$\int_{\text{planes}} L(Y \cap T)\,\mathrm{d}T = \frac{\pi^2}{2} S(Y), \tag{4.13}$$

where $L(Y \cap T)$ is the length of the curve $Y \cap T$ of intersection between the surface Y and the plane T, cf. Figure 4.7. Applications of this formula allow the area of a surface to be statistically estimated from the lengths of plane section curves. It is not difficult to show (4.13) if Y is a union of planar elements. The general result can be obtained by approximation.

Figure 4.7 *Crofton formula in three dimensions. Left: $m = 2, k = 2$, equation (4.13). Middle: $m = 2, k = 1$, equation (4.14). Right: $m = 1, k = 2$, equation (4.17).*

Let Y be a (one-dimensional) space curve in $\mathbf{R}^3$ of finite length $L(Y)$, satisfying certain regularity and rectifiability conditions. Then

$$\int_{\text{planes}} N(Y \cap T)\,\mathrm{d}T = \pi L(Y), \tag{4.14}$$

where $N(Y \cap T)$ is the number of intersections between the curve Y and the plane T, cf. Figure 4.7. Using this formula, we can estimate the length of a curved filament from the number of crossings it makes with plane sections. A proof of (4.14) can be obtained by first considering a piecewise linear curve Y.

4.1.4 Lines in $\mathbf{R}^3$

Instead of two-dimensional plane sections, we may use probes of other types. A one-dimensional straight line probe can be used to estimate the area of a surface (Figure 4.7, middle) or the volume of a solid.

To determine the position of a one-dimensional infinite straight line T in $\mathbf{R}^3$ we specify its direction ω (a unit vector on S^2_+ parallel to T) and its vector displacement t from the origin. We will take t to be perpendicular to ω so that it lies in the two-dimensional plane $\omega^\perp$ perpendicular to ω. The line with orientation ω and position t is denoted $T_{\omega,t}$. See Figure 4.8.

The appropriate uniform measure for integration over all lines in $\mathbf{R}^3$ turns out to be

$$\mathrm{d}T = \mathrm{d}t\,\mathrm{d}\omega \tag{4.15}$$

where $\mathrm{d}\omega$ is the element of uniform measure on the hemisphere as in (4.10), and $\mathrm{d}t$ is the element of area (Lebesgue measure) on the plane $w^\perp$.

Notice that the set of all pairs (ω, t) with $t \in \omega^\perp$ is not a Cartesian product, so (4.15) only makes sense when we integrate over $t \in \omega^\perp$ with ω fixed in the innermost integral.

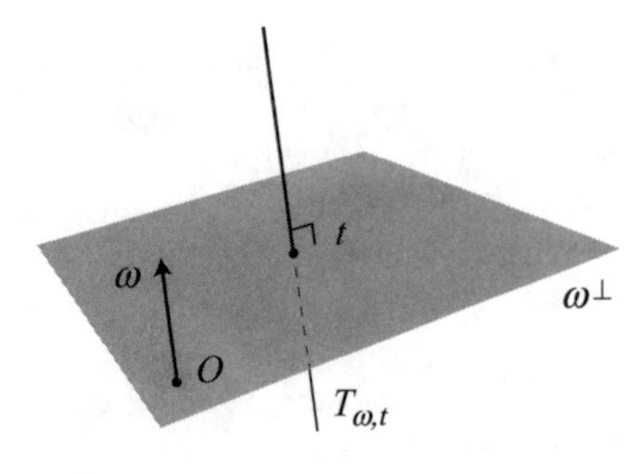

Figure 4.8 *Parametrization of lines in space.*

The volume $V(Y)$ of a measurable subset $Y \subset \mathbf{R}^3$ can be recovered from

$$\int_{\text{lines}} L(Y \cap T)\,\mathrm{d}T = 2\pi V(Y), \tag{4.16}$$

where $L(Y \cap T)$ is the length of the intersection $Y \cap T$ between the feature Y and the line probe T. This is a straightforward consequence of Fubini's theorem

$$\int_{\omega^\perp} L(Y \cap T_{\omega,t})\,\mathrm{d}t = V(Y)$$

for all ω, see also (4.18) below. Applications of (4.16) allow volumes to be statistically estimated from intersections with systems of test lines.

Let Y be a two-dimensional surface in $\mathbf{R}^3$ of finite surface area $S(Y)$, satisfying certain regularity and rectifiability conditions. Then

$$\int_{\text{lines}} N(Y \cap T)\, \mathrm{d}T = \pi S(Y), \tag{4.17}$$

where $N(Y \cap T)$ is the number of intersection points between the surface Y and the line probe T, cf. Figure 4.7. Applications of this formula allow the area of a surface to be statistically estimated from its intersections with systems of test lines. Details are given in the following chapters, in particular in Chapters 7 and 8.

4.1.5 *Fixed orientation*

A counterpart of Crofton's formula (4.2) holds for lines or planes of fixed orientation in the case $k = d$ only, that is, only when Y is a full-dimensional subset of $\mathbf{R}^d$. For example, (2.6) is the case $k = d = 3$ and $m = 2$.

Consider all m-dimensional planes T in $\mathbf{R}^d$ that are parallel to a fixed direction, say, parallel to a fixed plane T_0 through the origin. All such planes may be obtained by shifting the plane T_0

$$T = h + T_0$$

by a vector h belonging to the orthogonal complement $T_0^\perp$ of T_0. See Figure 4.9. This establishes a 1–1 correspondence between planes T parallel to T_0 and vectors h in $T_0^\perp$, a subspace of dimension $d - m$.

The fixed-orientation Crofton formula states that, for a d-dimensional subset Y of $\mathbf{R}^d$,

$$\int_{T_0^\perp} V_m(Y \cap (h + T_0))\, \mathrm{d}h = V_d(Y), \tag{4.18}$$

where $\mathrm{d}h$ denotes integration with respect to Lebesgue measure in $T_0^\perp$. This is a simple application of Fubini's theorem.

The practically relevant cases are $(k,d,m) = (3,3,2)$, $(3,3,1)$ and $(2,2,1)$. For example, the case $k = d = 3$, $m = 1$ states that the volume of a three-dimensional object Y can be recovered by integrating the length $L(Y \cap T)$ of its intersection with a one-dimensional line probe T over all parallel displacements of T.

It is important in practice that there is no counterpart of (4.18) for $k < d$, using planes of fixed orientation. Consider the urban legend (sometimes called the 'breadcrust theorem') that the surface area of a curved surface Y in $\mathbf{R}^3$ can be determined by integrating the length $L(Y \cap T)$ of intersection with all planes parallel to a fixed plane:

$$S(Y) \overset{\text{FALSE}}{=} \int_{-\infty}^{\infty} L(Y \cap (h + T_0))\, \mathrm{d}h. \tag{4.19}$$

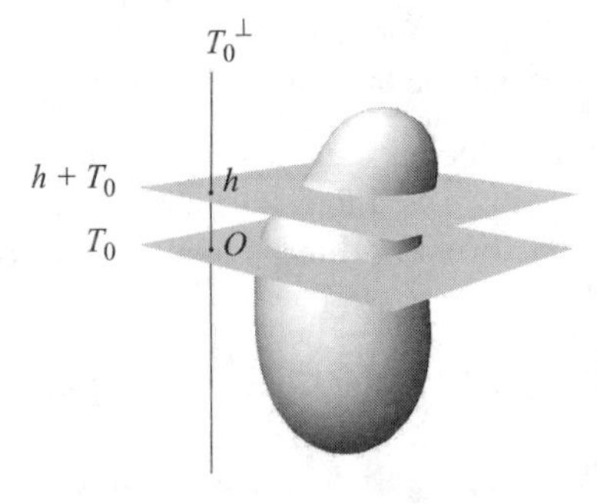

Figure 4.9 *Representation of planes of fixed orientation.*

A counterexample is provided by taking Y to be a flat surface (i.e. contained in a two-dimensional plane in $\mathbf{R}^3$), for which we can easily calculate that

$$\int_{-\infty}^{\infty} L(Y \cap (h + T_0))\, \mathrm{d}h = S(Y)\, |\sin \alpha|$$

where α is the angle between the normal to the plane of Y and the normal to T_0. For example if Y is parallel to T_0, the integral is zero. For a piecewise differentiable surface Y,

$$\int_{-\infty}^{\infty} L(Y \cap (h + T_0))\, \mathrm{d}h = \int_Y |\sin \alpha(x)|\, \mathrm{d}x$$

where $\alpha(x)$ is the angle between the unit vector normal to Y at the point x and the normal to T_0. This is strictly less than $S(Y)$ for every closed surface. Thus the 'breadcrust theorem' gives an underestimate of the surface area of any object — even a perfectly rectangular loaf of bread.

4.2 Quadrats

This section describes another type of geometric identity known as the kinematic formula. The kinematic formula also has the general form (4.1) but the probe T is now a compact set, of fixed shape and size. For example, the probe may be a rectangle of fixed dimensions but variable position and orientation.

Such probes are called 'quadrats' in stereology and often denoted by Q instead of T. This derives from the technique of 'quadrat sampling' in ecology, in which the investigator throws a rigid rectangular frame onto a meadow and examines the vegetation types visible within the frame where it lands.

Stereologists use the term 'quadrat' for a bounded set Q of any shape and any dimension, placed at a random position in space. This embraces some kinds of stereological test systems: for example a bounded array of test line segments printed on a transparent sheet can be treated as a quadrat.

The kinematic formula (or 'principal kinematic formula' in [500, 509]) expresses the k-dimensional volume of a fixed k-dimensional subset Y in $\mathbf{R}^d$ as an integral over all positions of an m-dimensional quadrat Q, provided $0 \leq d-k \leq m \leq d$. It states that

$$\int_{m\text{-quadrats}} V_{m+k-d}(Y \cap Q)\,\mathrm{d}Q = c_{k,m,d} V_k(Y) V_m(Q), \tag{4.20}$$

where the integral is with respect to all positions and orientations of Q, $\mathrm{d}Q$ is 'uniform integration' and $c_{k,m,d}$ is a known constant. See [500, eq. (15.16)]. We explain $\mathrm{d}Q$ below.

For quadrats Q of full dimension ($m = d$) or objects Y of full dimension ($k = d$) there is a counterpart of (4.20) for quadrats of fixed orientation.

4.2.1 *Kinematic density* dQ

We first explain 'uniform integration' $\mathrm{d}Q$ for a quadrat in two dimensions. Suppose Q_0 is a fixed, bounded, closed set in $\mathbf{R}^2$. Any set Q congruent to Q_0 can be obtained by first rotating and then translating Q_0:

$$Q = Q_{\theta,u} = u + R_\theta Q_0,$$

where R_θ denotes rotation by an angle $\theta \in [0, 2\pi)$ anticlockwise about the origin, and u is a vector in $\mathbf{R}^2$. See Figure 4.10.

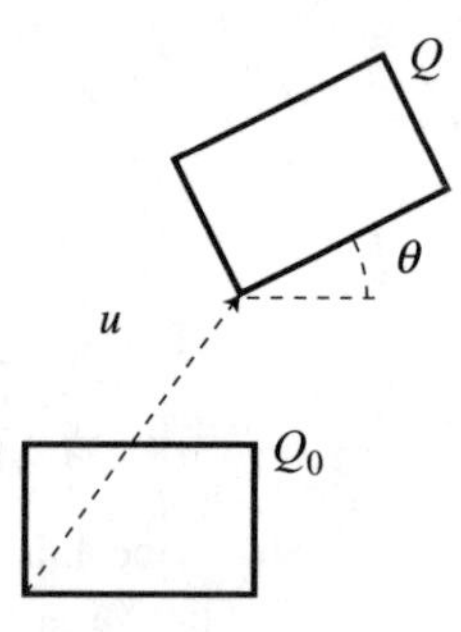

Figure 4.10 *The position of a quadrat Q in the plane is determined by a rotation θ and translation u relative to its original position Q_0.*

The appropriate uniform measure for quadrats in $\mathbf{R}^2$ is the 'kinematic density'

$$\mathrm{d}Q = \mathrm{d}u\ \mathrm{d}\theta \tag{4.21}$$

where $\mathrm{d}u$ is area (Lebesgue) measure. The term 'kinematic' comes from the fact that $\mathrm{d}Q$ represents uniform integration over all Euclidean rigid motions (rotations and translations).

A subtlety is that Q may have symmetries; for example, a rectangle is symmetric under a rotation of 180 degrees around its centre. In such cases, several different (θ, u) pairs will yield the same set Q. To avoid such complications, we should strictly define $\mathrm{d}Q$ as a measure on (θ, u) coordinates, and interpret the integral on the left side of the kinematic formula (4.20) as an integral over (θ, u) space. In geometrical terms, imagine a Cartesian coordinate system ('coordinate frame') rigidly attached to the moving set Q. Then u is the location of the origin of this coordinate frame, and θ is the angle between the x-axis in this frame and in the standard coordinate system.

In higher dimensions, a similar representation holds [500, Chap. 15]. Let Q_0 be a fixed, compact set of dimension m in $\mathbf{R}^d$ for $0 \le m \le d$. Any set Q congruent to Q_0 can be obtained by subjecting Q_0 to a rotation followed by a translation:

$$Q = u + RQ_0$$

where R is a d-dimensional rotation (a linear map represented by an orthonormal matrix with determinant +1) and u is a vector in $\mathbf{R}^d$. The kinematic measure is

$$\mathrm{d}Q = \mathrm{d}u \; \mathrm{d}R$$

where $\mathrm{d}u$ is d-dimensional Lebesgue measure and $\mathrm{d}R$ is the invariant measure for rotations. A detailed description of $\mathrm{d}R$ would occupy too much space here. See [500, pp. 197–199] and [425, 509].

4.2.2 *Case of full dimension*

Here we consider the kinematic formula (4.20) for quadrats Q of full dimension $(m = d)$ or objects Y of full dimension $(k = d)$.

In these cases there is also a counterpart of (4.20) for fixed orientations. It is convenient to discuss this first. If $Q = u + Q_0$ where Q_0 is a fixed reference quadrat, we have

$$\int_{\mathbf{R}^d} V_{m+k-d}(Y \cap (u + Q_0))\,\mathrm{d}u = V_k(Y)V_m(Q) \tag{4.22}$$

if either $m = d$ or $k = d$.

Consider the case $m = k = d$. Then, (4.22) is proved by a simple change of variables. Write the integrand of (4.22) as

$$V_{m+k-d}(Y \cap (u + Q_0)) = \int_{u+Q_0} \mathbf{1}\{x \in Y\}\,\mathrm{d}x.$$

Observe that a point x belongs to $u+Q_0$ if and only if $x-u$ belongs to Q_0. Changing variables from (x,u) to $(y,u)=(x-u,u)$ we get

$$\int_{\mathbf{R}^d}\int_{u+Q_0}\mathbf{1}\{x\in Y\}\,\mathrm{d}x\,\mathrm{d}u=\int_{\mathbf{R}^d}\int_{Q_0}\mathbf{1}\{y+u\in Y\}\,\mathrm{d}y\,\mathrm{d}u$$

since the Jacobian of the inverse transformation $(y,u)\mapsto(y+u,u)$ is 1. Exchanging the order of integration, the last expression becomes

$$\int_{Q_0}\int_{\mathbf{R}^d}\mathbf{1}\{y+u\in Y\}\,\mathrm{d}u\,\mathrm{d}y.$$

Since $y+u\in Y$ if and only if $u\in Y-y$, the inner integral is simply the volume $V_d(Y-y)$ of the shifted set $Y-y$, which is equal to $V_d(Y)$. The double integral reduces to $V_d(Q_0)V_d(Y)$. This proves (4.22) when $m=k=d$. The general case is similar.

For later use, we spell out this formula for 2-dimensional quadrats in $\mathbf{R}^2$ of fixed orientation:

$$\int_{\mathbf{R}^2}A(Y\cap(u+Q_0))\,\mathrm{d}u = A(Y)A(Q) \tag{4.23}$$

$$\int_{\mathbf{R}^2}L(Y\cap(u+Q_0))\,\mathrm{d}u = L(Y)A(Q) \tag{4.24}$$

$$\int_{\mathbf{R}^2}N(Y\cap(u+Q_0))\,\mathrm{d}u = N(Y)A(Q), \tag{4.25}$$

for objects $Y\subset\mathbf{R}^2$ of dimension $k=2$, 1 and 0, respectively. Here we have abused notation by writing $A(Q)$ for $A(Q_0)$.

The kinematic formula (4.20) is obtained from the formula for fixed orientation (4.20) by integrating both sides over all rotations $\theta\in[0,2\pi)$.

For example, for 2-dimensional quadrats in $\mathbf{R}^2$, the three formulae listed above yield

$$\int A(Y\cap Q)\,\mathrm{d}Q = 2\pi A(Y)A(Q) \tag{4.26}$$

$$\int L(Y\cap Q)\,\mathrm{d}Q = 2\pi L(Y)A(Q) \tag{4.27}$$

$$\int N(Y\cap Q)\,\mathrm{d}Q = 2\pi N(Y)A(Q). \tag{4.28}$$

4.2.3 *Curves in the plane*

Here we consider one-dimensional quadrats in the two-dimensional plane, that is, Q is a union of piecewise smooth curves in $\mathbf{R}^2$ with finite total length. The position and orientation of Q can again be specified as

$$Q=Q_{\theta,u}=u+R_\theta Q_0,$$

where Q_0 is a reference position of the curve array, R_θ is the rotation through angle $\theta \in [0, 2\pi)$ anticlockwise, and $u \in \mathbf{R}^2$ indicates translation, cf. Figure 4.11.

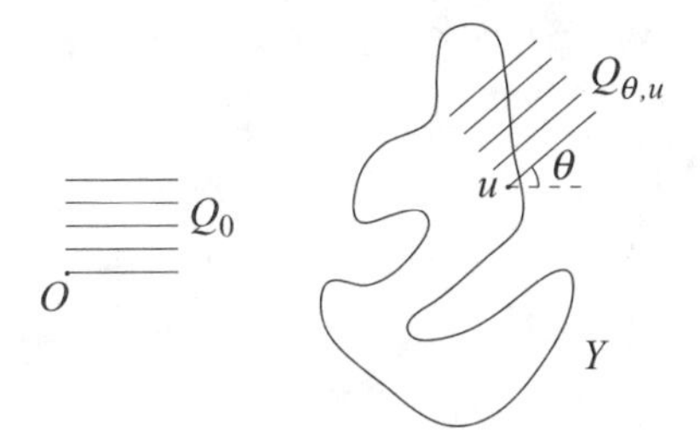

Figure 4.11 *Parametrization of positions of a 1-dimensional quadrat Q.*

In the plane, we have two versions of the kinematic formula for 1-quadrats, viz.

$$\int_0^{2\pi} \int_{\mathbf{R}^2} L(Y \cap Q_{\theta,u}) \, \mathrm{d}u \, \mathrm{d}\theta = 2\pi A(Y) L(Q) \tag{4.29}$$

for any measurable set $Y \subset \mathbf{R}^2$ with finite area, and

$$\int_0^{2\pi} \int_{\mathbf{R}^2} N(Y \cap Q_{\theta,u}) \, \mathrm{d}u \, \mathrm{d}\theta = 4 L(Y) L(Q) \tag{4.30}$$

for any rectifiable curve $Y \subset \mathbf{R}^2$.

The equation for area (4.29) is easily shown. Writing $Q_\theta = R_\theta Q_0$, we have for fixed θ

$$\begin{aligned}
&\int_{\mathbf{R}^2} L(Y \cap (u + Q_\theta)) \, \mathrm{d}u \\
&= \int_{\mathbf{R}^2} \int_{u+Q_\theta} \mathbf{1}\{z \in Y\} \, \mathrm{d}z \, \mathrm{d}u \\
&= \int_{\mathbf{R}^2} \int_{Q_\theta} \mathbf{1}\{u + v \in Y\} \, \mathrm{d}v \, \mathrm{d}u \\
&= \int_{Q_\theta} A(Y - v) \, \mathrm{d}v \\
&= A(Y) L(Q).
\end{aligned}$$

The inner integral of (4.29) has constant value $A(Y)L(Q)$ and (4.29) follows immediately.

To prove the equation for curve length (4.30) we may take the case where both Y and Q are polygonal curves. By linearity, it suffices to consider the case where both Y and Q are line segments. Now for two line segments Y and Q_0 of fixed orientation, the set of all translation vectors u under which the translated segment $u + Q$ will intersect Y, is a quadrilateral (cf. Figure 4.12) with area

$|\sin\alpha|L(Y)L(Q)$ where α is the angle between the two segments. That is, for line segments Y and Q,

$$\int_{\mathbf{R}^2} N(Y \cap (u+Q))\,\mathrm{d}u = |\sin\alpha|\,L(Y)L(Q). \tag{4.31}$$

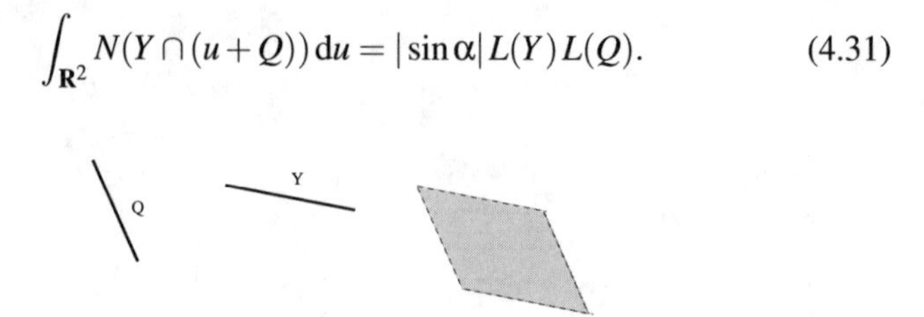

Figure 4.12 *Area of all translations of a line segment Q that intersect a line segment Y.*

Integrating over all rotations yields

$$\begin{aligned}\int_0^{2\pi}\int_{\mathbf{R}^2} N(Y \cap (u+R_\theta Q))\,\mathrm{d}u\,\mathrm{d}\theta &= L(Y)L(Q)\int_0^{2\pi}|\sin\alpha|\,\mathrm{d}\theta \\ &= 4L(Y)L(Q).\end{aligned}$$

The general result (4.30) holds by linearity and approximation. Note again that this requires uniform ('isotropic') integration over all rotations θ.

4.2.4 *Plane curves of fixed orientation*

Curves of fixed orientation in the plane can also be used in some stereological techniques. For polygonal curves Y and Q consisting of line segments $Y_1,\ldots,Y_n$ and $Q_1,\ldots,Q_m$, respectively, we get from (4.31) that

$$\int_{\mathbf{R}^2} N(Y \cap (u+Q))\,\mathrm{d}u = \sum_i\sum_j |\sin(\beta(Y_i)-\beta(Q_j))|\,L(Y_i)L(Q_j) \tag{4.32}$$

where $\beta(s)$ denotes the signed angle between the y-axis and the line segment s.

Define the 'orientation distribution' of the polygonal curve Y as the probability distribution on $[0,\pi)$ which takes the value $\beta(Y_i)$ with probability $L(Y_i)/L(Y)$. This is the probability distribution of the random angle between the y-axis and the curve Y at a random point that is uniformly distributed on Y. Then (4.32) can be rewritten

$$\int_{\mathbf{R}^2} N(Y \cap (u+Q))\,\mathrm{d}u = L(Y)\,L(Q)\,\mathbf{E}\left[|\sin(\Theta-\Psi)|\right] \tag{4.33}$$

where Θ, Ψ are independent random variables whose probability distributions are the orientation distributions of Y and of Q, respectively. This result holds more generally for piecewise differentiable curves Y and Q.

An interesting consequence of (4.33) is that we **can** determine curve length

$L(Y)$ from a system of test curves Q with fixed orientation, provided the orientation distribution of the test curves is uniform. If Ψ is uniformly distributed on $[0,\pi)$ then $\mathbf{E}[|\sin(\Theta-\Psi)|] = 1/2$ so that

$$\int_{\mathbf{R}^2} N(Y \cap (u+Q))\,\mathrm{d}u = \frac{1}{2}\,L(Y)\,L(Q). \tag{4.34}$$

For example, a test system Q consisting of circles has uniform orientation distribution, and satisfies (4.34). See Section 7.2.2.

4.2.5 *2-quadrats in* $\mathbf{R}^3$

Here we consider two-dimensional quadrats in three-dimensional space, that is, Q is a surface or union of surfaces with finite total area $S(Q)$. In principle, Q may be of any shape.

A coordinate frame may be attached to Q, with its origin at a point u in Q, such that its third axis ('z-axis') is normal to Q at the point u. The position and orientation of Q are then specified by (ω, τ, u) where $\omega \in S^2_+$ is the direction of the third axis (i.e. the normal vector to Q at u) and $\tau \in [0, 2\pi)$ is the angle of rotation of the coordinate frame around the normal (defined as the angle of rotation around the third axis which would map the first axis of the coordinate frame into a direction in the horizontal plane in standard coordinates). See Figure 4.13. The uniform measure in this situation is $\mathrm{d}Q = \mathrm{d}u\,\mathrm{d}\tau\,\mathrm{d}\omega$.

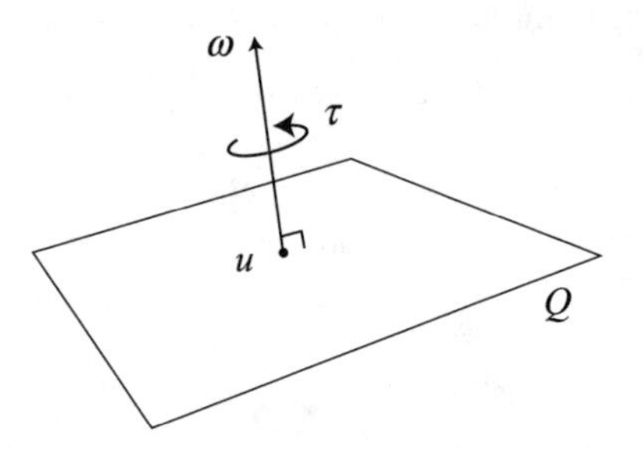

Figure 4.13 *Parametrization of 2-quadrats.*

For 2-quadrats in $\mathbf{R}^3$, the kinematic formula takes the following forms

$$\int_{\text{2-quadrats}} A(Y \cap Q)\,\mathrm{d}Q = (2\pi)^2 V(Y) A(Q) \tag{4.35}$$

$$\int_{\text{2-quadrats}} L(Y \cap Q)\,\mathrm{d}Q = \frac{\pi}{4}(2\pi)^2 S(Y) A(Q) \tag{4.36}$$

$$\int_{\text{2-quadrats}} N(Y \cap Q)\,\mathrm{d}Q = \frac{1}{2}(2\pi)^2 L(Y) A(Q), \tag{4.37}$$

depending on the dimension of Y. Proofs of these formulae are discussed in Exercise 4.7.

4.3 Curvature

Fundamental geometric identities also exist for certain quantities related to curvature.

4.3.1 Curvature quantities

We define the quantities of interest for differentiable curves and surfaces, and then give the corresponding definitions for polygonal curves and polyhedral surfaces.

Curves in the plane

Consider a differentiable curve Y in the plane. Using classical differential geometry (e.g. [532, 547]), we may approximate the curve locally (in the neighbourhood of any point s) by a quadratic. Adopting coordinates x', y' with their origin at the point s, and with the y'-axis normal to the curve, the approximating parabola is $y' = \frac{1}{2}\kappa(x')^2$ where the coefficient $\kappa = \kappa(s)$ is called the (local) *curvature* of Y at the point s.

A circle of radius r has local curvature $\kappa(s) = 1/r$ at every point on the circle. In general the local curvature $\kappa(s)$ depends on the location of the point s. If the curve is traversed by a point moving at unit speed, then $\kappa(s)$ is the instantaneous rate of rotation of the unit normal vector.

The *total* curvature of Y is

$$C(Y) = \int_Y \kappa(s)\,\mathrm{d}s \tag{4.38}$$

the integral of local curvature with respect to arc length along the curve. Equivalently $C(Y)$ is the total angle of rotation undergone by the unit normal when the curve is traversed once.

Total curvature is a dimensionless quantity, and is topologically invariant. For any simple closed curve (either differentiable or polygonal) we have $C(Y) = 2\pi$. It is also closely related to counting the number of components in a set. If Y consists of n separate components, each of which is a simple closed curve, then $C(Y) = 2\pi n$.

For polygonal curves, the corresponding definition of $C(Y)$ is the total exterior angle. The two definitions of $C(Y)$ are compatible in the sense that, under approximation of a differentiable curve Y by a sequence of polygonal curves Y_n, we will have $C(Y_n) \to C(Y)$. Similarly for approximation of a polygonal curve by differentiable curves.

Surfaces in space

Consider a surface Y in three-dimensional space which is twice continuously differentiable. The neighbourhood of any point s on the surface may be approximated by a quadratic surface. Adopting coordinates (x', y', z') with their origin at the point s, with z'-axis aligned with the outward unit normal to the surface at s, and with an appropriate alignment of the x'- and y'-axes, we may express the approximating surface as the graph of

$$z' = \frac{1}{2}\kappa_1 x'^2 + \frac{1}{2}\kappa_2 y'^2$$

where the coefficients $\kappa_1 = \kappa_1(s)$, $\kappa_2 = \kappa_2(s)$ are called the *principal curvatures*. See Figure 4.14.

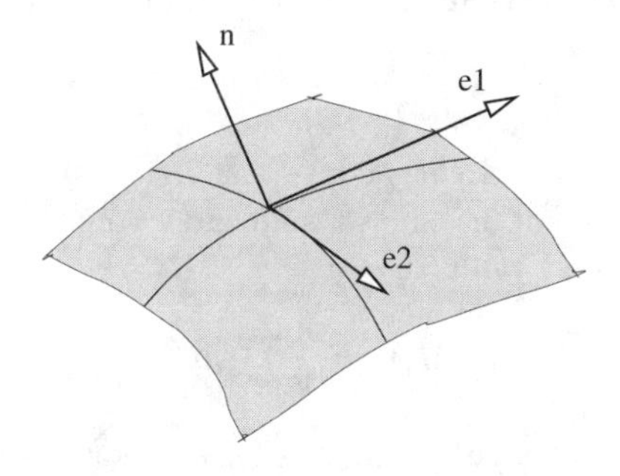

Figure 4.14 *Principal axes and local curvatures of a surface.*

The principal curvatures are functions of the location s on the surface. A sphere of radius r has principal curvatures $\kappa_1 = \kappa_2 = 1/r$ at every point. A circular cylinder of radius r has principal curvatures $\kappa_1 = 1/r$ and $\kappa_2 = 0$ at every point on its curved surface. The saddle surface $z = 5x^2 - 7y^2$ has principal curvatures $\kappa_1 = -14$ and $\kappa_2 = +10$ at the origin. A surface is locally convex (in a neighbourhood of s) if $\kappa_1, \kappa_2 \geq 0$; it is concave if $\kappa_1, \kappa_2 \leq 0$; and it is locally a saddle surface if $\kappa_1 \kappa_2 < 0$. Of course it is flat if $\kappa_1 = \kappa_2 = 0$.

Two important quantities are obtained by integrating functions of the principal curvatures. The *mean curvature* at point s is

$$m(s) = \frac{1}{2}(\kappa_1 + \kappa_2) \tag{4.39}$$

and the *Gaussian curvature* at s is

$$k(s) = \kappa_1 \kappa_2. \tag{4.40}$$

Integrating these functions over the surface Y we obtain the *integral of mean curvature*

$$M(Y) = \int_Y m(s)\,\mathrm{d}s \tag{4.41}$$

and the *integral of Gaussian curvature*

$$K(Y) = \int_Y k(s)\,\mathrm{d}s. \tag{4.42}$$

The integral of Gaussian curvature $K(Y)$ is dimensionless and is a topological invariant. For any closed convex surface Y, or any surface which may be deformed into a sphere, $K(Y) = 4\pi$.

The integral of mean curvature $M(Y)$ has dimensions length^1 and is closely associated with measures of length and diameter. For a sphere of radius r we have

$$M(Y) = \int_Y \frac{1}{r}\,\mathrm{d}s = 4\pi r^2 \times \frac{1}{r} = 4\pi r.$$

For the curved surface of a circular cylinder of radius r and length ℓ,

$$M(Y) = \int_Y \frac{1}{2r}\,\mathrm{d}s = 2\pi r\ell \times \frac{1}{2r} = \pi\ell.$$

For polyhedral surfaces Y, we may correspondingly define the total Gauss curvature $K(Y)$ as the sum of the exterior solid angles at all vertices of Y. Define $M(Y)$ for a polyhedron Y with edges $e_1, \ldots, e_n$ by

$$M(Y) = \frac{1}{2}\sum_i \ell_i \alpha_i$$

where ℓ_i is the length of edge e_i, and α_i is the exterior angle (subtended by the two faces which meet at e_i).

4.3.2 *Section formula for mean curvature*

Equation (4.14) has an analogue for the integral mean curvature M. For a surface Y in three-dimensional space (either twice-differentiable or polyhedral)

$$\int_{\text{planes}} C(Y \cap T)\,\mathrm{d}T = 2\pi M(Y), \tag{4.43}$$

where C is the total curvature of a plane curve and M is the integral of mean curvature, as defined above.

It is fairly straightforward to establish (4.43) for polyhedra. The curve of intersection $Y \cap T$ is polygonal, so $C(Y \cap T)$ is defined as the sum of exterior angles at the vertices of $Y \cap T$. Each vertex v_j is the intersection between the section plane T and an edge e_i of the polyhedral surface. The exterior angle ε_j observed in the section plane is related to the dihedral angle α_i of the polyhedral surface by $\varepsilon_j = \alpha_i \cos\beta$ where β is the angle between the section plane T and the edge e_i. We obtain

$$\int_{\text{planes}} \varepsilon_j \, dT = \pi \ell_i \alpha_i$$

yielding (4.43) by additivity. A similar argument applies in the differentiable case.

4.4 Exercises

Exercise 4.1 Consider all the lines which pass through a given point $v = (x, y)$ in the plane. Show that the invariant measure $\mu([v])$ of this set of lines is equal to zero.

Exercise 4.2 Rotation of the plane by an angle α around the origin transforms the line $T_{\theta,p}$ to $T_{\theta+\alpha,p}$. What is the effect on $T_{\theta,p}$ of a translation of the plane which maps the origin to the point (x, y)? Verify that the invariant measure $dT = dp\, d\theta$ for lines in the plane (4.3) is indeed invariant under rotations and translations.

Exercise 4.3 Verify (4.13) for a sphere Y of radius R.

Exercise 4.4 Show (4.14) for a piecewise linear curve Y.

> **Hint.** It is enough to show (4.14) for a single line segment Y of length ℓ, say, in which case (4.14) takes the form
>
> $$\int_{\text{planes}} \mathbf{1}\{Y \cap T \neq \emptyset\} \, dT = \pi \ell.$$
>
> Without loss of generality, it can be assumed that Y is parallel to the z-axis.

Exercise 4.5 Show (4.17) by combining (4.5) and (4.13) with the following decomposition result

$$\pi \, dT_1^3 = dT_1^2 \, dT_2^3,$$

where dT_1^3 is the element of the uniform measure on lines in $\mathbf{R}^3$ (Section 4.1.4), dT_2^3 is the element of the uniform measure on planes in $\mathbf{R}^3$ (Section 4.1.3) and dT_1^2 is the element of the uniform measure on lines in the plane T_2^3 (Section 4.1.2). Here, the lower index refers to the dimension of the probe and the upper index to the dimension of the containing space.

Exercise 4.6 1) Verify (4.43) for a sphere of radius R (for which the principal curvatures are $\kappa_1 \equiv \kappa_2 \equiv 1/R$).

2) Verify that, for a rectangular box Y of dimensions $a \times b \times c$, the integral of mean curvature is $M(Y) = \pi(a+b+c)$. Hence determine the measure of all planes that intersect the box.

Exercise 4.7 In this exercise, we discuss proofs of the kinematic formula for 2-quadrats in $\mathbf{R}^3$. We concentrate on (4.35) and (4.37).

1) Show (4.35) by first establishing the following result

$$\int_{\mathbf{R}^3} A(Y \cap Q_{\omega,\tau,u})\,\mathrm{d}u = V(Y)A(Q).$$

Here, it is useful to write $Q_{\omega,\tau,u} = u + Q_{\omega,\tau,0}$ and use the same techniques as described in Section 4.2.2.

2) Show (4.37) for the case where Y is a line-segment and Q is a circular disc. Then, $Q_{\omega,\tau,u}$ does not depend on τ and will be denoted $Q_{\omega,u}$. The formula (4.37) takes the form

$$\int_{S_+^2} \int_0^{2\pi} \int_{\mathbf{R}^3} \mathbf{1}\{Y \cap (u + Q_{\omega,0}) \neq \emptyset\}\,\mathrm{d}u\,\mathrm{d}\tau\,\mathrm{d}\omega$$
$$= \frac{1}{2}(2\pi)^2 L(Y)A(Q). \tag{4.44}$$

Start by showing that

$$\int_{\mathbf{R}^3} \mathbf{1}\{Y \cap (u + Q_{\omega,0}) \neq \emptyset\}\,\mathrm{d}u = V(Y \oplus Q_{\omega,0}),$$

where

$$Y \oplus Q_{\omega,0} = \{y + z : y \in Y, z \in Q_{\omega,0}\}.$$

Then, show that

$$V(Y \oplus Q_{\omega,0}) = L(p_\omega Y)A(Q),$$

where p_ω is the orthogonal projection onto a line with direction ω. Finally, show that

$$\int_{S_+^2} L(p_\omega Y)\,\mathrm{d}\omega = \pi L(Y).$$

At this final step, it can be assumed without loss of generality that Y is parallel to the z-axis. It is also useful to employ the decomposition (4.10) of the uniform measure ω in spherical coordinates.

4.5 Bibliographic notes

The subject of this chapter is properly called *integral geometry*. See Schneider and Weil [509], Santaló [500] and Gardner [192] for comprehensive accounts of this beautiful theory. The classic works of Blaschke [57] and Hadwiger [233] are also readable.

Precise regularity conditions for the validity of these identities are provided in geometric measure theory, e.g. [177]. The theory of invariant measures for geometrical elements such as lines and planes is an offshoot of the theory of Haar measures (e.g. [242, chap. XI]).

CHAPTER 5

Geometrical probability

Stereology works by reinterpreting the fundamental geometrical identities (like those of Chapter 4) in terms of a random sampling experiment. That is, we construct a probability model, in which either the probe T or the solid material Y is random. The geometrical identities then give us formulae for expected values under this probability model.

In fact, each geometrical identity has several different stochastic interpretations. In this chapter, we consider the conceptually simplest ones, in which the probe T is random, with a probability distribution analogous to the uniform distribution. This is called a **'uniformly random'** (UR) probe. Identities of the form (4.1) then give the expected value $\mathbf{E}[\alpha(Y \cap T)]$ of the 'size' of the intersection $Y \cap T$. These models were the main topic of classical geometrical probability [309, 526], because of their beautiful properties.

The acronyms IUR (isotropic, uniformly random) and FUR (fixed orientation, uniformly random) are also used, to indicate whether the orientation of the probe is random or fixed, respectively. In this chapter we elaborate the theory of FUR and IUR probes when the probe T is a section plane, line probe or quadrat.

These stereological equations are the easiest to derive, as they depend only on elementary probability concepts. They contain important lessons, especially about sampling design, and about sampling bias for particles. However, they are usually not the most useful formulae for direct application in stereology. Chapters 7 and 12 give related results which have more direct practical application.

Section 5.1 covers important properties of the uniform distribution. Sections 5.3 and 5.4 describe FUR planes and FUR quadrats respectively. Isotropic random directions are explained in Section 5.5, followed by IUR planes and IUR quadrats in Sections 5.6 and 5.7, respectively.

5.1 UR points

We start by recalling properties of the uniform distribution. A random variable U is uniformly distributed on the interval $[a,b]$, denoted $U \sim \mathsf{Unif}([a,b])$, if U has probability density

$$f_U(u) = \begin{cases} 1/(b-a) & \text{if } u \in [a,b] \\ 0 & \text{otherwise.} \end{cases}$$

A random point U in d-dimensional space may be defined as a vector $(U_1, \ldots, U_d)$ of coordinate random variables U_i. Define U to be *uniformly distributed* in a region $X \subset \mathbf{R}^d$ of finite positive volume $V_d(X)$, written $U \sim \mathsf{Unif}(X)$, if the joint probability density of the coordinates U_i is

$$f_U(u_1, \ldots, u_d) = \begin{cases} 1/V_d(X) & \text{if } (u_1, \ldots, u_d) \in X \\ 0 & \text{otherwise.} \end{cases}$$

Note carefully that the coordinates U_i are *jointly* uniform, but are not necessarily *marginally* uniform. They are not independent random variables, in general.

An exception occurs when X is a generalized rectangle with sides parallel to the coordinate axes,

$$X = [a_1, b_1] \times \cdots \times [a_d, b_d].$$

Then U is uniformly distributed in X iff $U_1, \ldots, U_d$ are independent and uniformly distributed, $U_i \sim \mathsf{Unif}([a_i, b_i])$, $i = 1, \ldots, d$. See Figure 5.1. This makes it easy to generate a UR point in a generalized rectangle. To generate UR points in other sets X, one may use special transformations (e.g. Exercise 5.4) or the rejection method (see below).

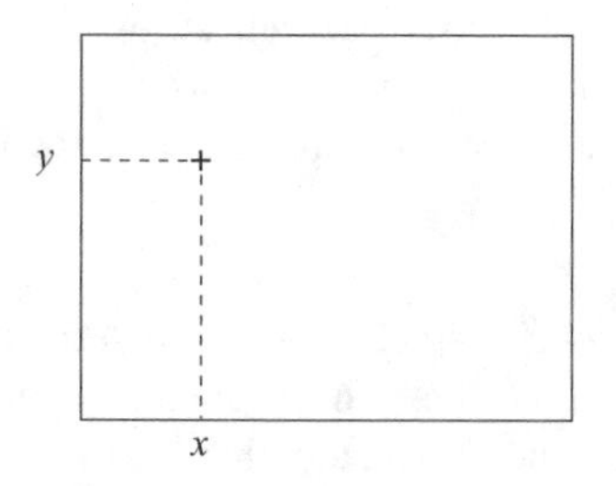

Figure 5.1 *A random point is uniformly distributed in a rectangle iff its Cartesian coordinates are independent and uniformly distributed random variables.*

A uniform random point samples space with probability *proportional to volume*: if $U \sim \mathsf{Unif}(X)$ and $Y \subset X$, then

$$\mathbf{P}[U \in Y] = V_d(Y)/V_d(X).$$

This property is in fact equivalent to the definition of the uniform distribution given above.

The uniform distribution has the following *conditional property*: if $U \sim \mathsf{Unif}(X)$ and $Y \subset X$, then the conditional distribution of U given $U \in Y$ is uniform in Y. This property is often exploited in constructing sampling designs and in analysing them.

The conditional property makes it possible to generate a UR point in any irregularly shaped set X, provided it is bounded, by *rejection sampling*. Enclose X in a larger set X^+, such as a generalized rectangle, in which it is easy to generate UR points. See Figure 5.2. Generate a sequence of random 'proposal' points $U^{(1)}, U^{(2)}, \ldots$ in X^+ which are independent and uniformly distributed in X^+. Inspect each successive proposal point $U^{(i)}$ and determine whether it falls inside X. 'Reject' any proposal point which falls outside X, and 'accept' the first proposal point which falls in X. Then the accepted point U (that is, $U = U^{(I)}$ where $I = \min\left\{i : U^{(i)} \in X^+\right\}$) is a UR point in X.

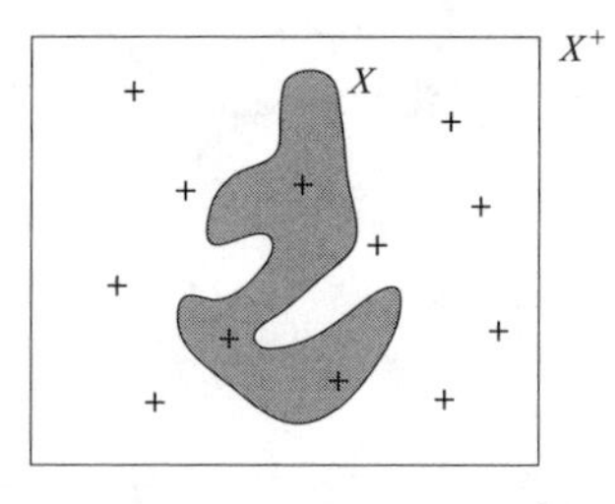

Figure 5.2 *Generation of UR points in X by the rejection method.*

The marginal and conditional probability densities of the Cartesian coordinates U_i of a UR point $U \sim \mathsf{Unif}(X)$ can be derived from the joint density by standard techniques. They have simple geometrical interpretations. For example, the marginal probability density of U_i is

$$f_i(u) = V_{d-1}(X \cap T_u(i))/V_d(X)$$

where $T_u(i)$ is the hyperplane $\left\{x \in \mathbf{R}^d : x_i = u\right\}$. A subset of coordinates $(U_i, i \in I)$ for $I \subset \{1, \ldots, d\}$ has marginal joint density

$$f_I(u) = V_{d-|I|}(X \cap T_u(I))/V_d(X)$$

where $u = (u_i, i \in I)$ and $T_u(I) = \{x \in \mathbf{R}^d : x_i = u_i, i \in I\}$. Thus, for a UR point in a three-dimensional solid X, the z coordinate has marginal density $f(u) = A(X \cap T_u(3))/V(X)$, proportional to the area of the horizontal plane section of X at height u. The x and y coordinates have marginal joint density $f(x,y) = L(X \cap T_{(x,y)}(1,2))/V(X)$ proportional to the length of the transect of X by the vertical line through $(x,y,0)$.

The Cartesian coordinates of a UR point $U \sim \mathsf{Unif}(X)$ are *conditionally uniform*, in the sense that the conditional distribution of $(U_i, i \in I)$ given $(U_j = u_j,\ j \notin I)$ is the uniform distribution on $X \cap T_{(u_j : j \notin I)}(I^c)$. For a UR point in a three-dimensional solid X, given the z coordinate, the conditional joint distribution of the (x,y) coordinates is uniform in the horizontal plane section $X \cap T_z(3)$. Given the x and y coordinates, the conditional distribution of the z coordinate is uniform in the interval $X \cap T_{(x,y)}(1,2)$.

In other coordinate systems, the coordinates will generally not be conditionally uniform. This occurs because the change of coordinates introduces a Jacobian which is not constant. This is a potential source of errors in sampling designs. See Exercise 5.4.

5.2 IUR and FUR lines in the plane

It is easiest to start by explaining the case of random lines in the two-dimensional plane. Suppose we want to generate a line at random through a domain Y in the plane $\mathbf{R}^2$. The line might be required to have a given, fixed orientation (FUR) or a random orientation (IUR). See Figure 5.3.

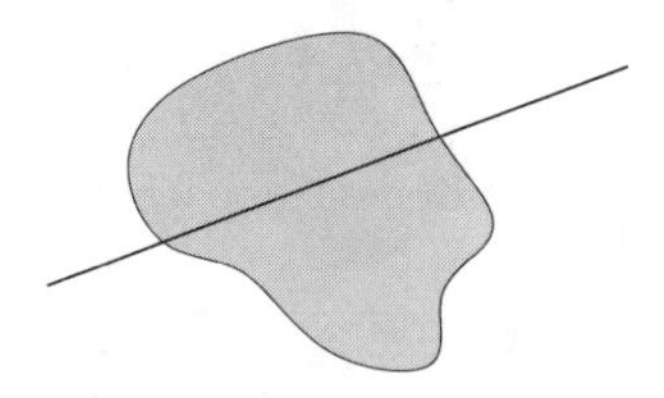

Figure 5.3 *Random line T through a domain Y.*

5.2.1 FUR lines

First we consider random lines of fixed orientation. Using the polar coordinates (θ, p) for lines (Section 4.1.2), a random line of fixed orientation θ is obtained by randomising the value of the coordinate p. In order that the random line may inherit some of the nice properties of the uniform distribution, the p coordinate should be uniformly distributed.

The appropriate range of values for p can be found by noticing that $T_{\theta,p}$ hits Y if and only if p lies in $p_\theta(Y)$, the orthogonal projection of X onto the line through the origin at an angle θ to the x-axis. Let $H_\theta(Y)$ be the length of this projected set $p_\theta(Y)$, i.e. $H_\theta(Y)$ is the length of the orthogonal projection of Y onto the line in the direction θ, variously called the 'caliper diameter' or '(projected) height' in direction θ. See Figure 5.4.

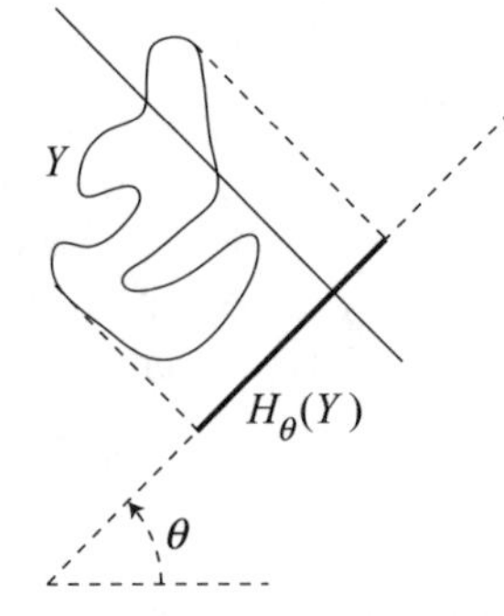

Figure 5.4 *Caliper diameter $H_\theta(Y)$ of an object Y in the plane.*

Then the random coordinate p has uniform probability density

$$f(p) = \begin{cases} \frac{1}{\mu} & \text{if } p \in p_\theta(Y) \\ 0 & \text{otherwise,} \end{cases}$$

where the normalising constant is $\mu = H_\theta(Y)$. A random line $T_{\theta,p}$ with θ fixed and p uniformly distributed as above is called a **FUR line hitting** Y.

Some geometric identities for lines of fixed orientation were discussed in Section 4.1.2. They can be translated into statements about mean values for FUR lines. Suppose T is a FUR line of orientation θ hitting a set X, and consider its intersection $Y \cap T$ with a subset $Y \subset X$. By equation (4.6) the expected length of the intersection is

$$\mathbf{E}[L(Y \cap T)] = \frac{A(Y)}{H_\theta(X)}.$$

5.2.2 *IUR lines*

Write

$$[Y] = \{T_{\theta,p} : Y \cap T_{\theta,p} \neq \emptyset\}$$

for the set of all lines that intersect (or 'hit') a given object Y in the plane. An **isotropic, uniformly random (IUR) line hitting** Y is defined as a random line $T_{\Theta,P}$ where the coordinates (Θ, P) are random variables that are *jointly* uniformly distributed on $[Y]$.

For example, if Y is a disc of radius r centred at the origin, then $T_{\theta,p}$ intersects Y if and only if $|p| \leq r$, so $[Y]$ can be identified with $[0,\pi) \times [-r,r]$. The coordinates (Θ, P) are jointly uniform in $[Y]$ if and only if Θ and P are independent and uniformly distributed, $\Theta \sim \mathsf{Unif}([0,\pi))$ and $P \sim \mathsf{Unif}([-r,r])$. In other words, we can generate an IUR line through the disc by choosing independent, uniformly distributed polar coordinates.

Note carefully that the polar coordinates Θ and P will be dependent and marginally non-uniform, in general. Recall the properties of jointly uniform random variables described in Section 5.1. For example, an IUR line through a rectangle does not have a uniformly distributed orientation Θ.

Explicit expressions can be found for the normalising constant, the invariant measure of all lines hitting Y,

$$\mu([Y]) = \int_{[Y]} \mathrm{d}p \, \mathrm{d}\theta.$$

By a similar argument to that in Section 5.2.1, we have

$$\mu([Y]) = \int_0^{\pi} H_\theta(Y) \, \mathrm{d}\theta = \pi \, \overline{H}(Y) \tag{5.1}$$

where $\overline{H}(Y)$ denotes the mean of $H_\theta(Y)$ over all θ, often termed the mean caliper diameter.

If Y is a line segment, we have already calculated (Section 4.1.2) that

$$\mu([Y]) = 2L(Y).$$

For a convex set $K \subset \mathbf{R}^2$ with non-empty interior,

$$\mu([K]) = \int_{[K]} \mathrm{d}p \, \mathrm{d}\theta = L(\partial K) \tag{5.2}$$

where ∂K is the boundary of K. This result is a simple consequence of (4.5) applied to $C = \partial K$, using the fact that almost all lines intersecting K meet ∂K in either 0 or 2 points, so that

$$\mathbf{1}\{K \cap T_{\theta,p} \neq \emptyset\} = \frac{1}{2} N(\partial K \cap T_{\theta,p}).$$

Comparing (5.1) and (5.2), it follows that the mean caliper diameter $\overline{H}(K)$ of a convex set K is equal to $L(\partial K)/\pi$.

If T is an IUR line through a set X, and if $Y \subset X$, then (a) the probability that T hits Y is

$$\mathbf{P}[T \text{ hits } Y] = \frac{\overline{H}(Y)}{\overline{H}(X)}; \tag{5.3}$$

and (b) given that T does hit Y, then T is an IUR line hitting Y.

The conditional property (b) is useful for generating an IUR line through an arbitrary set Y using the 'rejection method'. We enclose T in a disc X, then generate a sequence of independent IUR lines through X, until one of these lines actually hits Y. This line is IUR hitting Y.

In particular (5.3) has a nice form for convex sets. If T is an IUR line hitting a convex set K, and if $K' \subset K$ is also convex, then

$$\mathbf{P}\left[T \text{ hits } K'\right] = \frac{L(\partial K')}{L(\partial K)}, \tag{5.4}$$

the ratio of the perimeter lengths of the convex sets.

Integral geometric identities for lines under the invariant measure (Section 4.1.2) can be translated into mean values for IUR lines. Suppose T is an IUR line hitting X and consider the intersection $Y \cap T$ with a subset $Y \subset X$. The mean length of intersection is

$$\mathbf{E}\left[L(Y \cap T)\right] = \frac{\pi A(Y)}{\overline{H}(X)}. \tag{5.5}$$

For a rectifiable curve $Y \subset X$ the expected number of intersection points is

$$\mathbf{E}\left[N(Y \cap T)\right] = \frac{2L(Y)}{\overline{H}(X)}. \tag{5.6}$$

5.3 FUR planes

Next we consider random planes of fixed orientation in $\mathbf{R}^3$. Without loss of generality, we may take the planes to be horizontal. Write T_u for the horizontal plane with z-coordinate ('height') $u \in \mathbf{R}$. In the more elaborate notation of the previous section this was denoted $T_u(3)$. A random horizontal plane can be generated by letting the height coordinate u be a random variable U.

Typically we want to generate a random plane through an object X in three dimensions. The appropriate range of values for the random coordinate U can again be seen by noticing that T_u hits X iff $u \in p(X)$, where $p(X)$ is the orthogonal projection of X onto the z-axis. See Figure 5.5.

Therefore define a **fixed orientation uniform (FUR) random plane hitting** X to be a plane with random u-coordinate uniformly distributed in $p(X)$, i.e. the plane is distributed as T_U where U has probability density

$$f(u) = \begin{cases} \frac{1}{\mu([X])} & \text{if } X \cap T_u \neq \emptyset \\ 0 & \text{otherwise,} \end{cases}$$

The normalising constant is

$$\mu([X]) = \int_{-\infty}^{\infty} \mathbf{1}\{X \cap T_u \neq \emptyset\}\, \mathrm{d}h = L(p(X)) = H(X),$$

the projected height of X onto a line normal to the plane, also known as the caliper diameter. See equation (2.18) and Figure 2.11 in Section 2.3.2.

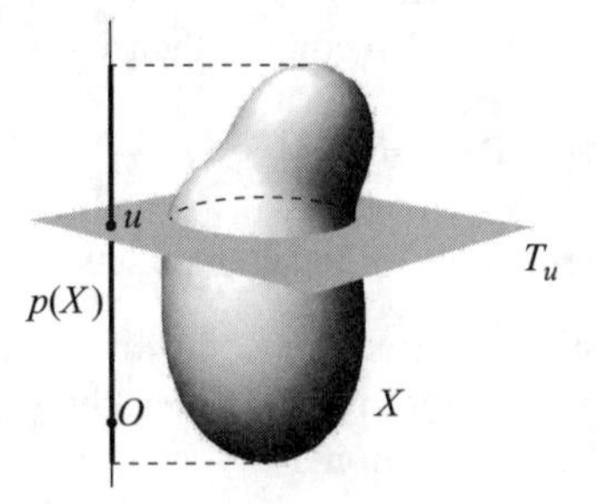

Figure 5.5 *A FUR plane T_u hitting X is generated by making its height u uniformly distributed in $p(X)$.*

A FUR plane has the following properties. If T is an FUR plane hitting X and if $Y \subset X$,

- *hitting probability:*

$$\mathbf{P}[T \text{ hits } Y] = \frac{H(Y)}{H(X)}. \tag{5.7}$$

- *conditional property:* given T hits Y, T is FUR hitting Y.

These are straightforward consequences of properties of the uniform distribution stated above.

The statement (5.7) about hitting probabilities has important consequences for stereology. For a finite population of particles $Y_1, \ldots, Y_N$ contained in X, the probability that a given particle Y_i will be hit by a FUR plane T is proportional to its height $H(Y_i)$. The sample of particles collected by taking a FUR plane section is therefore a biased sample of the population, in the sense that the particles do not have equal probability of being sampled. The sampling bias is proportional to particle height.

The conditional property is useful when generating a FUR plane hitting X, using the rejection method. Let an interval $[a,b]$ be chosen such that $p(X) \subset [a,b]$. Let $U \sim \mathsf{Unif}([a,b])$. Then, given that T_U hits X, conditionally T_U is a FUR plane hitting X.

A common fallacy of sampling designs is that a UR point in X can be selected by first generating a FUR plane T hitting X, then selecting a UR point in the plane section, $V \sim \mathsf{Unif}(X \cap T)$ conditional on T. The resulting random point V has probability density $f(v) = 1/[H(X)A(X \cap T)]$ for $v \in X$, and zero otherwise. This is clearly a nonuniform density if $A(X \cap T)$ depends on the position of T. In fact a FUR plane cannot in general be used to generate a UR point in X

by any such mechanism; a FUR plane has a uniformly distributed z-coordinate, and we know that the marginal distribution of the z-coordinate of a UR point in X is generally not uniform. See Exercise 5.9.

The volume of $Y \subset X$ can be estimated stereologically using a FUR plane section hitting X. The estimator

$$\widehat{V}(Y) = H(X)A(Y \cap T)$$

is an unbiased estimator of $V(Y)$. This is straightforward, since

$$\begin{aligned}
\mathbf{E}\widehat{V}(Y) &= H(X)\mathbf{E}[A(Y \cap T_U)] \\
&= H(X)\int_{p(X)} A(Y \cap T_u)\frac{\mathrm{d}u}{H(X)} \\
&= \int_{-\infty}^{\infty} A(Y \cap T_u)\,\mathrm{d}u \\
&= V(Y).
\end{aligned}$$

The concept of a FUR plane in $\mathbf{R}^3$ can be generalized to a FUR m-dimensional plane in $\mathbf{R}^d$. The particular case of a FUR line in $\mathbf{R}^3$ is treated in Exercise 5.8.

5.4 FUR quadrats in the plane

This section deals with random quadrats of fixed orientation (FUR quadrats) in the two-dimensional plane. The results presented below can be formulated in $\mathbf{R}^d$, but we concentrate on the planar case because it is the most important for applications.

5.4.1 Definition

Let Q_0 be any compact subset of the plane $\mathbf{R}^2$. Consider a translated copy $Q = u + Q_0$ where u is a vector in $\mathbf{R}^2$. We call Q a 'position' of the quadrat, and Q_0 its 'reference position'. The translation vector u determines the position Q. For example, if Q_0 is a rectangle with lower left corner at the origin, then $Q = u + Q_0$ has lower left corner at the point u.

Let X be a fixed set in the plane and consider all quadrat positions Q which intersect X. See Figure 5.6. The corresponding set of translations is

$$\begin{aligned}
\{u : X \cap (u + Q_0) \neq \emptyset\} &= \{u : x = u + q \text{ for some } x \in X,\ q \in Q_0\} \\
&= \{u : u = x - q \text{ for some } x \in X,\ q \in Q_0\} \\
&= X \oplus \check{Q}_0
\end{aligned}$$

where $\oplus$ denotes the Minkowski sum of two sets,

$$A \oplus B = \{a+b : a \in A, b \in B\}$$

and $\check{}$ denotes reflection through the origin

$$\check{A} = (-1)\, A = \{-a : a \in A\}.$$

In mathematical morphology, the set $X \oplus \check{Q}_0$ is called the *dilation* of X by Q_0, cf. [514, pp. 43–50], [361] and Figure 5.6.

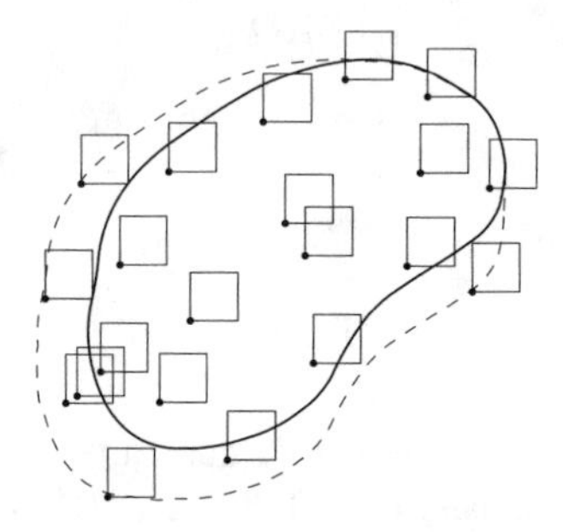

Figure 5.6 *Morphological dilation. The set of all positions of (the corner point of) the quadrat Q which intersect a fixed set X (solid lines) is equivalent to the dilation of X by Q_0 (dashed lines).*

We can then define a *fixed orientation, uniform random (FUR) quadrat hitting* X to be a random position $Q = U + Q_0$ where $U \sim \mathsf{Unif}(X \oplus \check{Q}_0)$, i.e. U is a random translation vector uniformly distributed in the dilation of X by Q_0. The probability density of U is

$$f(u) = \begin{cases} \frac{1}{\mu([X])} & \text{if } X \cap (u + Q_0) \neq \emptyset \\ 0 & \text{otherwise.} \end{cases}$$

where the normalising constant is the area of the dilation,

$$\mu([X]) = A(X \oplus \check{Q}_0).$$

The area of the dilation, $A(X \oplus \check{Q}_0)$, does not have a simple form in general, and depends on the relative orientation of the sets X and Q_0. For a convex set K and a disc $B(r)$ of radius r, *Steiner's formula* states that

$$A(K \oplus \check{B}(r)) = A(K) + rL(\partial K) + \pi r^2$$

see [500, pp. 7, 220]. What can be said in general is that

$$A(X \oplus \check{Q}_0) \geq \max\{A(X), A(Q_0)\}.$$

5.4.2 *Properties*

Many properties of FUR quadrats are inherited from the uniform distribution. Suppose Q is a FUR quadrat hitting X, and $Y \subset X$. Then we have

- **hitting probability:** Q hits Y with probability

$$\mathbf{P}[Q \text{ hits } Y] = \frac{A(Y \oplus \check{Q}_0)}{A(X \oplus \check{Q}_0)}.$$

- **conditional property:** given Q hits $Y \subset X$, Q is conditionally a FUR quadrat hitting Y.

Note that the hitting probability $\mathbf{P}[Q \text{ hits } Y]$ is the same for all translations of Y. Provided $Y' = Y + v$ is still a subset of X, we have $\mathbf{P}[Q \text{ hits } Y'] = \mathbf{P}[Q \text{ hits } Y]$. Hence we are justified in calling Q a uniformly random quadrat.

Using (4.23)–(4.25) we have the following mean value formulae for a FUR quadrat hitting X:

$$\mathbf{E}A(Y \cap Q) = \frac{A(Q_0)}{A(X \oplus \check{Q}_0)} A(Y) \tag{5.8}$$

$$\mathbf{E}L(Y \cap Q) = \frac{A(Q_0)}{A(X \oplus \check{Q}_0)} L(Y) \tag{5.9}$$

$$\mathbf{E}N(Y \cap Q) = \frac{A(Q_0)}{A(X \oplus \check{Q}_0)} N(Y) \tag{5.10}$$

for sets Y of dimension 2, 1 and 0, respectively. Unbiased estimators are

$$\widehat{A}(Y) = \frac{A(X \oplus \check{Q}_0)}{A(Q_0)} A(Y \cap Q) \tag{5.11}$$

$$\widehat{L}(Y) = \frac{A(X \oplus \check{Q}_0)}{A(Q_0)} L(Y \cap Q) \tag{5.12}$$

$$\widehat{N}(Y) = \frac{A(X \oplus \check{Q}_0)}{A(Q_0)} N(Y \cap Q). \tag{5.13}$$

5.4.3 *Alternative models*

Quadrats inside a region

An alternative is to constrain the quadrat position Q to lie strictly inside the region X. For example, if Q represents the rectangular field of view of a microscope and X is the two-dimensional section of material on the microscope slide, then constraining $Q \subset X$ ensures that the field of view lies within the material of interest.

The set of $u \in \mathbf{R}^2$ such that $Q \subset X$ is

$$\begin{aligned} \{u : u + Q_0 \subseteq X\} &= \{u : u + q \in X \text{ for all } q \in Q_0\} \\ &= (X^c \oplus \check{Q}_0)^c, \end{aligned}$$

cf. Figure 5.7. In mathematical morphology this is called the *erosion* of X by Q_0 and denoted $X \ominus \check{Q}_0$. See [514, pp. 43–50], [361]. The related concept of Minkowski subtraction is defined by $X \ominus Q_0 = (X^c \oplus Q_0)^c$.

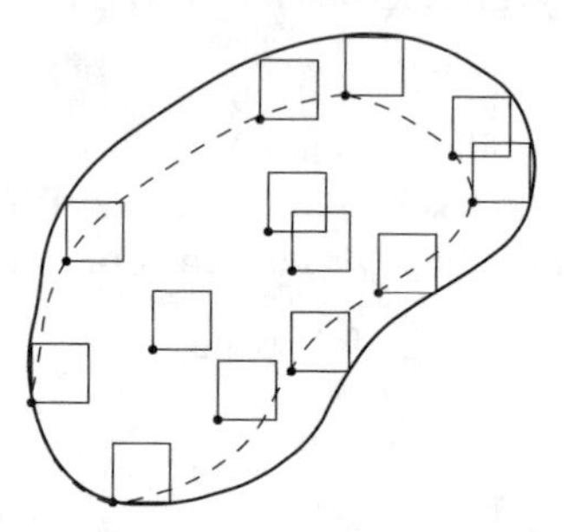

Figure 5.7 *The erosion of X by Q_0 (quadrat in reference position) is the set of positions of the quadrat such that the quadrat is contained in X.*

We may then define a *FUR quadrat Q inside X* as a random quadrat $Q = U + Q_0$ where $U \sim \mathsf{Unif}(X \ominus \check{Q}_0)$.

This model is complicated by 'edge effects'. Quantities such as $\mathbf{P}[Q \text{ hits } Y]$ and $\mathbf{E}A(Y \cap Q)$ now depend on the position of the subset Y within X. If Y is close to the boundary of X, then there are some quadrat positions Q which hit Y but do not lie inside X. Thus, for all $Y \subset X$ we can only say

$$\mathbf{P}[Q \text{ hits } Y] = \frac{A((Y \oplus \check{Q}_0) \cap (X \ominus \check{Q}_0))}{A(X \ominus \check{Q}_0)}$$

while under the stricter condition that Y avoids the edges,

$$Y \oplus \check{Q}_0 \subset X \ominus \check{Q}_0,$$

this reduces to

$$\mathbf{P}[Q \text{ hits } Y] = \frac{A(Y \oplus \check{Q}_0)}{A(X \ominus \check{Q}_0)}.$$

Hence, this model does not sample uniformly from X. The edge effects can be neglected only if

$$\frac{A(X \oplus \check{Q}_0)}{A(X \ominus \check{Q}_0)} - 1$$

is negligible.

Uniform reference point

An alternative distribution for Q, sometimes unintentionally implemented in sampling designs, is one in which the translation vector or reference point u is uniformly distributed in the target region X.

For example, if X represents the microscope section, we might randomly translate the slide so that the centre of the field of view is at a uniformly random point in X, then capture the field of view.

This design also suffers from 'edge effects' insofar as, for general $Y \subseteq X$,

$$\mathbf{P}[Q \text{ hits } Y] = \frac{A((Y \oplus \check{Q}_0) \cap X)}{A(X)};$$

if $Y \subseteq X \ominus Q_0$ then this reduces to

$$\mathbf{P}[Q \text{ hits } Y] = \frac{A(Y \oplus \check{Q}_0)}{A(X)}.$$

5.5 Isotropic directions

5.5.1 Isotropic directions in the plane

A direction in the plane can be indicated by a unit vector ω or equivalently by an angle θ with $\omega = (\cos\theta, \sin\theta)$. The direction is *isotropically random* (IR) if $\Theta \sim \mathsf{Unif}([0, 2\pi))$.

In many cases it is appropriate to identify opposite directions, so that $-\omega \equiv \omega$ and $\theta + \pi \equiv \theta$ represent 'unsensed directions' in the plane. Then the possible unit vectors ω are effectively those on the semicircle

$$S^1_+ = \{(x,y) : x^2 + y^2 = 1, y \geq 0\}$$

corresponding to angles $\theta \in [0, \pi)$. An *isotropically random* unsensed direction Ω corresponds to $\Theta \sim \mathsf{Unif}([0, \pi))$, cf. Figure 5.8.

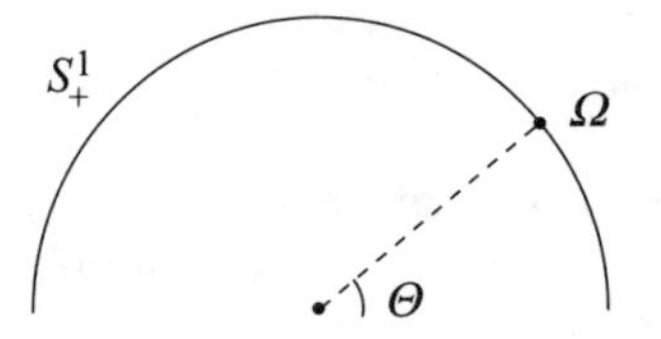

Figure 5.8 *An isotropic direction in the plane is a uniform point Ω on the semicircle S^1_+. The angle Θ is uniform in the interval* $[0, \pi)$.

An isotropic random line in the plane is a random line through O, containing an isotropic direction. In polar coordinates (Section 4.1.2) this is the line $T_{\Theta,0}$ where $\Theta \sim \mathsf{Unif}([0,\pi))$.

5.5.2 *Isotropic directions in space*

A direction in $\mathbf{R}^3$ is a unit vector ω corresponding to cartographic spherical coordinates θ, φ through

$$\omega = (\sin\theta\cos\varphi, \sin\theta\sin\varphi, \cos\theta)$$

with $\theta \in [0,\pi]$, $\varphi \in [0,2\pi)$. An *isotropic random* (IR) direction Ω is a random point uniformly distributed on the unit sphere. Its coordinates Θ, Φ are independent, with $\Phi \sim \mathsf{Unif}([0,2\pi))$ while Θ has probability density

$$f(\theta) = \frac{1}{2}\,|\sin\theta|,\ 0 \le \theta \le \pi.$$

If opposite directions are identified, so that $-\omega \equiv \omega$ and $(\pi-\theta, \varphi+\pi) \equiv (\theta, \varphi)$, then ω is effectively a point on the hemisphere

$$S^2_+ = \{(x,y,z) : x^2+y^2+z^2 = 1, z \ge 0\}$$

while $\theta \in [0, \frac{\pi}{2})$ and $\varphi \in [0, 2\pi)$. An isotropically random unsensed direction in $\mathbf{R}^3$ is a uniformly random point Ω on the hemisphere S^2_+, with independent coordinates Θ, Φ, with $\Phi \sim \mathsf{Unif}([0,2\pi))$ while Θ has probability density

$$f(\theta) = |\sin\theta|,\ 0 \le \theta \le \frac{\pi}{2}. \tag{5.14}$$

An isotropic line in $\mathbf{R}^3$ is a random line through O, containing an isotropic direction (in either sense). An isotropic plane in $\mathbf{R}^3$ is a random plane through O whose normal direction is an isotropic random direction.

An isotropic line through O in $\mathbf{R}^3$ can also be generated by a two-step procedure, by first generating an isotropic plane through O in $\mathbf{R}^3$, and then an isotropic line in this plane [500, eq. (12.54), p. 207].

5.5.3 *Isotropic planes and the 'orientator'*

An isotropic random plane through the origin in three dimensions is obtained by choosing the plane's normal vector to be uniformly distributed on the hemisphere S^2_+.

One interesting practical technique for cutting an isotropic random plane in a physical material is the 'orientator' method [373], illustrated in Figure 5.9. First an arbitrary plane C is cut in the material (panel a). The material is then placed on the laboratory bench with the plane C downward (panel b) and is rotated uniformly randomly about the vertical axis. This could be done formally by placing the object on a graduated circle, choosing a random angle $\Phi \sim \mathsf{Unif}([0, 2\pi])$ and applying the corresponding rotation. A second plane V is then cut perpendicular to the laboratory bench at an arbitrary position (panel c). The material is now placed with the plane V downward (panel d). A third and final plane T is cut perpendicular to the laboratory bench (panels e–f). The random angle Θ between V and C must have the sin-weighted probability density (5.14). With a little three-dimensional thinking it can be seen that the resulting plane T has isotropic orientation.

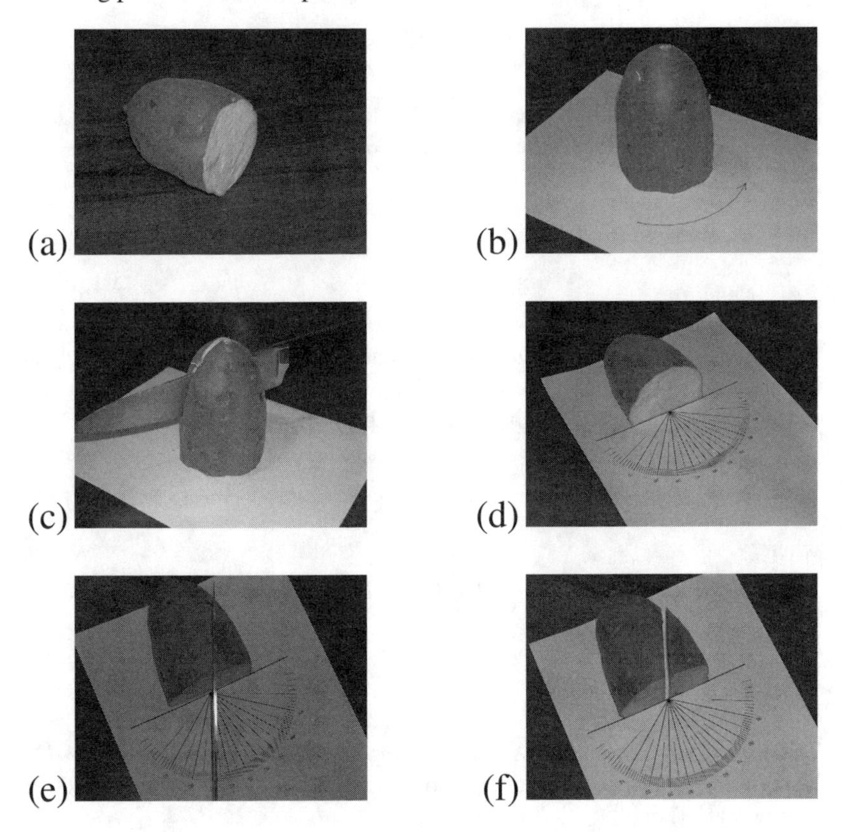

Figure 5.9 *The orientator technique applied to a yam. See text.*

To generate sin-weighted random angles in Figure 5.9, panels d–f, we used the "sin-weighted protractor" shown in Figure 5.10. This is a semicircle with markings from 1 to 100 at coordinates (x_i, y_i) where $x_i = 1 - i/200$ and $y_i = \sqrt{1 - x_i^2}$ for $i = 1, \ldots, 100$. The random angle Θ generated by selecting a uniform random number in the range $[0, 100]$ has probability density $f(\theta) = (1/2)\sin\theta$ on $[0, \pi]$.

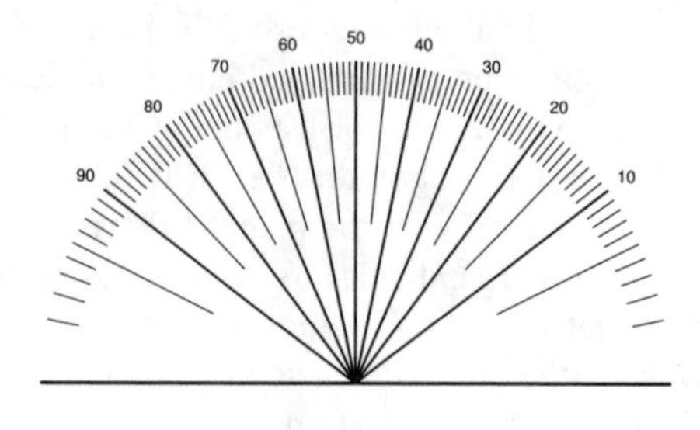

Figure 5.10 *Sin-weighted protractor. See text.*

An alternative, useful with computers, is to let $\Theta = \cos^{-1}(U)$ where $U \sim \mathsf{Unif}([-1,1])$. Then Θ has the density (5.14).

5.6 IUR planes

5.6.1 Definition

An *isotropic, uniformly random* (IUR) plane T hitting a fixed set X in $\mathbf{R}^3$ can be defined in terms of the polar coordinates (ω, s) (cf. Section 4.1.3) as the plane $T = T_{\Omega,S}$ where (Ω, S) have the joint probability density

$$f(\omega,s) = \begin{cases} \frac{1}{\mu([X])} & \text{if } X \cap T_{\omega,s} \neq \emptyset \\ 0 & \text{otherwise,} \end{cases}$$

where

$$\mu([X]) = \int_{S_+^2} \int_{-\infty}^{\infty} \mathbf{1}\{X \cap T_{\omega,s} \neq \emptyset\}\, \mathrm{d}s\, \mathrm{d}\omega.$$

5.6.2 Normalising constant

The normalizing constant $\mu([X])$ has a nice geometric interpretation. Let $p_\omega(X)$ be the orthogonal projection of X onto the line through O with direction ω, cf. Figure 5.11. Then,

$$\begin{aligned} \mu([X]) &= \int_{S_+^2} \int_{\mathbf{R}} \mathbf{1}\{X \cap T_{\omega,s} \neq \emptyset\}\, \mathrm{d}s\, \mathrm{d}\omega \\ &= \int_{S_+^2} L(p_\omega(X))\, \mathrm{d}\omega \\ &= 2\pi \bar{H}(X), \end{aligned}$$

where $\bar{H}(X)$ denotes the *mean projected height* of X, i.e. the average over all orientations of the total length of the orthogonal projection of X onto a line. For convex X, $\bar{H}(X) = M(X)/(2\pi)$ where $M(X)$ is the integral of mean curvature of X, see Section 4.1.3.

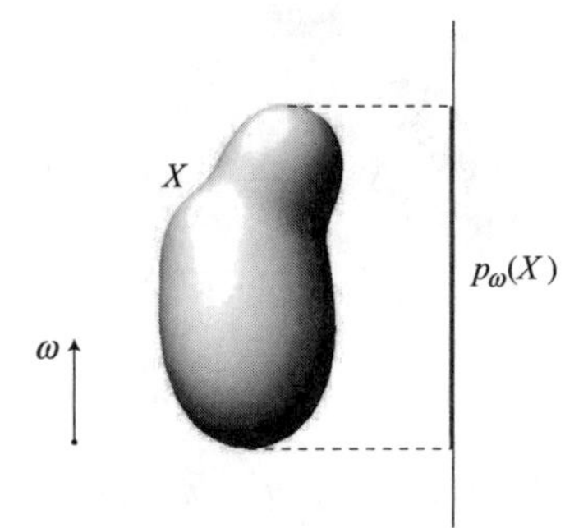

Figure 5.11 *Illustration of the orthogonal projection of X onto a line through O with direction* ω.

5.6.3 Distributional properties

For an IUR plane T hitting X, we have the following properties:

- *hitting probability:*

$$\mathbf{P}[T \text{ hits } Y] = \frac{\bar{H}(Y)}{\bar{H}(X)}, \tag{5.15}$$

- *conditional property:* given T hits Y, T is an IUR plane hitting Y.

The result (5.15) implies that the probability of observing a particle Y_i from a particle population on an IUR plane is proportional to its mean linear projection $\bar{H}(Y_i)$.

Note carefully that the coordinates (Ω, S) are *jointly* uniform. They are not *marginally* uniform, nor independent, in general. Statements about the conditional and marginal distributions of the coordinates (Ω, S) carry over from Section 5.1. In particular, an IUR plane does *not* usually have a marginally uniformly distributed orientation ω. Indeed the marginal density of ω with respect to the uniform measure on S_+^2 is

$$\frac{1}{2\pi\bar{H}(X)} L(p_\omega(X)), \quad \omega \in S_+^2.$$

Thus, the term 'isotropic, uniformly random' may be slightly misleading, in that an IUR plane is not marginally isotropic.

Figure 5.12 *Enclosing an arbitrary object Y in a sphere X to generate IUR planes through Y.*

The conditional distribution of signed distance s given orientation ω is uniform over $p_\omega(X)$. That is, an IUR plane conditioned on its orientation becomes an FUR plane.

An important practical application of the conditional property is that we may generate an IUR plane through an arbitrary object Y by enclosing the object in a sphere X and generating IUR planes through the sphere until we hit Y. See Figure 5.12. Note again that the orientation of the resulting section is not isotropic in general, because of the effect of conditioning.

A closely related technique, for generating IUR serial sections, is the 'isector' method [441] in which a small piece of biological tissue is embedded in a spherical mould before sectioning. See Section 12.4.1.

5.6.4 Section formulae

The Crofton formula gives that

$$\int_{\text{planes}} \gamma(Y \cap T)\,\mathrm{d}T = 2\pi\beta(Y),$$

where corresponding values of (β, γ) are

$$(\beta,\gamma) = (V,A), (S, \frac{4}{\pi}L), (L, 2N), \tag{5.16}$$

see also (4.12), (4.13) and (4.14). It follows that if T is an IUR plane hitting X, then for $Y \subseteq X$

$$\mathbf{E}\gamma(Y \cap T) = \frac{\beta(Y)}{\bar{H}(X)}.$$

An unbiased estimator of $\beta(Y)$ is therefore

$$\widehat{\beta}(Y) = \bar{H}(X)\gamma(Y \cap T)$$

assuming $\bar{H}(X)$ is known.

5.7 IUR quadrats

We will concentrate on the case of 2-dimensional quadrats Q in $\mathbf{R}^3$. We will assume that the position and orientation of Q can be specified by the coordinates (ω, τ, u), as explained in Section 4.2.5 and Figure 4.13.

We will assume that the density of (ω, τ, u) with respect to $du\, d\tau\, d\omega$ is

$$f(\omega, \tau, u) = \begin{cases} \frac{1}{\mu([X])} & \text{if } X \cap Q_{\omega,\tau,u} \neq \emptyset \\ 0 & \text{otherwise.} \end{cases} \tag{5.17}$$

The corresponding random plane probe is called an IUR quadrat or an IUR bounded plane probe hitting X. When Q is IUR hitting X, the probability that Q hits Y does not depend on the position of Y inside X and given Q hits Y, Q is IUR hitting Y.

Using (4.35)–(4.37), we get for an IUR quadrat Q

$$\mathbf{E}\gamma(Y \cap Q) = \frac{(2\pi)^2}{\mu([X])} A(Q_0)\beta(Y), \tag{5.18}$$

where corresponding values of (β, γ) are given in (5.16). We can then unbiasedly estimate $\beta(Y)$ by

$$\widehat{\beta}(Y) = \frac{1}{(2\pi)^2} \frac{\mu([X])}{A(Q_0)} \gamma(Y \cap Q).$$

5.8 Advice to consultants

Subtle fallacies about probability may lead to biased sampling protocols. Suppose it is required to choose a sampling location at random, uniformly distributed in a region of two- or three-dimensional space. The coordinates identifying this location must be *jointly* uniformly distributed; they are generally not *marginally* uniform or independent.

There are many instances of this. For example, suppose we want to choose a location at random in a potato. We slice the potato thinly, choose one of the slices at random with equal probability, and choose a point at random in this slice with uniform probability density in the slice. This is *not* a valid way to select a point with uniform probability density throughout the potato, as we saw in Section 5.3. See Exercise 5.9.

A fairly common error in sampling design is to generate a single IUR plane by first choosing an isotropic random direction in space and then placing the plane perpendicular to this direction at a random position. An IUR plane hitting a specimen X does *not* have isotropic random orientation, in general. See Section 5.6.3 and Exercise 5.7.

5.9 Exercises

Practical exercises

Exercise 5.1 Draw five independent IUR lines across (a) a discarded compact disc; (b) a business card or credit card.

Exercise 5.2 Implement the orientator technique on a potato.

Exercise 5.3 Discuss the methodological problems which may arise in sampling a two-dimensional region if the experimental technique requires us to place the microscope field-of-view entirely inside the region.

Theoretical exercises

Exercise 5.4 Suppose U is a uniformly random point in a disc in the two-dimensional plane. Say $U \sim \mathsf{Unif}(X)$ where X is the disc of radius s centred at the origin in $\mathbf{R}^2$. Let R and Θ be the polar coordinates of U, i.e. $U = (R\cos\Theta, R\sin\Theta)$.

1. show that R and Θ are independent, Θ is uniformly distributed, and R^2 is uniformly distributed, by
   (a) calculating the joint probability density of R and Θ using the representation of area in polar coordinates $\mathrm{d}x\,\mathrm{d}y = r\,\mathrm{d}r\,\mathrm{d}\theta$; **or**
   (b) using the fact that a UR point samples the plane with probability proportional to area.
2. verify that the following algorithm generates a UR point U in the disc X. If V_1, V_2 are independent and uniformly distributed in $[0,1]$, let the point $U = (U_1, U_2)$ be determined by $U_1 = s\sqrt{V_1}\cos(V_2)$ and $U_2 = s\sqrt{V_1}\sin(V_2)$.

Exercise 5.5 Let U be a UR point in any region $X \subset \mathbf{R}^2$ with positive finite area. Let (R, Θ) be the polar coordinates of U.

1. Show that the joint density of (R, Θ) is

$$f_{(R,\Theta)}(r,\theta) = \frac{r}{A(X)}\mathbf{1}\{(r\cos\theta, r\sin\theta) \in X\}.$$

2. Hence show that the marginal density of R is

$$f_R(r) = \frac{rL(X_r)}{A(X)}, \quad r \geq 0,$$

where

$$X_r = \{\theta \in [0, 2\pi) : (r\cos\theta, r\sin\theta) \in X\}.$$

The marginal distribution of R is therefore non-uniform, unless $L(X_r) \propto r^{-1}$.

3. In particular, R is not uniformly distributed in the case where X is a circular disc. See Figure 5.13. For example, a sampling protocol for the iris or retina of the eye, which places the microscope field of view at a uniformly random distance from the centre of the eye, is not uniformly random.

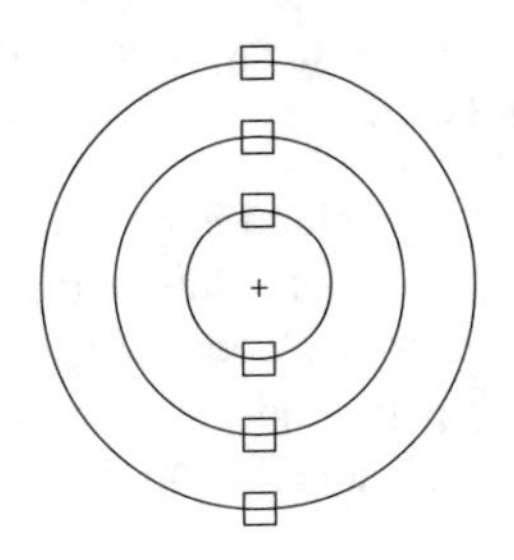

Figure 5.13 *A UR point in a disc does not have a uniformly distributed distance from the centre.*

4. Find an expression for the marginal density of Θ. Show that Θ is uniform if X is a circular disc with center at the origin. In general, Θ is not uniform. It follows that a sampling scheme which begins by making the polar coordinate θ uniformly distributed, cannot yield a uniformly random point. See Figure 5.14.

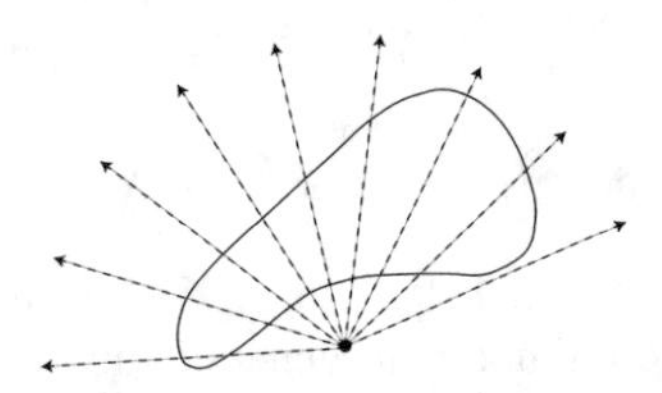

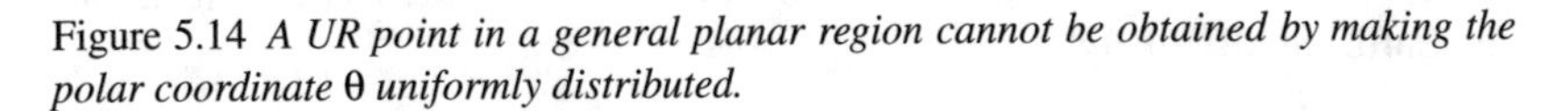

Figure 5.14 *A UR point in a general planar region cannot be obtained by making the polar coordinate θ uniformly distributed.*

Exercise 5.6 Suppose T is an IUR line intersecting a line segment X in the two-dimensional plane. Show that the angle of incidence between T and X is not uniformly distributed on $[0, \pi)$. Find its distribution and explain why it is not uniformly distributed.

Exercise 5.7 An experimenter wishes to generate an IUR plane through a rectangular box B, using software. The lab's programmer reads Section 5.6.3 but decides that the rejection method (Figure 5.12) would be wasteful. Instead, the programmer's code first chooses an isotropic random orientation ω in three dimensions, then chooses a uniformly random coordinate s in the range of values that lead to an intersection with the box ($T_{\omega,s} \cap B \neq \emptyset$), giving a random plane $T_{\omega,s}$. Explain why this code is incorrect.

Exercise 5.8 A FUR line T_U hitting a set X in three-dimensional space is a random line with fixed orientation and uniform random position U chosen among those positions for which T_U hits X. Develop a theory for FUR lines in three dimensions, corresponding to the theory for FUR planes presented in Section 5.3. In particular, this involves finding an expression for the normalizing constant of the density and deriving the hitting probability and the conditional property.

Exercise 5.9 A patient's adrenal gland is removed surgically and cut end-to-end into parallel slices of constant thickness 100 microns. A lab technician selects one of the slices at random with equal chance, and sends it to the pathology lab. Within this slice, the pathologist chooses a point uniformly at random, and investigates the tissue at this location under a microscope at high magnification.

Explain why this sampling design does **not** select a location inside the pituitary gland with uniform probability. In what way is it biased? How can this be fixed?

Exercise 5.10 Let X and Y be solid spheres in $\mathbf{R}^3$ with centres at the origin O and radius R and r, respectively, $R \geq r$. Let us consider N independent FUR planes hitting X and let $\widehat{V}_i(Y)$ and $\widehat{V}_i(X)$ be the volume estimates based on the ith section. Find, by simulation, the distribution of $\frac{1}{N}\sum_{i=1}^{N} \widehat{V}_i(Y)$, of $\frac{1}{N}\sum_{i=1}^{N} \widehat{V}_i(X)$ and of

$$\sum_{i=1}^{N} \widehat{V}_i(Y) / \sum_{i=1}^{N} \widehat{V}_i(X)$$

for $N = 1$, 10 and 100 and selected values of r/R.

Exercise 5.11 A well-travelled stereologist decides to choose a holiday destination at random. The method is to generate random longitude and latitude coordinates which are independent and uniformly distributed. Are all locations on Earth equally likely?

Exercise 5.12 This exercise concerns generation of an IUR plane hitting a bounded subset $X \subset \mathbf{R}^3$. We suppose that X is contained in a ball $B_3(O,R)$ centred at the origin O of radius R.

1) Show that if $T_{\omega,s}$ is an IUR plane hitting $B_3(O,R)$, then ω and s are independent, ω has density $\frac{1}{2\pi}$ with respect to $d\omega$ and s is uniform in $[-R,R]$.

2) Suppose $T_{\omega,s}$ is an IUR plane hitting $B_3(O,R)$. Show that, given $T_{\omega,s}$ hits X, then $T_{\omega,s}$ is an IUR plane hitting X.

3) Give practical procedures for generating ω and s with the distribution mentioned in 1).

Exercise 5.13 Which of the following is true for an IUR plane through a cube?

(a) the orientation of the plane is marginally isotropic;

(b) the plane is conditionally FUR, given its orientation;

(c) the location and orientation are jointly uniform.

Exercise 5.14 An incorrect simulation algorithm for generating FUR quadrats runs as follows. Let X and Q_0 be bounded sets. Generate independent random vectors $A \sim \mathsf{Unif}(X)$ and $B \sim \mathsf{Unif}(Q_0)$, then calculate $U = A - B$, then output $Q = U + Q_0$. Prove that this is invalid, i.e. that U is not uniformly distributed in $X \oplus \check{Q}_0$.

5.10 Bibliographic notes

Classical geometrical probability originated in the eighteenth century with the famous 'needle problem' and related problems of Buffon [68]. Important early work includes [37, 142, 157]. The monographs of Kendall and Moran [309] and Solomon [526] are very accessible introductions, along with some chapters of [500]. Literature reviews on geometrical probability include [421, 423, 345, 24]. Geometrical probability is now viewed as part of stochastic geometry [48, 299, 388, 543].

The abbreviations IUR and FUR are due to Miles and Davy.

CHAPTER 6

Statistical formulations of stereology

Statistical inference in stereology needs to be formulated carefully, as R.E. Miles [401] first demonstrated. Miles distinguished three kinds of statistical model for a stereological experiment, and showed that the choice of model has important practical consequences: it affects the meaning of parameters, the appropriateness of sampling techniques, the validity of stereological estimators, and the nature of inferences.

This chapter describes a general template for a statistical model of a stereological experiment, developed principally by Miles, Cruz-Orive and Weibel. We give a few illustrative examples; more detailed examples are laid out in the chapters to follow.

Section 6.1 introduces general terms; Section 6.2 defines the three basic types of statistical model. Parameters are defined in Section 6.3. Sections 6.5–6.7 explain the basis for statistical inference in each of the three types of model. Finally Section 6.4 describes some modifications of the basic theory.

6.1 Terminology

Miles [401] introduced the following generic description of a stereological experiment, sketched in Figure 6.1.

The **'feature of interest'** Y is the microscopic structure which is our primary focus of study. It must be a well-defined subset of three-dimensional space (although its size and shape are not known).

For example, in a study of normal human lung, the feature of interest might be the alveolar gas exchange surface. This must be unambiguously defined in anatomical terms (in *three* dimensions, not on plane sections) and would usually be idealised as a curved two-dimensional surface (consisting of many disconnected pieces of surface) in three-dimensional space.

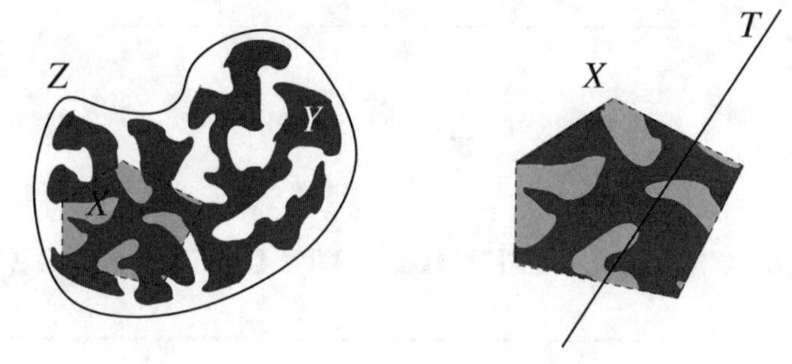

Figure 6.1 *Miles' schematic description of a stereological experiment.* Left: *the feature of interest Y is contained in the reference space Z. The specimen X is taken from Z.* Right: *We generate the probe T through the specimen X and observe its intersection with Y inside X.*

The **'reference space'** Z is a well-defined region of space which serves three purposes:* it delimits the feature of interest; it contains the feature of interest; and the size of Y may be expressed with reference to the size of Z.

For example, in a study of neurons in the neocortex of the human brain, the feature of interest Y may be the union of all neurons in the neocortex, and the reference space Z may be the entire neocortex. One quantity of interest may be the numerical density $N_V = N(Y)/V(Z)$ of neurons per unit volume of neocortex. Since there are neurons in the hippocampus as well, this study is unambiguous and reproducible only if we have clearly defined the boundary between neocortex and hippocampus in order to delimit Y ('neurons in neocortex') and to define Z ('neocortex').

The **specimen** X is the experimental material initially available to us. It may be the whole reference space Z, or it may be a proper subset of Z. For example, a pathologist studying human liver Z might investigate the entire liver from a cadaver ($X = Z$), or might obtain a needle biopsy of liver tissue from a clinical patient ($X \subset Z$). The specimen X contains a subset $Y \cap X$ of the feature of interest. The relation between X and Z has a strong influence on the choice of stereological technique, as we discuss below.

Finally, we generate a **probe** T intersecting X and we are able to observe the intersections of the probe with X and $Y \cap X$. The probe may be a section plane, a line probe, a stack of parallel section planes, an array of test lines, etc. The observable information is the pair of sets $(Y \cap X \cap T, X \cap T)$ or some measurements performed on them.

* Refinements of the concept of a reference space are discussed in Section 6.4.

6.2 Generic models

The very strong analogy between stereology and survey sampling was noted in Section 3.2.

In the general theory of sample surveys there are two broad approaches, depending on whether sampling variation arises from the random selection of the sample (a *design-based* or 'randomisation' approach) or from intrinsic randomness in the population (a *model-based* or 'superpopulation' approach).

In stereology, things are slightly more complicated, because of the two-stage process in which we first take the specimen X, then generate the probe T through X. Miles [401] therefore distinguishes three kinds of inference:

'Restricted deterministic case': the specimen X available for examination is the whole reference space Z. The feature Y and reference space Z are non-random, arbitrary, bounded sets which are the main objects of interest (e.g. $X = Z$ is a whole human kidney, and Y is a structure within the kidney). The probe T is randomised by the experimental sampling procedure. Statistical inference relies on the randomness of the sampling design. See Figure 6.2.

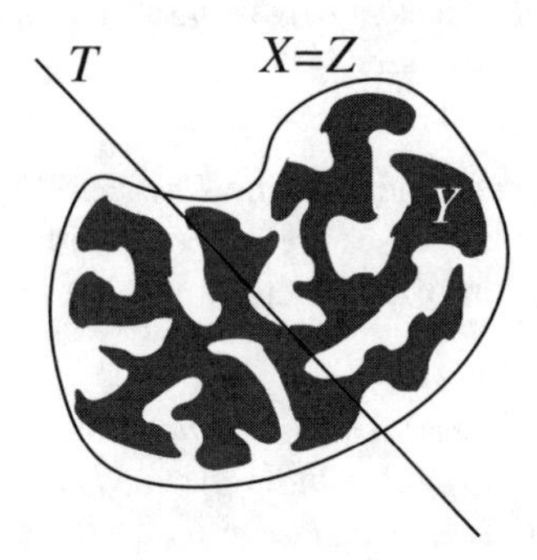

Figure 6.2 *Illustration of the 'restricted deterministic' case where the specimen X is the whole reference space. The spatial structure (X,Y) is non-random and arbitrary, while the probe T is randomised, following a specified distribution.*

'Extended deterministic case': the specimen X is but a portion taken from a much larger reference space Z of unknown extent (e.g. X is a rock sample from a large rock outcrop Z). The feature Y and reference space Z are non-random, bounded sets. The position of the specimen X inside Z was randomly chosen, in some fashion which allows us to regard its contents as 'statistically homogeneous'. The probe T may also be randomised by the experimenter. Statistical inference relies on the assumed randomness of X, possibly combined with the randomness of the probe T. Figure 6.1 is appropriate if we imagine Z is much larger than X.

'Random case': the specimen X and reference space Z are arbitrary fixed sets, and the feature Y is a random set (e.g. X is a 1-cm cube of steel, sampled from a continuous roll of steel formed under given conditions, and Y consists of all the grain boundary surfaces in the steel). Typically we assume $Y = Y^* \cap Z$ where Y^* is a stationary and isotropic random closed set. The probe T may have a fixed position relative to the specimen X. Statistical inference relies entirely on the randomness of the feature Y. See Figure 6.3.

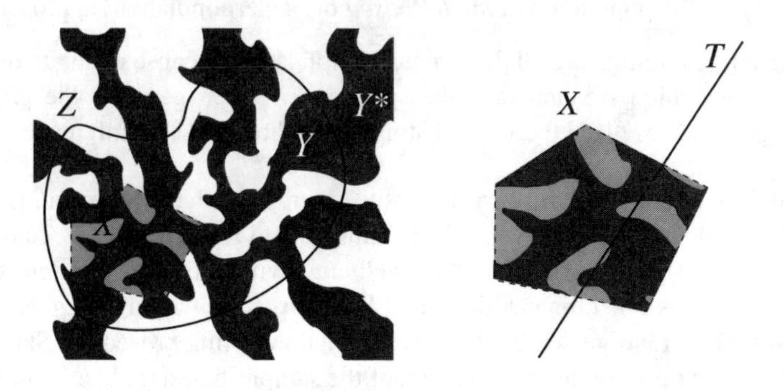

Figure 6.3 *Illustration of the* Random *case where Y is a random set, usually a partial realisation of a stationary random set Y^*.*

For brevity we shall refer to these as the *Restricted*, *Extended*, and *Random* cases, respectively. The three cases are different with regard to their sampling requirements, the definition of parameters, and the inferences which may be drawn from the sample.

Example 6.1 (Restricted case) *During kidney transplantation, the recipient's kidney is removed intact. A renal pathologist needs to estimate the total length of glomerular tubules in the excised kidney.*

It would not be prudent to assume the kidney exhibits any kind of spatial homogeneity. The estimate of total tubule length must refer to the specific kidney under investigation. Hence, this is an instance of the Restricted *case. The reference space Z is the excised kidney; the feature of interest Y is the assembly of all glomerular tubules in the excised kidney.*

In this case, inference depends on the randomisation of the sampling probe T. A random sampling design must be followed.

Notice that, in this example, there would be no basis for statistical inference about the whole kidney, unless the whole kidney were available for sampling — at least potentially available. If excision of the kidney is not complete, or if we only have access to a subset of the reference space, selected by an arbitrary, nonrandom procedure, the rationale of the *Restricted* case does not apply.

Example 6.2 (Extended case) *Evaluation of the quality of marble at a proposed quarry site involves selecting a 'representative block' of marble from the seam, cutting a polished plane section through the block, and applying stereological techniques.*

The Extended *case applies if Z is the entire seam of marble, X is the 'representative block', and T is the section plane. In order that X may be regarded as 'representative', its position (at least) must be random, and uniformly distributed within the seam Z. The orientation of the section plane T will also need to be randomized if the orientation of X was not. In this case, inference depends on the randomisation of X and possibly also on the randomisation of T.*

Example 6.3 (Random case) *A materials scientist develops a composite abrasive material and wishes to relate its wear resistance to its microstructure. Sample rods of the material are provided for stereological estimation of quantities such as the surface area density S_V of tungsten carbide particle boundaries.*

The parameter of interest is the average *composition of the material, and we are willing to assume the material is statistically homogeneous. So this is an example of the* Random *case. Inference is based on the assumption of homogeneity.*

Each of the three cases involves a different probabilistic argument and results in a different kind of statistical inference. In the jargon of sampling theory (see Appendix A), the *Restricted* case corresponds to finite population survey inference, the *Extended* case roughly to infinite population inference, and the *Random* case to superpopulation inference. These differences are explained further in Sections 6.5–6.7.

It is worth noting again that Miles' list does not include an experiment in which the specimen X is an *arbitrary* proper subset of the reference space Z, and X, Y, Z are all fixed sets which are not assumed to exhibit statistical homogeneity. In such an experiment there would be no basis for extrapolating from X to Z, so stereological techniques are invalid. We like to call this the *'Miraculous case'*.

6.3 Parameters

6.3.1 Types of parameters

The goal of stereological estimation is typically one of the following three types of parameters.

absolute size: the 'size' $\beta(Y)$ of the feature of interest Y, where β is some appropriate geometrical measure of size. Biological examples are the total volume $V(Y)$ of neocortex Y in a human brain Z, available for sampling at autopsy, and the surface area $S(Y)$ of the gas exchange surface Y in a gazelle lung Z. Absolute parameters are only meaningful in the *Restricted* case.

relative size: in the *Restricted* and *Extended* cases, relative size is the ratio $\beta(Y)/V(Z)$ of 'sizes' of the feature of interest Y to the reference space Z. Examples are the volume fraction $V_V = V(Y)/V(Z)$ of neocortex Y in brain Z, and the surface area per unit volume $S_V = S(Y)/V(Z)$ of the gas exchange surface Y in a gazelle lung Z. In the *Random* case, relative size parameters are defined as (expected) densities. Examples are the volume fraction $V_V = \mathbf{E}[V(Y \cap X)]/V(X)$ of air space Y in foam plastic, and the surface area density $S_V = \mathbf{E}[S(Y \cap X)]/V(X)$ of grain boundaries Y in steel. We need to assume that Y is stationary so that these quantities do not depend on X and can genuinely be interpreted as average densities.

particle average size: if Y consists of discretely identifiable objects or 'particles' $Y_1, Y_2, \ldots$, then the parameter of interest may be the average of the individual particle values $\beta(Y_i)$ of a measure of 'size' β, such as the particle volumes $V(Y_i)$, surface areas $S(Y_i)$, etc. and their higher moments if possible. In the *Restricted* and *Extended* cases, the particles are a finite population $Y_1, \ldots, Y_N$. Examples are the mean volume $\mathbf{E}_N[V] = \sum_i V(Y_i)/N$ and the volume-weighted mean volume $\mathbf{E}_V[V] = \sum_i V(Y_i)^2 / \sum_i V(Y_i)$ of glomeruli in the kidney. In the *Random* case, particles are assumed to be random sets with a common probability distribution (their 'size-shape distribution'), and the particle population mean size $\mathbf{E}_N[V]$ is then defined as the expectation of individual particle volume with respect to this distribution. See Chapter 11.

Other possible measures of 'size' include length L, number of components N, integral of mean curvature M, Euler-Poincaré characteristic χ, second-order moments and so on.

6.3.2 Meaning of parameters

In applications it is very important to distinguish carefully the three types of parameters listed above, as their practical interpretations are quite different.

Under the analogy with survey sampling (Appendix A), measures of absolute size $\beta(Y)$ are analogous to population totals, while ratios $\beta(Y)/V(Z)$ are analogous to population means, where the 'population' is the reference space Z. Particle averages are also analogous to population means, but with respect to the population of discrete particles. Note that the total number N of particles in a finite population of particles $Y_1, \ldots, Y_N$ is regarded as an 'absolute' quantity under this classification.

In biology, absolute quantities often have an interpretation as the amount of cellular processing capacity. For example, the number of neurons in a region of the brain, the total surface area of the alveolar gas exchange surface in the lung, and the total glomerular filtration surface area in the kidney, are overall measures of the 'competence' or 'endowment' of these organs. Relative quantities sometimes have a useful intuitive interpretation relating to the 'packing' of tissues. For example the reciprocal of the numerical density N_V of glial cells in the brain is the quantity $V_N = 1/N_V$, the average volume of brain tissue served by one glial cell. Particle population averages often have a useful biological interpretation relating to the 'typical' particle. For example the mean volume of the cell nuclei in a tumour biopsy seems to have some value for cancer diagnosis and prognosis.

In most areas of materials science, absolute quantities are not of direct interest, since most materials have indefinite extent. Relative quantities (V_V, S_V, L_V, N_V) have a direct interpretation relating to the composition of a material. Particle population averages ($\mathbf{E}_N[V]$, $\mathbf{E}_N[S]$, ...) are very important descriptions of mineral grains, crystalline metal grains, etc.

Weighted distributions and weighted means of particle size are important in stereology and image analysis [514, pp. 326–330, 345–346, 368–369]. We saw in Section 2.4.2 that plane sections of a particle population give us easier access to the volume-weighted distribution than to the ordinary (unweighted or 'number-weighted') distribution of particles. These two distributions are different: every stereologist should know the recipe for horse and lark pie [514, p. 368]. The volume-weighted distribution of particle sizes is also more natural in connection with the sampling of mineral particles by sieving (in which particles are segregated using a series of sieves with decreasing mesh openings, and the sieve fractions are measured by total weight). Weighted size distributions are also natural when the larger particles are more important or more valuable, for example, when economic value is proportional to particle size. In many cancers, enlarged cell nuclei are an important indication of cancer progression, and the volume-weighted mean volume $\mathbf{E}_V[V]$ of cell nuclei has been found to have diagnostic value [17, 52, 533].

6.3.3 Methodological errors: reference trap

Changes or group differences in parameters are liable to misinterpretation, a problem related to Simpson's paradox. For relative parameters such as N_V, this problem is known as the 'reference trap' or reference space artefact and was described in Section 3.2.5. A decrease in the N_V of cells in a developing foetal organ does not necessarily imply that the organ is losing cells, but may alternatively be due to an increase in the reference volume (organ volume) with no change in cell number. Similarly the numerical density of pores (enclosed air bubbles) in a neoprene wetsuit will be increased when the neoprene is compressed at depth underwater, but this does not indicate that new pores have formed.

For particle population mean parameters there is a corresponding reference space artefact. A decrease in the mean volume of bubbles in an unstable liquid foam may arise because individual bubbles shrink, but may also arise because larger bubbles pop at a faster rate than smaller bubbles. An increase in the mean volume of hepatocytes in liver disease may be due to enlargement of individual hepatocytes, or may occur because smaller hepatocytes die at a faster rate than larger hepatocytes.

Weighted and unweighted means have different interpretations. The volume-weighted mean volume of cell nuclei $\mathbf{E}_V[V]$ is related to the unweighted mean volume $\mathbf{E}_N[V]$ by

$$\mathbf{E}_V[V] = \frac{\mathbf{E}_N[V^2]}{\mathbf{E}_N[V]} = \frac{\mathsf{var}[V]}{\mathbf{E}_N[V]} + \mathbf{E}_N[V].$$

A change in $\mathbf{E}_V[V]$ may occur without a change in $\mathbf{E}_N[V]$.

6.4 Reference spaces

Some useful refinements to Miles' general scheme are noted here. They affect the definition of the reference space Z.

6.4.1 Reference space and containing space

Gundersen and Baddeley [223] make the further distinction between a *containing space* and a *reference space*. A *containing space* C serves simply as a 'container' for the feature of interest Y. We randomly sample the containing space in order to obtain a random sample of the feature of interest Y. This is often the case when we are estimating an absolute geometrical quantity. The containing space C is required to completely contain Y, but its size and extent need not be known exactly. For example, in order to estimate the total volume of a tumour, it suffices to remove enough tissue from the patient to ensure that the entire tumour has been removed, and then to sample (correctly!) from this excised tissue.

In contrast, a *reference space* is a well-defined object Z of known or measurable size, whose sole purpose is to allow us to standardise the size of the feature of interest Y (i.e. to express the size of Y relative to the size of Z). The reference space Z need not contain the feature Y, although it often does. Indeed Z may be lower-dimensional (a curved surface, a space curve, or a finite set of points). The size of Z will be measured by an appropriate absolute quantity $\phi(Z)$. Then the absolute size $\beta(Y)$ of the feature of interest Y is related to the size of Z using the ratio $\beta(Y)/\phi(Z)$.

This simply recognises the existing practice in stereology. Weibel and Knight [589] determined the mean thickness $\bar{\tau}$ of the air-blood barrier in the lung by writing $\bar{\tau} = 2V(Y)/S(Z)$ where Y is the air-blood barrier (a three-dimensional region of tissue) and Z is the air exchange surface. In dermatology it is common and desirable to express stereological parameters of a skin layer (a three-dimensional region) relative to the area of the outer surface of skin [73]. For tubular structures it is common to define the average cross-sectional area as the total volume divided by total length.

6.4.2 *Variable or fixed sample size*

Mayhew and Cruz-Orive [380], Cruz-Orive [109] and Cruz-Orive & Weibel [140] introduced a further distinction between the two situations sketched in Figure 6.4. In Model I, the bounded plane probe Q is almost surely included in the reference space Z, while in Model II, part of Q may not belong to the reference space.

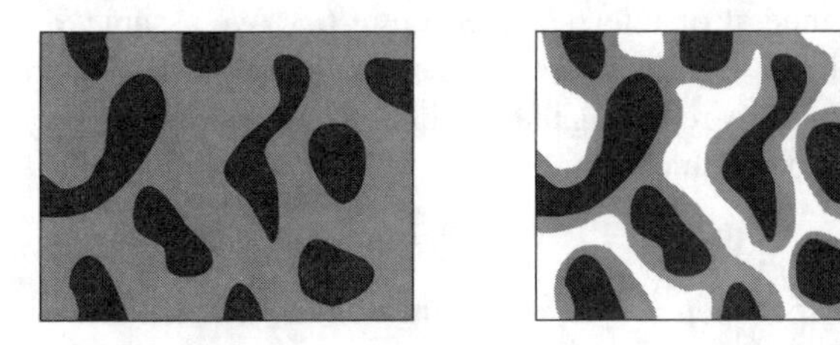

Figure 6.4 *Two cases distinguished by Cruz-Orive. In a sample of material (rectangle) we observe part of the feature of interest Y (black) and part of the reference space Z (grey and black). In Model I, shown at left, the sample is completely contained in the reference space. In Model II, at right, not all of the sample lies inside the reference space.*

The key difference is that in Model I, the sample size (the size of $Z \cap Q$) is fixed and may be controlled by the design, while in Model II it is random. The behaviour of ratio estimators will be different in the two cases [109].

Miles' generic scheme effectively assumes Model I in the *Random* and *Extended* cases. It can be modified to incorporate Model II by introducing a containing space C and treating the reference space Z as if it were a 'feature of interest'. In the *Extended* case the specimen X is a portion sampled from the much larger containing space C. The feature of interest Y and reference space Z are non-random, bounded subsets of C. We observe $Y \cap X$ and $Z \cap X$. In the *Random* case we assume the specimen X and the containing space C are arbitrary fixed sets, the feature of interest is $Y = C \cap Y^*$ where Y^* is a stationary and isotropic random closed set, and the reference space is $Z = C \cap Z^*$ where Z^* is also a stationary and isotropic random closed set. We observe $Y \cap X$ and $Z \cap X$.

6.5 Estimation in the *Random* case

Having defined the statistical model and the parameters to be estimated, we now explain the basis for statistical inference about these parameters. Each of Miles' three cases requires a different probabilistic argument, and leads to a different kind of statistical inference. We start with the *Random* case.

6.5.1 Example: Delesse's principle

The theory for the *Random* case is easiest to grasp in the simplest example of Delesse's principle.

Inside a nominated reference region of rock Z, consider the sub-region Y occupied by the mineral of interest. See Figure 6.3. We assume Y is random, specifically that $Y = Y^* \cap Z$ where Y^* is a 'first-order stationary' random set. This may be defined by requiring that the indicator $I(x) = \mathbf{1}\{x \in Y^*\}$ for $x \in \mathbf{R}^3$ is a random variable with constant mean,

$$\mathbf{E}[I(x)] = \mathbf{P}[x \in Y^*] = p \quad \text{for all } x.$$

For an arbitrary specimen $X \subseteq Z$, define the volume fraction

$$V_V = \mathbf{E}[V(Y \cap X)]/V(X)$$

provided $0 < V(X) < \infty$. In particular, for $X = Z$, we have

$$V_V = \mathbf{E}[V(Y)]/V(Z).$$

Noting that

$$V(Y \cap X) = \int_X I(x)\,\mathrm{d}x$$

and taking the expectation we obtain

$$\mathbf{E}[V(Y \cap X)] = \int_X \mathbf{E}[I(x)]\,\mathrm{d}x = pV(X)$$

so that $V_V = p$. Hence the definition of V_V does not depend on the choice of X

(or Z) and it is meaningful to speak of a volume density. Similarly define the area fraction on a section plane T by

$$A_A = \mathbf{E}[A(Y \cap X \cap T)]/A(X \cap T)$$

provided $0 < A(X \cap T) < \infty$. By a similar argument it follows that $A_A = p$. Hence the definition of A_A does not depend on the choice of section plane and specimen. In particular, we have

$$A_A = \mathbf{E}[A(Y \cap T)]/A(Z \cap T).$$

Delesse's relation $V_V = A_A$ follows immediately. Note that the only assumption required was that the indicator process $I(x)$ has constant mean.

6.5.2 *Statistical inference in the* Random *case*

In Delesse's principle, for example, the parameter to be estimated is the volume fraction V_V. It should be remembered that A_A is also a parameter in this setting, rather than an estimator. Our *estimator* of V_V will be the *empirical* area fraction

$$\widehat{A_A} = \frac{A(Y \cap X \cap T)}{A(X \cap T)}.$$

observed on a plane section T. Inference is based on the fact that $\widehat{A_A}$ is an **unbiased estimator** of the true volume fraction V_V,

$$\mathbf{E}\left[\frac{A(Y \cap X \cap T)}{A(X \cap T)}\right] = \frac{\mathbf{E}[A(Y \cap X \cap T]}{A(X \cap T)} = V_V. \qquad (6.1)$$

The estimation is unbiased in the 'superpopulation' sense; **E** denotes expectation with respect to the random process, that is, over hypothetical realisations of the contents of the material.

Note also that the target parameter V_V is a 'superpopulation' quantity: it is the *expected* volume fraction of the feature of interest, averaged over hypothetical realisations of the random contents of the material.

A superpopulation parameter cannot be observed directly, even if we know the entire contents of the reference space Z, since it is a characteristic of the random process that generated the rock.

Under the extra assumption that the random process is ergodic, we may invoke limit theorems which state that the stereological estimate (e.g. observed volume fraction $V(Y)/V(Z)$) over a large domain Z converges to its superpopulation mean (V_V). In this case the stereological technique is a *consistent* estimator. Some accounts of stereology have adopted consistent estimation, rather than unbiased estimation, as the goal (e.g. [78]).

In summary, in the *Random* case, the contents of the specimen X are assumed to be 'statistically homogeneous'. The parameters of interest are defined as expectations over the random contents of the specimen X. Stereological estimators of these parameters are unbiased in the sense that their expected value, over all realisations of the random contents of the specimen, equals the true parameter value. They can only be interpreted as unbiased *predictors* of the contents of the actual specimen X (which is not usually of direct interest).

6.5.3 *Fundamental Formulae in the* Random *case*

Here we prove the Fundamental Formulae of Stereology in the *Random* case. These concern the estimation of densities such as the volume fraction V_V, surface area fraction S_V and length fraction L_V of a feature of interest Y. When the probe T is a plane section, the relevant formulae are

$$\begin{aligned} V_V &= A_A \\ S_V &= \frac{4}{\pi} L_A \\ L_V &= 2N_A, \end{aligned}$$

which all have the generic form

$$\beta_V = \gamma_A, \tag{6.2}$$

where β and γ represent geometrical functionals satisfying (5.16).

In the *Random* case, the specimen X and the reference space Z are fixed bounded subsets of $\mathbf{R}^3$, while $Y \subseteq Z$ is considered as the part inside Z of a random closed set Y^*, i.e. $Y = Z \cap Y^*$, cf. Figure 6.3. The interpretations of the fundamental formulae of stereology rely on a measure theoretic framework in this case. The reader is referred to [543]. Here, we will just indicate some of the ideas.

It will be assumed that Y^* is 'first-order stationary and isotropic' in the sense that

$$\beta_V = \frac{\mathbf{E}\beta(Y \cap X)}{V(X)} \tag{6.3}$$

does not depend on the specimen X and

$$\gamma_A = \frac{\mathbf{E}\gamma(Y \cap X \cap T)}{A(X \cap T)} \tag{6.4}$$

does not depend on X and the position and orientation of the plane T. These assumptions are in particular fulfilled if the distribution of Y^* is invariant under translations and rotations.

The Crofton formula states that

$$\beta(Y \cap X) = \frac{1}{2\pi} \int_{S^2_+} \int_{-\infty}^{\infty} \gamma(Y \cap X \cap T_{\omega,s})\, \mathrm{d}s\, \mathrm{d}\omega. \tag{6.5}$$

Combining (6.3)–(6.5), we find that

$$\begin{aligned}
\beta_V V(X) &= \mathbf{E}\beta(Y \cap X) \\
&= \frac{1}{2\pi}\int_{S^2_+}\int_{-\infty}^{\infty} \mathbf{E}\gamma(Y \cap X \cap T_{\omega,s})\,\mathrm{d}s\,\mathrm{d}\omega \\
&= \frac{1}{2\pi}\int_{S^2_+}\int_{-\infty}^{\infty} \gamma_A A(X \cap T_{\omega,s})\,\mathrm{d}s\,\mathrm{d}\omega \\
&= \gamma_A V(X)
\end{aligned}$$

or

$$\beta_V = \gamma_A.$$

6.6 Estimation in the *Restricted* case

Suppose that both the reference space Z and its contents Y are fixed, non-random, arbitrary regions in three dimensions. We shall take a plane section T through Z. This is the *Restricted* case, depicted in Figure 6.2.

6.6.1 Example: estimation of absolute volume

In the *Restricted* case it is simpler to estimate absolute quantities like the volume $V(Y)$ of the feature of interest. Recall the equation expressing the volume of a solid as the integral of its plane section areas:

$$\int A(Y \cap T_u)\,\mathrm{d}u = V(Y), \tag{6.6}$$

where T_u denotes the horizontal plane at height u.

Let us take a fixed orientation, uniform random (FUR) plane through the reference region Z. This was defined in Section 5.3. Notice that T_u has a nonempty intersection with Z if and only if u belongs to $p(X)$, the projection of X onto the z-axis. We select the section plane T at random by choosing a uniformly distributed random height, $T = T_U$, where $U \sim \mathsf{Unif}(p(X))$. The probability density of U is

$$f(u) = \begin{cases} \frac{1}{\ell} & \text{if } X \cap T_u \neq \emptyset \\ 0 & \text{otherwise} \end{cases} \tag{6.7}$$

where $\ell = L(p(X))$ is the length of $p(X)$.

It follows immediately that

$$\mathbf{E}\left[\ell\, A(Y \cap T)\right] = V(Y) \tag{6.8}$$

so that $\widehat{V}(Y) = \ell\, A(Y \cap T)$ is an **unbiased estimator** of the true volume $V(Y)$.

We stress this is an elementary example for illustrative purposes.

6.6.2 Statistical inference in the Restricted *case*

A key point here is that statistical inference relies entirely on the randomisation of the section plane T. The estimator $\widehat{V}(Y)$ is unbiased *without any assumptions about the material* Y. There is instead a very strict requirement on the sampling protocol, namely that the probability distribution of the random section plane T be exactly the uniform distribution specified in (6.7).

Recall that, in design-based survey sampling, estimators of population parameters are unbiased by virtue of the randomisation of the sample, without any need for assumptions about the population structure (cf. [555, p. 3]). Similarly, in design-based stereology, estimators of three-dimensional quantities are unbiased under appropriate randomisation of the section plane, *regardless* of the spatial arrangement of the material. This vastly increases the scope and power of stereological methods, since we no longer need to assume the material is 'spatially homogeneous' in any sense at all.

In the *Restricted* case, we do not need to assume anything about the spatial configuration of the feature Y and the reference space Z. These are arbitrary, fixed subsets of three-dimensional space. However, the randomised probe T must follow a specific probability distribution. Stereological estimators obtained under these conditions give unbiased estimates of properties of the specimen X itself. An unbiased estimator in this context is one whose average over all possible positions of the probe T is equal to the true parameter value.

6.6.3 Fundamental Formulae in the Restricted *case*

In the *Restricted* case, the specimen X and the feature Y are non-random, bounded subsets in $\mathbf{R}^3$ which are the sole objects of interest, cf. Figure 6.2. The basic type of random planar section of X is an IUR plane T hitting X, cf. Section 5.6. An unbiased estimator of $\beta(Y)$ is

$$\widehat{\beta}(Y) = \bar{H}(X)\gamma(Y \cap T),$$

where $\bar{H}(X)$ is the mean linear projection of X. In particular,

$$\widehat{V}(X) = \bar{H}(X)A(X \cap T)$$

is an unbiased estimator of $V(X)$. A stochastic interpretation of (6.2) is therefore

$$\frac{\mathbf{E}\gamma(Y \cap T)}{\mathbf{E}A(X \cap T)} = \frac{\beta(Y)}{V(X)}.$$

In Chapter 9 we discuss how this result can be used in statistical inference for ratio parameters.

6.7 Estimation in the *Extended* case

Suppose again that both the reference space Z and its contents Y are fixed, non-random, arbitrary regions in three dimensions. We now take a specimen $X \subset Z$ and generate a section plane T through X. This is the *Extended* case, cf. Figure 6.1.

6.7.1 Example: Delesse's principle

We apply the theory for FUR quadrats. In Section 5.4, the case of FUR quadrats in the plane has been treated in detail. We will here use the corresponding theory in three dimensions.

Suppose that X is a uniform random position of a fixed region X_0 hitting Z. That is,

$$X = u + X_0$$

where u is a random point, uniformly distributed in $Z \oplus \check{X}_0$. Recall that $Z \oplus \check{X}_0$ is the set of u such that $u + X_0$ intersects Z, while $Z \ominus \check{X}_0$ is the set of u such that $u + X_0$ lies entirely inside Z. In the *Extended* case Z is typically 'much larger' than X_0 in the sense that

$$V(Z \oplus \check{X}_0) \approx V(Z) \approx V(Z \ominus \check{X}_0). \tag{6.9}$$

This implies that X may be assumed to lie entirely inside Z for geometrical arguments.

Now suppose T_0 is an *arbitrary* fixed plane through X_0, and $T = u + T_0$ is the corresponding plane through X. Then the section $X \cap T = u + (X_0 \cap T_0)$ is a translated copy of the section $X_0 \cap T_0$. That is, $Q = X \cap T$ is a random two-dimensional quadrat in space (visualise a playing card that is free to move in space but not to rotate). Using (4.22) with $d = k = 3$ and $m = 2$, we find with $Q_0 = X_0 \cap T_0$

$$\mathbf{E}[A(Y \cap Q)] = \frac{V(Y)A(Q_0)}{V(Z \oplus \check{X}_0)}$$

and

$$\mathbf{E}[A(Z \cap Q)] = \frac{V(Z)A(Q_0)}{V(Z \oplus \check{X}_0)}.$$

Accordingly,

$$\mathbf{E}[\frac{A(Y \cap Q)}{A(Z \cap Q)}] = \frac{V(Y)}{V(Z)}. \tag{6.10}$$

This is a version of Delesse's principle in the *Extended* case.

We have derived this statement for a fixed relative position of T and X (i.e. for a fixed plane T_0 intersecting X_0). In practice we would first select X, and subsequently select a plane T through X. The statement (6.10) remains true in this situation, provided the conditional distribution of T given X depends only on the relative positions of T and X (and not, for example, on the location of X or on $Y \cap X$).

6.7.2 *Statistical inference in the* Extended *case*

We have seen that the empirical area fraction $A(Y \cap Q)/A(Z \cap Q)$ is **a ratio unbiased estimator** of the true volume fraction $V(Y)/V(Z)$ of the reference space. Recall that $Q = X \cap T$. Expectation in (6.10) is taken over all possible positions of the specimen X, and possibly also over randomised positions of the section plane T through X. The primary basis for statistical inference is *the randomisation of the specimen.*

The target parameter $V(Y)/V(Z)$ is a ratio of sizes of the fixed sets Y and Z. *No assumptions about Y are required.*

The *Extended* and *Random* cases have certain similarities. The contents of the specimen X are 'statistically homogeneous' in both cases. In the *Extended* case, this is a consequence of randomizing the specimen X inside Z in a specified way; in the *Random* case the feature Y is explicitly assumed to be a random process. In both *Extended* and *Random* cases, the parameters of interest can be written as expectations with respect to the random contents of the specimen X.

However, the similarities end there. In the *Random* case the parameters of interest are densities (V_V, S_V etc) originally defined as expectations for the random process Y which is assumed to be statistically homogeneous. In the *Extended* case the parameters of interest are ratios ($V_V = V(Y)/V(Z)$, etc) defined for the non-random feature of interest Y and reference space Z about which no assumptions are made.

In the *Extended* case, estimators are unbiased with respect to the randomisation of the specimen X and possibly also of the randomisation of the probe T.

6.7.3 *Fundamental Formulae in the* Extended *case*

In the *Extended* case, a stochastic interpretation of (6.2) can be given, using an IUR 2-quadrat Q in $\mathbf{R}^3$ hitting Z, cf. Section 5.7.

Using the results in Section 5.7, we find that

$$\widehat{\beta}(Y) = c\gamma(Y \cap Q) \text{ and } \widehat{V}(Z) = cA(Z \cap Q)$$

are unbiased estimators of $\beta(Y)$ and $V(Z)$, respectively, where

$$c = \frac{1}{(2\pi)^2} \frac{\mu(Z)}{A(Q_0)}$$

and $\mu(Z)$ is the normalization constant in the density of the IUR quadrat. The stochastic interpretation of the Fundamental Formulae of Stereology is therefore of the same type as in the *Restricted* case

$$\frac{\mathbf{E}\gamma(Y \cap Q)}{\mathbf{E}A(Z \cap Q)} = \frac{\beta(Y)}{V(Z)}.$$

Statistical inference on ratios based on this result is taken up in Chapter 9.

6.8 Advice to consultants

In summary, the formulation of a stereological experiment has a great bearing on the validity and interpretation of the experimental findings. It should be pondered at an early stage of experimental design.

A useful point to begin discussion with the client is to focus on the final objective. We should define the features of interest in three dimensions, rather than by the operational criteria for recognising these features on a plane section. After discussing the scientific questions in this context, it is usually easier to identify the appropriate target parameters.

A clear definition of the features of interest must resolve issues such as the approximation of thick membranes or layers by ideal two-dimensional surfaces (e.g. we may nominate the upper/outer surface of the membrane as the ideal surface); the spatial extent of the features of interest (e.g. deciding whether we count all neurons in the brain or only in the neocortex); and resolution effects (e.g. the 'fractal' increase in apparent surface area with increasing magnification).

The quantitative aspects of the study should be discussed at length, without fear of appearing naive, until it is absolutely clear precisely what three-dimensional quantities are desired, and what two-dimensional quantities are measured.

Great care needs to be taken with the specification of the parameters that are to be estimated. The distinction between absolute and relative parameters should be discussed in the context of the experiment. Sobering examples of the misinterpretation of parameters are cited in Sections 2.3.2 and 2.3.3. 'Reference space artefacts' analogous to Simpson's paradox hamper the interpretation of changes or differences in ratios.

The choice of appropriate statistical formulation and sampling design depends on many factors. See Chapter 12 for further discussion.

6.9 Exercises

Exercise 6.1 An experimental protocol for a proposed study of the optic nerve defines the cells of interest as those whose nuclei are stained blue in the section. Explain why this cell population is not a well-defined 'feature of interest Y' in the sense of Miles. Discuss the possible consequences if the objective is (a) to estimate the volume fraction occupied by these cells, (b) to count these cells.

Exercise 6.2 Explain the difficulty in reconciling two studies which have estimated the numerical density N_V of neurons in the CA1 layer of the human hippocampus using different anatomical/histological definitions of the CA1 layer.

Exercise 6.3 A needle biopsy of the iliac crest (rim of the hip bone) is taken by inserting a hollow cylindrical needle into the patient's upper hip at an arbitrary position (determined of course by clinical considerations).

(a) Identify Z and X in this experiment.

(b) Decide which of Miles' three kinds of stereological inference definitely **cannot** be used here.

Exercise 6.4 Material from kidney biopsies from diabetic patients was analysed stereologically to yield estimates, for each patient, of the surface area density S_V (cm^2/cm^3) of the glomerular basement membrane per unit volume of kidney. These numbers were subsequently converted to estimates of absolute surface areas (cm^2) by multiplying them by the (notional) volume of a typical human kidney, 270 cm^3.

(a) What statistical interpretation, if any, can be placed on these estimates?

(b) Calculated in this way, absolute surface area seems to decrease with the progression of type II diabetes. Comment on this interpretation.

Exercise 6.5 When an organ removed from a clinical patient requires investigation, standard practice at your hospital is to cut the organ end-to-end into a few thick slices, and to send one slice to each of the hospital departments involved (histology, toxicology, pathology etc). If the slices are allocated to departments in a fixed order (say, slices from left to right are sent to departments in alphabetical order; or the toxicology department always asks for the distal end) explain why *none* of the departments can validly use stereological methods to infer information about the entire kidney.

6.10 Bibliographic notes

Concerns about the fundamental validity of statistical inference in classical stereology were expressed throughout the classical literature. Delesse [156] discussed the extent to which various rocks may be treated as homogeneous. The inhomogeneity of biological structures was discussed extensively in the 1960's, notably by Weibel [575, p. 15], who proposed systematic sampling as a remedy. See also [579]. Weibel et al [586] summarise the limitations of classical techniques.

Progress was impeded by disagreement about what should constitute 'validity'. While the basic theory of stereology provides unbiased estimators, some writers (e.g. [78, 72]) required that stereological estimators be consistent, that is, that in large samples the parameter estimates should converge to the correct parameter value. Consistency requires a much more explicit stochastic model.

Rigorous statistical formulations of some experiments in stereology can be found in early studies of the performance of stereological estimators by Moran [420, 422] and others. Problems with existing theory were raised throughout the 1970's. Mayhew and Cruz-Orive [380] drew attention to the problem of variable sample size in estimating stereological ratios.

Thompson [557, p. 22] mentions random sampling of an organ to estimate its absolute and relative characteristics. This appears to be the first mention of a design based approach in stereology.

The seminal work of R.E. Miles [399, 398, 401, 403, 404] drew attention to the statistical aspects of stereology and to the need for proper model specification. See also Moran [424]. Stereological theory for design-based (*Restricted* and *Extended* case) experiments was first elaborated by P.J. Davy [146] and by Davy and Miles [147, 402, 408, 409, 410]. It was extended by L.M. Cruz-Orive and E.R. Weibel [109, 112, 119, 140]. Serra [514, chap. VIII] describes five "sampling situations" related to stereology and image analysis. See also Coleman [89, 90].

Matheron [366] developed the theory of random closed sets and showed that it provided a rigorous formulation for 'model-based' stereology [364, 367]. About 1980, J. Mecke, D. Stoyan and their collaborators [387, 389, 390, 433, 540] provided a rigorous probabilistic formulation of stereology in the model-based context, using J. Mecke's theory of stationary random measures. For modern accounts see [543, Chap. 10], [510]. See also [299].

The integral-geometric basis for stereology was elucidated by Giger and Hadwiger [196, 197, 198], Moran [424] and Santaló [500, Chap. 16]. See [509, 510] for modern accounts.

CHAPTER 7

Uniform and isotropic uniform designs

This chapter describes sampling designs that can be used for estimating an absolute quantity $\beta(Y)$ such as the volume, surface area or length of the feature of interest Y.

It is assumed that Y is a fixed and arbitrary set, the entire reference space or containing space Z is available for sampling, and the probe T is randomised; thus we are in the *Restricted* case of stereological inference, see Sections 6.2 and 6.6. The estimation of absolute quantities is also of interest in the *Extended* case, but then the estimation is usually performed via intermediate ratios. That topic is treated in Chapter 9.

All sampling designs considered in this chapter use systematic sampling. A regular array of probes is used, such as a stack of equally spaced, parallel sections; a spatial point grid; or a systematic array of quadrats in the plane. The position of the array is randomised, with a uniformly distributed starting location, and in some cases also with isotropic random orientation.

7.1 Estimation of volume and area

Here we describe the estimation of volume in $\mathbf{R}^3$, and estimation of area in $\mathbf{R}^2$, from systematic samples with fixed orientation. These techniques are the easiest to explain, because we do not need to randomise the orientation of the probe.

7.1.1 *Serial section plane design*

The volume $V(Y)$ of a subset Y of $\mathbf{R}^3$ can be estimated using a stack of equally spaced, parallel plane sections. Fix a direction ω in three-dimensional space and consider planes $T_{\omega,s}$ normal to this direction (using the polar coordinates ω, s defined in Section 4.1.3.) A systematic sample of planes is an infinite set

$$\mathcal{T} = \{T_{\omega,u+mt} : m \in \mathbf{Z}\}$$

of parallel planes at constant spacing $t > 0$. See Figure 7.1.

The coordinate u may be taken to satisfy $0 \le u < t$, specifying the location of the 'initial' plane, and this determines the location of the entire array of planes. The array $\mathcal{T}$ is a *systematic random sample* of parallel planes when the starting location U is uniformly random, $U \sim \mathsf{Unif}([0,t))$.

In stereology this is sometimes known as the 'egg slicer design'. A household egg slicer does indeed take a systematic sample of a boiled egg — provided the slicer has enough cutting wires to reach beyond both ends of the egg. It is a systematic random sample provided that the position of the slicer relative to the egg is properly randomised. Thus, the first slicing plane should not be positioned at the end of the egg (tangent to the surface) or at any other fixed place.

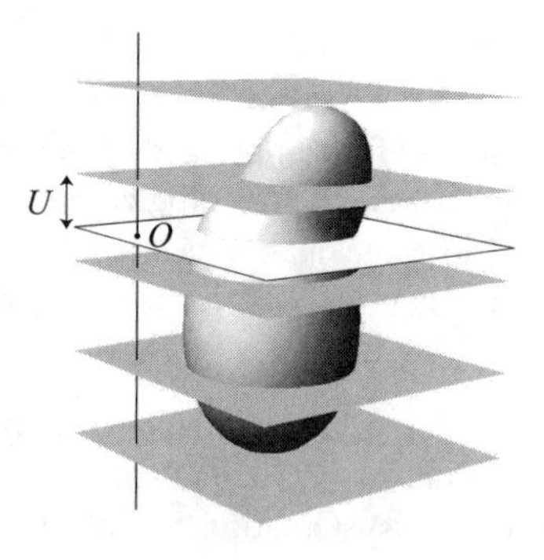

Figure 7.1 *The serial section plane design. The distance between neighbour planes (grey) is t. The parallel white plane is a reference plane, passing through the origin O.*

Many practical procedures for section preparation in microscopy are equivalent to taking a systematic random sample of plane sections. A microtome cuts physical slices of embedded tissue of approximately constant thickness s. We may subsample every kth slice from a microtome, giving us a systematic random sampling with spacing $t = ks$. In three-dimensional images obtained from laser confocal microscopy, each horizontal plane in the image corresponds to a different position of the optical focal plane, and a systematic random sample of these planes may easily be taken.

The information available from the systematic sample $\mathcal{T}$ is the collection of plane sections

$$Y \cap \mathcal{T} = \{Y \cap T_{\omega,U+mt} : m \in \mathbf{Z}\}.$$

An estimator of $V(Y)$ based on this information is

$$\widehat{V}(Y) = t \sum_{m} A(Y \cap T_{\omega,U+mt}) \tag{7.1}$$

where the sum is over all signed integers m. Since Y is bounded, there are only finitely many nonzero terms in the sum.

To justify this estimator, recall that the volume of Y can be expressed by Fubini's theorem as the integral of the sectional areas

$$V(Y) = \int_{-\infty}^{\infty} A(Y \cap T_{\omega,h}) \, \mathrm{d}h. \tag{7.2}$$

The estimator (7.1) can be viewed simply as a finite sum approximation to the integral in (7.2) of the function $f(h) = A(Y \cap T_{\omega,h})$ over the real line, based on a sequence of equally-spaced sample points $U + mt, \ m \in \mathbf{Z}$.

Note that the position of the sample is randomised, by using a uniform random variable U. Under these conditions the sample points constitute a systematic *random* sample on the real line. The estimator (7.1) is then *unbiased* because

$$\begin{aligned}
\mathbf{E}[\widehat{V}(Y)] &= \int_0^t t \sum_m A(Y \cap T_{u+mt}) \frac{1}{t} \, \mathrm{d}u \\
&= \sum_m \int_0^t A(Y \cap T_{u+mt}) \, \mathrm{d}u \qquad (7.3) \\
&= \sum_m \int_{mt}^{(m+1)t} A(Y \cap T_h) \, \mathrm{d}h \\
&= \int_{-\infty}^{\infty} A(Y \cap T_h) \, \mathrm{d}h \\
&= V(Y). \qquad (7.4)
\end{aligned}$$

This method is a direct analogue of systematic sampling in finite populations (Appendix A). This connection was noted by Moran [419].

In applied stereology, the estimation of volume from a systematic set of planes is called "estimation of volume by Cavalieri's principle" or loosely "the Cavalieri estimator". Cavalieri's principle is the statement that two solid objects which have equal cross-sectional areas on all horizontal planes must have equal volumes. This is as a consequence of (7.2) first stated by Cavalieri * in the two-dimensional case. Hence we could say that it is Cavalieri's principle which enables us to estimate volume from cross-sectional areas without reconstructing the three-dimensional geometry [341].

The variance of $\widehat{V}(Y)$ depends on the geometry of Y. There is no simple closed-form expression for the variance, but there are simple asymptotic approximations as $t \to 0$. Usually, but not always, systematic sampling yields a lower variance than simple (independent) random sampling. Systematic sampling

* Francesco Buonaventura Cavalieri (1598–1647), professor of mathematics at Bologna from 1629, invented the method of indivisibles (1635) that foreshadowed integral calculus.

is asymptotically 'superefficient' as $t \to 0$. Variances are discussed further in Chapter 13.

The two-dimensional version of the Cavalieri estimator gives an estimate of the area of a region in the plane from measurements of its intercept lengths with a grid of parallel lines. This is treated in Exercise 7.5.

7.1.2 *Grids of test points*

Planar point grid

Crofton [105] first pointed out that the area of a planar region Y can be estimated using a regular lattice of test points, randomly superimposed on the plane.

For simplicity we take the case of a rectangular grid with horizontal spacing r and vertical spacing s,

$$\mathcal{G} = \{U + (kr, ls) : k, l \in \mathbf{Z}\},$$

where the vector U determines the position of the entire grid $\mathcal{G}$. We may take U to lie in the 'basic tile'

$$D_0 = [0, r) \times [0, s).$$

If U is uniformly random, $U \sim \mathsf{Unif}(D_0)$, then $\mathcal{G}$ is a systematic random probe.

The area of any measurable plane set Y of finite area can be estimated by

$$\widehat{A}(Y) = a\, N(Y \cap \mathcal{G}) \tag{7.5}$$

where $a = rs$ is the area of the basic tile.

In the limit as the lattice becomes increasingly fine, this estimator is the standard pixel-counting rule for measuring the area of a region in a binary image. However, most practitioners would regard this as a very rough approximation when the grid spacing is large. Our point is that for *any* grid spacing, $\widehat{A}(Y)$ is an *unbiased estimator* of $A(Y)$, provided the grid location is uniformly randomised. It is also surprisingly accurate in practice.

One way to prove unbiasedness is sketched in Figure 7.2. We may write $N(Y \cap \mathcal{G})$ as a sum of indicators

$$N(Y \cap \mathcal{G}) = \sum_{k,l} \mathbf{1}\{U_{k,l} \in Y\}$$

where $U_{k,l} = U + (kr, ls)$. Observe that $U_{k,l}$ is a uniformly random point in the rectangle

$$D_{k,l} = D_0 + (kr, ls) = [kr, (k+1)r) \times [ls, (l+1)s).$$

Therefore,

$$\begin{aligned} EN(Y \cap \mathcal{G}) &= \sum_{k,l} P[U_{k,l} \in Y] \\ &= \sum_{k,l} \frac{A(Y \cap D_{k,l})}{A(D_{k,l})} \\ &= \frac{1}{rs} \sum_{k,l} A(Y \cap D_{k,l}) \\ &= \frac{1}{rs} A(Y). \end{aligned}$$

where the last line follows because the rectangles $D_{k,l}$ tessellate $\mathbf{R}^2$.

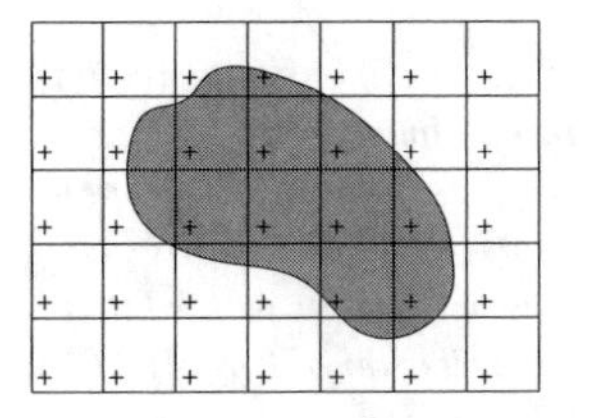

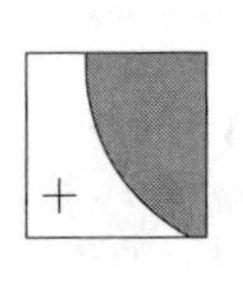

Figure 7.2 *Proof of unbiasedness of the point counting estimator of area, (7.5).* Left: *the point count is the sum of indicators from experiments performed in disjoint tiles.* Right: *each point is uniform in the corresponding tile.*

The variance of $\widehat{A}(Y)$ is not known analytically. It has been the subject of research for half a century, as described in Chapter 13.

Spatial point grid

Similarly, the volume of a three-dimensional region Y can be estimated by placing a regular lattice of test points in three-dimensional space and counting the number of test points hitting the region. Let

$$\mathcal{G} = \{U_{k,l,m} : k, l, m \in \mathbf{Z}\},$$

where $U_{k,l,m} = U + (kr, ls, mt)$ and U is uniform in

$$D_0 = [0, r) \times [0, s) \times [0, t).$$

This is the *spatial point grid design*. Then an unbiased estimator of $V(Y)$ is

$$\widehat{V}(Y) = vN(Y \cap \mathcal{G}), \tag{7.6}$$

where $v = rst$ is the volume of the basic parallelepiped D_0 of the grid. A proof of unbiasedness may be constructed as in the case of a planar point grid. It is also possible to prove unbiasedness by combining expectation (over random starting point) and summation (over points in the grid) into a single integral over $\mathbf{R}^3$, see Exercise 7.7.

An important practical implementation of the spatial grid occurs in confocal and transmission optical microscopy. Instead of cutting thin sections of tissue, we are able to visualise an *'optical section'* (an image of the focal plane only) situated inside a thick section. Images of a series of equally spaced optical sections are taken, and the same planar point grid is used on all optical sections.

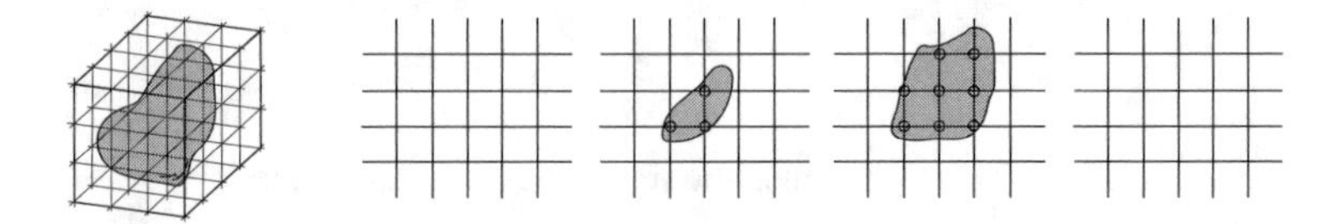

Figure 7.3 *Estimating the volume of a three-dimensional region using a spatial grid of test points. First panel: intended three-dimensional situation. The solid region of interest (shaded) is intersected by a grid of test points (located at the intersections of the lines). Subsequent panels: implementation on plane sections (usually optical sections). Four horizontal plane sections are shown. Test points which hit the region are indicated by thickened dots. Reproduced by kind permission of Prof. H.J.G. Gundersen.*

As in the case of the serial section plane design, the variance of $\widehat{V}(Y)$ is not known analytically, but asymptotic approximations are known. See Chapter 13.

7.1.3 Systematic subsampling in the plane

Systematic subsampling of the sampled material is an essential technique for reducing effort in microscopy. One important case concerns the measurement of area in the plane (usually a plane section of a solid material). In some experiments, it would be too laborious to measure areas exactly. Instead they are estimated from a subsample, using a systematic array of quadrats. We let Y denote a bounded subset of the plane, whose area $A(Y)$ is to be estimated. The set Y may either be a plane section of a three-dimensional solid, cf. e.g. Section 7.1.1, or an intrinsically two-dimensional region.

In order to describe systematic subsampling in the plane in sufficient generality, we introduce the concept of a lattice of fundamental regions in the plane $\mathbf{R}^2$. Let D_0 be a bounded subset of $\mathbf{R}^2$ and $\{u_j\}$ a sequence in $\mathbf{R}^2$, and let $D_j = u_j + D_0$. The sets $\{D_j\}$ are then called a lattice of fundamental regions in $\mathbf{R}^2$ if they tile the plane, i.e.

$$\begin{aligned} \cup_j D_j &= \mathbf{R}^2 \\ D_{j_1} \cap D_{j_2} &= \emptyset, \quad j_1 \neq j_2. \end{aligned}$$

We will from now on assume that $\{D_j\}$ is such a lattice of fundamental regions.

Now let $Q_0 \subseteq D_0$ be a set of dimension d where d may take the values 0, 1 or 2. Depending on the value of d, Q_0 is either a finite set of points ($d = 0$), a finite set of smooth curves with finite lengths ($d = 1$) or a domain, i.e. a set with positive finite area ($d = 2$). We let α denote number, length or area, depending on the value of d. Let $Q_j = u_j + Q_0$. The idea is now to randomly place this set, i.e. to consider $\{U + Q_j\}$ where U is uniform random in D_0 and use observation on this set to estimate the area of Y. The set $\{U + Q_j\}$ is called a uniform random *test system*.

An example of the set-up is shown in Figure 7.4. The set D_0 is the quadrat with partly stippled edges. The lattice of fundamental regions $\{D_j\}$ is not indicated on the figure but can be obtained by translating D_0 with vectors u_j having coordinates equal to integer multiples of the side lengths of D_0. The set Q_0 is a quadrat. In Figure 7.4, the quadrat with lower left corner U is $U + Q_0$. The systematic set of quadrats $\{U + Q_j\}$ is also shown.

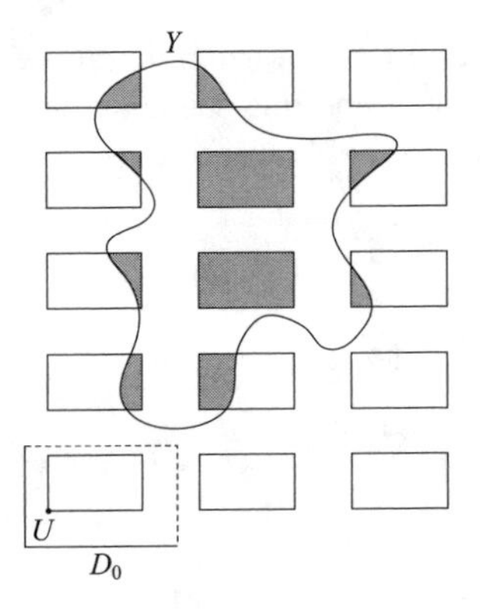

Figure 7.4 *Estimating the area of the planar set Y using a systematic set of quadrats.*

We can unbiasedly estimate the area of Y by

$$\widehat{A}(Y) = \frac{A(D_0)}{\alpha(Q_0)} \sum_j \alpha(Y \cap (U + Q_j)).$$

Recall that α is either number, length or area, depending on the dimension of the uniform random test system $\{U + Q_j\}$. In the case where $\alpha = A$ there also exists a discrete version of this set-up without random shifts.

To prove the unbiasedness, first recall that

$$\int_{\mathbf{R}^2} \alpha(Y \cap (u + Q_0))\,\mathrm{d}u = A(Y)\alpha(Q_0), \tag{7.7}$$

cf. (4.22) with $k = d = 2$ and $m = 0$, 1 or 2. On the other hand, we have

$$\int_{\mathbf{R}^2} \alpha(Y \cap (u + Q_0))\,\mathrm{d}u$$

$$
\begin{aligned}
&= \sum_j \int_{D_j} \alpha(Y \cap (u + Q_0))\, \mathrm{d}u \\
&= \sum_j \int_{D_0} \alpha(Y \cap (u + Q_j))\, \mathrm{d}u \\
&= \int_{D_0} \sum_j \alpha(Y \cap (u + Q_j))\, \mathrm{d}u
\end{aligned} \tag{7.8}
$$

Combining (7.7) and (7.8), we get

$$
\begin{aligned}
\mathbf{E}\widehat{A}(Y) &= \frac{A(D_0)}{\alpha(Q_0)} \int_{D_0} \sum_j \alpha(Y \cap (u + Q_j)) \frac{\mathrm{d}u}{A(D_0)} \\
&= \frac{1}{\alpha(Q_0)} \int_{\mathbf{R}^2} \alpha(Y \cap (u + Q_0))\, \mathrm{d}u \\
&= A(Y).
\end{aligned}
$$

Other examples of this set-up are shown in Figure 7.5. Here, Q_j may be either a two-dimensional quadrat, some line segments or a finite set of points (for instance, the endpoints of a set of line segments). See Exercises 7.9 and 7.10.

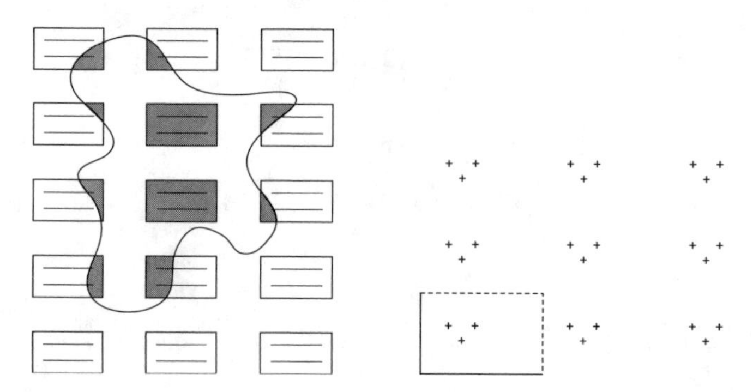

Figure 7.5 *Area estimation using a systematic set of quadrats, line segments or points. For details, see the text.*

7.2 Length and surface area estimation

In contrast to volume estimation, length and surface area estimation require isotropic random encounters between the probe and the structure. The key mathematical tool here is the Crofton formula (Chapter 4). Length and surface area estimation can be performed using isotropic uniform, vertical uniform, or local random sampling designs. This chapter deals with isotropic designs, which are the classical ones, while the next chapter presents vertical and local designs.

7.2.1 Length in the plane

The length of a curve in the plane can be estimated by superimposing a grid of test lines and counting the number of intersection points. See Figure 7.6.

Figure 7.6 *Estimating the length of a curve by counting the number of intersections it makes with an IUR grid of test lines.*

Using the polar coordinates (θ, p) for lines in the plane (Section 4.1.2), a grid of parallel lines with constant spacing h and orientation θ consists of the lines $T_{\theta,u+mh}$ for all signed integers m, where u is the displacement of the grid from the origin. An **isotropic uniform random (IUR)** position of the test grid is obtained if θ and u are randomised, so that the grid is

$$\mathcal{T} = \left\{T_{\Theta,U+mh} : m \in \mathbf{Z}\right\}$$

where Θ and U are independent random variables which are uniformly distributed, $\Theta \sim \mathsf{Unif}([0,\pi))$ and $U \sim \mathsf{Unif}([0,d])$.

An unbiased estimator of the length $L(Y)$ of a rectifiable curve Y in the plane is then

$$\widehat{L}(Y) = \frac{\pi}{2} h \sum_m N(Y \cap T_m) \tag{7.9}$$

where $T_m = T_{\Theta,U+mh}$. This follows directly from equation (4.5). See Exercise 7.10. Although this result has its roots in the famous Buffon needle problem, this estimation technique was first suggested by H. Steinhaus [534, 535, 536] and we shall call it the Steinhaus estimator.

Note carefully that isotropic random rotation of the test grid is required in order that $\widehat{L}(Y)$ be unbiased for $L(Y)$. The variance may be high when the curve Y has strongly preferential orientation. To improve accuracy, Steinhaus [535] proposed that the grid of parallel lines be repeatedly superimposed on the curve at n equally-spaced orientations $\theta_j = \Psi + j\pi/n$ for $j = 1,\ldots,n$ where $\Psi \sim \mathsf{Unif}([0,\pi/n])$. For example if $n = 2$ this may be achieved with a transparency bearing a square grid of test lines (two sets of parallel test lines, perpendicular to each other). The performance of this estimator was analysed by Steinhaus [535] and Moran [420]. See Section 13.2.6.

7.2.2 *Length estimation in the plane, without isotropy*

It was shown in Section 4.2.4 that curve length in the plane can be determined using an array of test curves, *without rotating the test system*, provided the orientation distribution of the test curves is isotropic. Examples of such arrays are shown in Figure 7.7.

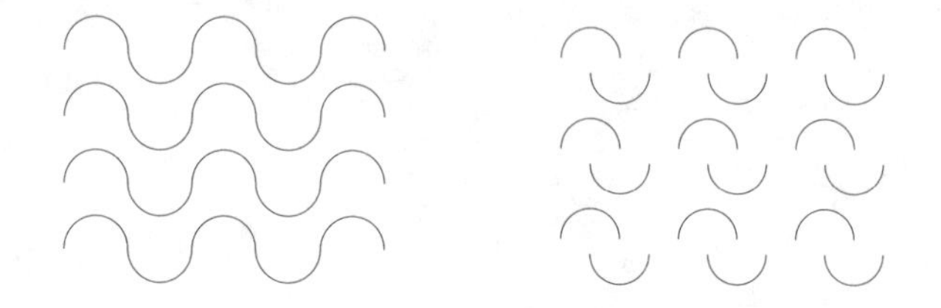

Figure 7.7 *Test systems consisting of circular arcs. The test system on the left is called the Merz grid [392].*

Consider a uniformly random test system (Section 7.1.3) consisting of translated copies Q_j of a curve Q_0 with isotropic orientation distribution (Section 4.2.4). By equation (4.34) we have, for any rectifiable plane curve Y,

$$\mathbf{E}\left[\sum_j N(Y \cap Q_j)\right] = \frac{\frac{1}{2}\, L(Y)L(Q_0)}{A(D_0)} \tag{7.10}$$

so that an unbiased estimator of curve length is

$$\widehat{L}(Y) = 2\frac{A(D_0)}{L(Q_0)} \sum_j N(Y \cap Q_j). \tag{7.11}$$

It should be noted carefully that this technique, when applied to plane sections of a surface, gives an unbiased estimator of the profile length of the surface on the section, but does not compensate for a lack of three-dimensional isotropic random orientation of the section planes. We still need to randomise the orientation of the section planes if surface area is to be estimated.

7.2.3 *Length and surface area in space from IUR serial sections*

The prime design is that of IUR serial section planes in $\mathbf{R}^3$

$$\mathcal{S} = \{T_{\Omega, U+mt} : m \in \mathbf{Z}\},$$

where $T_{\omega,u}$ denotes the plane with polar coordinates ω, u as defined in Section 4.1.3, and where Ω and U are independent and uniformly distributed over S^2_+ and $[0,t)$, respectively. The random set of planes $\mathcal{S}$ are called an isotropic, uniform random (IUR) serial section stack. Figure 7.1 is an appropriate illustration, except that the common orientation of the planes is no longer fixed but isotropically random.

The identities (4.12)–(4.14) yield, respectively,

$$\mathbf{E}[\sum_m A(Y \cap T_m)] = \frac{1}{t} V(Y) \tag{7.12}$$

$$\mathbf{E}[\sum_m L(Y \cap T_m)] = \frac{\pi}{4t} S(Y) \tag{7.13}$$

$$\mathbf{E}[\sum_m N(Y \cap T_m)] = \frac{1}{2t} L(Y) \tag{7.14}$$

in the same notation, where $T_m = T_{\Omega, U+mt}$. Thus, an IUR serial section stack enables us to estimate volume, surface area and length, knowing only the spacing t between the sections. Thus, unbiased estimators of the volume, surface area and length of spatial features Y are

$$\widehat{V}(Y) = t \sum_m A(Y \cap T_m) \tag{7.15}$$

$$\widehat{S}(Y) = \frac{4t}{\pi} \sum_m L(Y \cap T_m) \tag{7.16}$$

$$\widehat{L}(Y) = 2t \sum_m N(Y \cap T_m). \tag{7.17}$$

It is surface area and length estimation that are of principal interest, since volume can be estimated without randomising the orientation, as shown in Section 7.1. The set-up is illustrated in Figure 7.8.

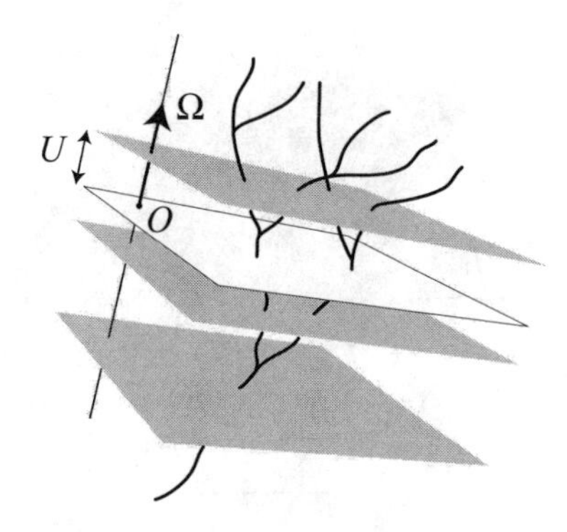

Figure 7.8 *Estimation of length from an IUR serial section stack.*

In some applications, it is not practical to analyse the entire section $Y \cap T_m$. A typical example comes from microscopy where the magnification needed for identifying Y may be so high that $Y \cap T_m$ is too large for complete observation. In such situations, subsampling is needed.

We have already been dealing with subsampling procedures for estimating $A(Y \cap T_m)$ in Section 7.1.3 on systematic subsampling using test systems. Here, we will concentrate on procedures for estimating $L(Y \cap T_m)$, the case of $N(Y \cap T_m)$ is dealt with in Exercise 7.11. For simplicity, we will denote $Y \cap T_m$ by Y which from now on is regarded as a planar curve in $\mathbf{R}^2$. Let us consider a uniform random test system $\{U + Q_j\}$ where U is uniform in a fundamental region D_0, $Q_j = u_j + Q_0$ and $Q_0 \subset D_0$. There are here two cases to consider, the case where Q_0 is a domain and the case where Q_0 is a set of curves. In the first case, $L(Y)$ can be estimated unbiasedly by

$$\widehat{L}(Y) = \frac{A(D_0)}{A(Q_0)} \sum_j L(Y \cap (U + Q_j)). \tag{7.18}$$

The unbiasedness can be proved, using (4.22) with $m = d = 2$ and $k = 1$. In order to estimate $L(Y)$ in the second case, we also need to randomise the orientation of the test system $\{U + Q_j\}$. Let R_Θ be a rotation with the uniform angle Θ in $[0, 2\pi)$. The unbiased estimator of $L(Y)$ becomes

$$\widehat{L}(Y) = \frac{\pi}{2} \frac{A(D_0)}{L(Q_0)} \sum_j N(Y \cap R_\Theta(U + Q_j)). \tag{7.19}$$

The unbiasedness is in this case a consequence of (4.30).

7.2.4 Estimating surface area using lines in space

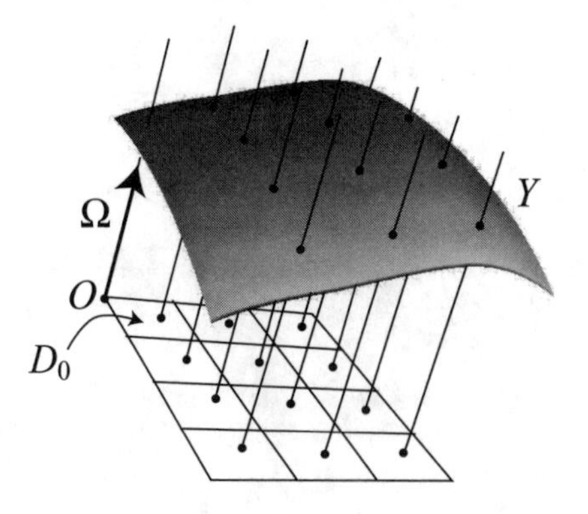

Figure 7.9 *Estimation of surface area from an IUR line grid in* $\mathbf{R}^3$.

Surface area can also be estimated using a systematic array of parallel lines in $\mathbf{R}^3$ (a 'bed of nails' or 'fakir's bed'). See Figure 7.9.

Geometry for lines in three-dimensional space was discussed in Section 4.1.4, where we introduced the coordinates (ω, t). There is the complication that the coordinates are not a Cartesian product: the displacement vector t must belong to the plane $\omega^\perp$ perpendicular to the direction ω. See Figure 4.8.

To construct a grid of parallel lines we first fix an orientation θ in the upper hemisphere S^2_+, then choose a regular grid of points $\{t_j\}$ in $\omega^\perp$, obtaining the line grid $\{T_{\omega,t_j} : j \in \mathbf{Z}^2\}$. To generate an **isotropic, uniformly random (IUR) grid of parallel lines** the common direction Ω of the lines should be isotropic random, i.e. uniformly random over S^2_+, and given Ω, the positions t_j of the lines should constitute a uniformly random grid (Section 7.1.2) of points in the plane $\Omega^\perp$, whose fundamental region D_0 is congruent to a fixed domain in $\mathbf{R}^2$.

Using an IUR grid of parallel lines, an unbiased estimator of the surface area $S(Y)$ of a curved surface in space is

$$\widehat{S}(Y) = 2A(D_0) \sum_j N(Y \cap T_j) \tag{7.20}$$

where $T_j = T_{\Omega,t_j}$. The sum on the right hand side is the total number of times the grid lines 'pierce' the surface Y.

Again, an important condition for unbiasedness of the estimator (7.20) is that the orientation Ω be isotropically random. The variance may be high if the surface has a strong preferential orientation.

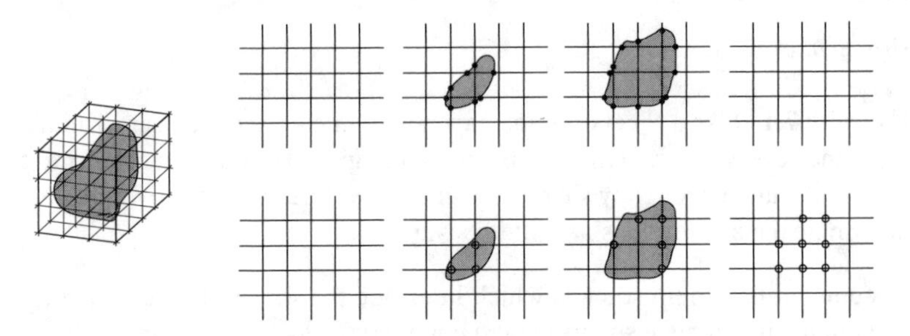

Figure 7.10 *Estimating the surface area (upper row) and volume (lower row) of a three-dimensional region using the spatial grid.*

One way to reduce the variance due to orientation is to use a test system containing test lines at several different orientations. Sandau [497, 498] proposed using a 'spatial grid' consisting of lines in three orthogonal directions. This has an elegant practical implementation in optical microscopy, as shown in Figure 7.10. The optical focal plane is positioned at successive depths $d, d+h, d+2h, \ldots$ in a systematic random sample of period h. At each position of the focal plane, a two-dimensional grid of lines is superimposed on the image, and we count the number of intersections between these lines and the profile of the surface Y. This yields the intersection count for two of the three sets of parallel test lines. The third set of test lines is traced out by the crossing-points between the two-dimensional test grid lines as we move the focal plane continuously up and down. Thus, as the focal plane is moved *between* the successive sample depths, we count any situations where the profile of the surface coincides with one of the grid crossing-points. This yields the intersection count for the third set of test lines. If all three systems of lines were arranged at the same spacing h then the unbiased estimator of surface area is $\frac{2}{3}h^2N$ where N is the total intersection count. See also [326, 372].

7.2.5 *Estimating length in space using test surfaces*

It is also possible to estimate the length of a spatial curve using a curved surface as the test system. This is analogous to the use of test curves to estimate curve length in the plane (Section 7.2.2).

Mouton *et al.* [428] investigated the case where the test curves are spheres. This method can be implemented in optical microscopy using a computer to draw the profile of the test sphere at each depth of the focal plane.

7.3 Exercises

Practical exercises

Exercise 7.1 Take a sheet of graph paper, with fine lines ruled every 1 mm and bold lines every 5 mm. Trim off the blank margins. Turn the paper over to the blank side, and cut out any shape you wish. Now turn the paper over again and determine the area of the shape in two ways:

(a) count all the 1-mm squares which lie inside the shape. (Decide what to do with the incomplete squares at the boundary.)

(b) count the number of crossing points of the bold lines which fall in the shape. Multiply this by 25.

Compare the accuracy of the two measurements, and compare the time they took.

Exercise 7.2 Photocopy onto a transparent (overhead projector) sheet a rectangular grid of horizontal and vertical lines, equally spaced at 1 cm in each direction. Use it to measure the area of one of the following: (a) a lake on a map (small enough for the earth's curvature to be negligible); (b) a patch of colour on the fur of a domestic pet; or (c) a slice of bread.

Exercise 7.3 Take a piece of cotton thread about 10–20 cm long and measure its length when it is held straight without stretching. Now place the thread in any curved position on a sheet of white paper. Hold it down with transparent tape. Placing the transparency from Exercise 7.2 over the page, estimate the length of the thread stereologically. Compare the estimate to the truth. Repeat.

Exercise 7.4 Hard-boil an egg, and remove the shell carefully. Measure the egg volume using water displacement. Now slice the egg evenly using an egg slicer, and estimate the volume of the egg and the volume of the yolk. Compare the two estimates of egg volume. Discuss sources of non-sampling error (such as the water meniscus, deflection of the egg slicer wires, etc).

Theoretical exercises

Exercise 7.5 This exercise concerns estimation of area from observation on a grid of lines. Let Y be a bounded subset of $\mathbf{R}^2$ and let

$$\mathcal{T} = \{T_{U+ks} : k \in \mathbf{Z}\},$$

where $T_h = \{(x,y) : y = h\}$ and U is uniform random in $[0,s)$, cf. Figure 7.11. Show that

$$\widehat{A}(Y) = s \sum_k L(Y \cap T_{U+ks})$$

is an unbiased estimator of $A(Y)$ where L denotes length.

Figure 7.11 *Estimation of area in* $\mathbf{R}^2$ *from lengths of intersection with a uniform line grid.*

Exercise 7.6 Let Y be a bounded subset of $\mathbf{R}^3$ and let

$$\mathcal{J} = \{J_{U+mt} : m \in \mathbf{Z}\},$$

where J_{U+mt} is the slab of thickness d bounded by $T_{\omega,U+mt}$ and $T_{\omega,U+d+mt}$, $\omega \in S^2_+$ is a fixed direction, $0 < d < t$ and $U \sim \mathsf{Unif}([0,t))$. Show that

$$\widehat{V}(Y) = \frac{t}{d} \sum_m V(Y \cap J_{U+mt})$$

is an unbiased estimator of $V(Y)$.

[*Hint:* use $V(Y \cap J_u) = \int_u^{u+d} A(Y \cap T_{\omega,v})\,\mathrm{d}v$.]

Exercise 7.7 Let the situation be as described in Section 7.1.2. Construct a proof of unbiasedness of $\widehat{V}(Y)$ in (7.6) by using that

$$\widehat{V}(Y) = v \sum_{k,l,m} \mathbf{1}\{U_{k,l,m} \in Y\}$$

and that $\mathbf{E}\widehat{V}(Y)$ can be expressed as a single integral over $\mathbf{R}^3$.

Exercise 7.8 Show the mean-value results (7.12)–(7.14).

Exercise 7.9 This exercise concerns Buffon's needle problem. Let us consider a grid of parallel lines in the plane with distance h between neighbour lines. The needle is a line-segment of length $l \leq h$. Buffon considered a fixed line grid onto which the needle was dropped with random position and orientation. He found that the probability that the needle crosses a grid line is $2l/(\pi h)$.

We will here show this result but with the randomness reversed. The needle is now a fixed line-segment Y while the line grid is randomly placed. More specifically, the line grid is denoted $\mathcal{T}$ and given by

$$\mathcal{T} = \{T_{\Theta,U+mh} : m \in \mathbf{Z}\},$$

where Θ and U are independent random variables, $\Theta \sim \mathsf{Unif}([0,\pi))$ and $U \sim \mathsf{Unif}([0,h))$, cf. Figure 7.12.

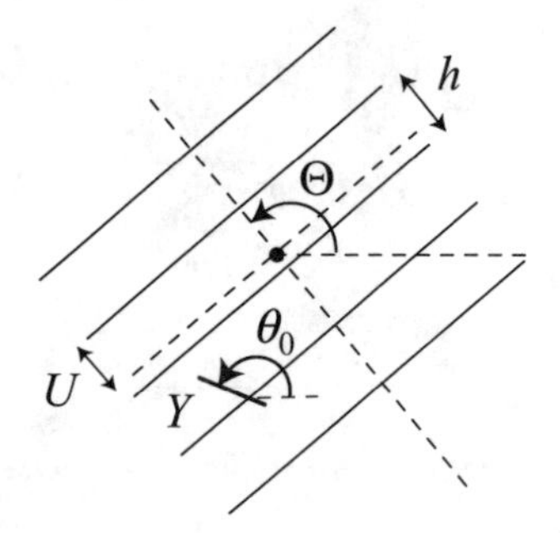

Figure 7.12 *The Buffon needle problem.*

1) Show that

$$P(Y \text{ is hit by } \mathcal{T} \,|\, \Theta = \theta) = \frac{l}{h}|\cos(\theta - \theta_0)|,$$

where $\theta_0 \in [0,\pi)$ is the angle that Y makes with the x-axis.

2) Show, using 1), that the unconditional probability is

$$P(Y \text{ is hit by } \mathcal{T}) = \frac{2l}{\pi h}.$$

Consider now n independent placements of the line grid, $(\Theta_i, U_i), i = 1, \ldots, n$. Let $X_i = 1$ if the ith line grid hits the needle and $= 0$, otherwise.

3) Show that

$$\widehat{l} = \frac{\pi h}{2} \frac{1}{n} \sum_{i=1}^{n} X_i$$

is an unbiased and consistent estimator of l.

4) Find the distribution of $\hat{l}$ by simulation, for $n = 10, 100$ and 1000 and selected values of $l \nearrow h$.

Exercise 7.10 This exercise concerns the estimation of length in the plane, using a line grid with random position and orientation. The exercise is a continuation of Exercise 7.9. Recall that the line grid is given by

$$\mathcal{T} = \{T_{\Theta, U+mh} : m \in \mathbf{Z}\},$$

where Θ and U are independent and uniform in $[0, \pi)$ and $[0, h)$, respectively. We let C be a planar curve of finite length.

1) Show, using (4.5), that

$$\widehat{L}(C) = \frac{\pi}{2} h \sum_{m} N(C \cap T_{\Theta, U+mh})$$

is an unbiased estimator of $L(C)$.

2) Show that

$$\mathbf{E}[\widehat{L}(C)|\Theta] = \check{L}(C),$$

where

$$\check{L}(C) = \frac{\pi}{2} \int_{-\infty}^{\infty} N(C \cap T_{\Theta, u}) \, \mathrm{d}u.$$

3) From 2), it follows that $\check{L}(C)$ is an alternative *unbiased* estimator of $L(C)$. Show that

$$\mathrm{Var}\check{L}(C) \leq \mathrm{Var}\widehat{L}(C).$$

Replication is performed using a systematic set-up with n line grids at angles

$$\Theta_i = \Theta + \frac{i\pi}{n}, \quad i = 0, 1, \ldots, n-1,$$

where $\Theta \sim \mathsf{Unif}([0, \frac{\pi}{n}))$. For $n = 2$, two orthogonal line grids are used.

Exercise 7.11 Let Y be a finite set of points in $\mathbf{R}^2$ and let $\{U+Q_j\}$ be a uniform random test system, cf. Section 7.1.3. Thus, U is uniform in a fundamental region D_0, $Q_j = u_j + Q_0$ and $Q_0 \subseteq D_0$ is a set with positive finite area.

- Construct an unbiased estimator of the number $N(Y)$ of points in Y based on observation in $\{U+Q_j\}$.

In Figure 2.8 of Chapter 2, a section is shown where this kind of procedure is appropriate. The set Y is here the fine dots in the right region of the image. The dots represent profiles of microvilli. It is clearly tedious to count *all* dots.

Exercise 7.12 In this exercise, we will develop a method of estimating the length of a spatial curve Y from isotropic projections.

Let Ω be an isotropic direction (a uniform point on S^2_+) and consider the plane $\Omega^\perp$, passing through the origin O with Ω as normal vector. Let $\mathcal{T}_\Omega$ be a random line grid in $\Omega^\perp$ with distance h between neighbour planes, as described in Exercise 7.10. Let $p_{\Omega^\perp}$ be the orthogonal projection onto $\Omega^\perp$. We will show in this exercise that

$$\widehat{L}(Y) = 2hN(p_{\Omega^\perp}Y \cap \mathcal{T}_\Omega)$$

is an unbiased estimator of $L(Y)$. For simplicity, we assume that Y is a piecewise linear curve.

1. Show that
$$\mathbf{E}[\widehat{L}(Y)|\Omega] = \frac{4}{\pi}L(p_{\Omega^\perp}Y).$$

Let Y_i be one of the line segments, constituting Y.

2. Show that
$$L(p_{\Omega^\perp}Y_i) = \sin\Theta_i \cdot L(Y_i),$$
where Θ_i is the angle between Ω and Y_i.
3. Show that
$$\mathbf{E}L(p_{\Omega^\perp}Y_i) = \frac{\pi}{4}L(Y_i).$$
4. Combine the results in 1.)–3.).

7.4 Bibliographic notes

Many of the fundamental principles of measuring volume were known to Archimedes, and to Cavalieri [75]. The "Cavalieri" volume estimator (7.1) seems to have been used in many practical contexts in the 19th century (cf. [557, p. 21]). The connection between Cavalieri's principle and the stereological volume estimator (7.1) was mentioned as early as 1935 in [341].

The first *probabilistic* statements about estimation of length, area and volume seem to be those of Buffon [68], Crofton [105] and Steinhaus [534, 535]. See also the reviews [115, 123]. Test systems in stereology originate with Rosiwal [486], Thomson [559] and Glagolev [200]. See Underwood [562] for the early history of test system methods. Parallel research on estimation of area in cartography is surveyed in [350]. Important theoretical work on the performance of test systems, and their relation to systematic sampling, is due to Moran [419, 420, 422] and others cited in Chapter 13. See also [302].

The spatial grid originates with Sandau [497, 498] and has been developed by many others including [326, 372]. 'Virtual' test surfaces were developed in [273, 342, 428]. The groundwork for these techniques was laid in stochastic geometry, e.g. [387, 433].

CHAPTER 8

Vertical and local designs

This chapter describes sampling designs in which the section plane is not isotropically oriented, or is not uniformly positioned in the experimental material. *Vertical sections* are plane sections parallel to a fixed axis. *Axial sections* are constrained to contain a fixed axis. *Local sections* are constrained to pass through a fixed point.

These special orientations and positions of the section plane are often used in microscopy, to reveal structural information, to see structures which are not visible on other sections, or for practical convenience. However, they are not suitable for most classical stereological techniques, because they are preferential samples which introduce sampling bias. Indeed axial and local sections are data-dependent (their position depends on the contents of the material).

It was long believed that there is no basis for stereological inference from these sections (without making assumptions about the material), since the classical stereological techniques require "isotropic uniform random encounters between the structure and the probe".

Stereological methods for these sections were discovered in the 1950's and 60's but were not widely understood. Materials scientists A.G. Spektor [531] and J.E. Hilliard [260] discovered that surface density S_V could be estimated from vertical sections, in a model-based setting, if the surface is assumed to be isotropic about the vertical axis. Geologist H. Röthlisberger [487] developed an estimator of mean particle size in which particles are sampled using test points and a line is drawn through each of these points (now called a 'point-sampled intercept'). G. Matheron and J. Serra, pioneers of mathematical morphology, developed the related concept of star volume for measuring the sizes of pores in rock [231, 232], [513, pp. 98], [514, pp. 325, 332, 347–350]. Of these three techniques, only the star volume gained a substantial following in stereology at the time.

With the development of design-based stereology in the 1980's it was discovered that there are indeed stereological principles for vertical, axial and local sections, which are valid on the same general terms as stereology for IUR sections [25, 33, 293]. These principles include the three techniques mentioned above. This chapter presents both the design-based and model-based formulations, with particular emphasis on design-based systematic sampling methods.

8.1 Vertical sections

8.1.1 Motivation

'Vertical' sections are plane sections which are constrained to be parallel to a fixed direction (nominated as the 'vertical' direction). Equivalently, they are sections constrained to be perpendicular to a fixed plane that is nominated as the 'horizontal'. See Figure 8.1. Examples are sections of a cylinder parallel to its central axis; sections of a flat slab normal to the surface of the slab; and sections of an object sitting on a table obtained by cutting sections perpendicular to the table surface.

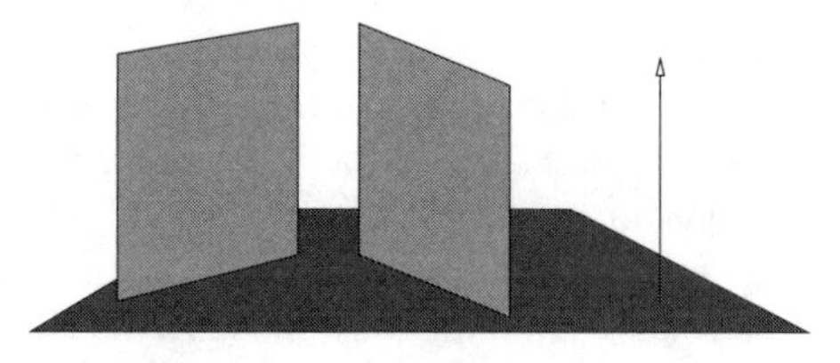

Figure 8.1 *Vertical sections (light grey) are plane sections that are parallel to the vertical direction (arrow) or equivalently are perpendicular to the horizontal plane (dark grey).*

We stress that the choice of 'vertical' direction is quite arbitrary so far as the theory is concerned. The vertical axis may or may not have any special meaning in relation to the experimental material.

We may choose to align the vertical direction with a special orientation in the experimental material. Examples of vertical sections of this kind occur when

- muscle tissue is cut longitudinally to the main axis of the muscle;
- skin is cut perpendicularly to the macroscopic plane of the skin surface;
- layered materials (sedimentary rock, synthetic materials) are cut perpendicularly to the interfaces between the layers;
- materials under stress (load-bearing bone, magnetised materials, concrete foundations) are cut parallel to the direction of stress;
- metal fracture surfaces are cut perpendicularly to the macroscopic plane of fracture;
- a cylindrical core sample of Antarctic ice is cut longitudinally.

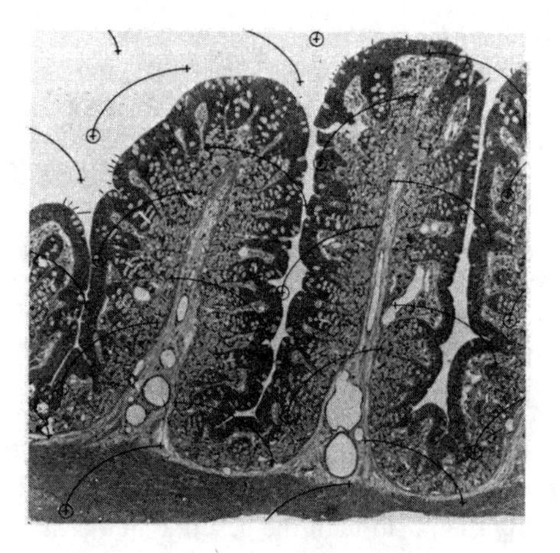

Figure 8.2 *Vertical section of rat colon. See text. Reproduced from [33] by kind permission of Blackwell Publishing. Copyright © 1986 Blackwell Publishing.*

For example, Figure 8.2 shows a vertical section of tissue from the wall of the colon (large intestine). The section is oriented so that the villi — finger-shaped features extending from the colon wall — are sectioned longitudinally. (The test system in the figure is explained further below.) We emphasise again that this type of section does not satisfy the requirements of any stereological method covered in the book up to this point, with the exception of estimators of volume.

Alternatively the vertical direction may be completely arbitrary, and chosen for convenience in the sampling design. Examples of this kind occur when:

- a needle biopsy of the hip bone (yielding a narrow cylinder of bone) is cut longitudinally;
- a biological organ is placed on a table and sliced with the blade perpendicular to the table surface;
- samples of biological tissue are cut into elongated bars ('French fries'), rolled around their long axis, then sliced longitudinally.

We emphasise that vertical planes have two degrees of freedom: rotation about the vertical axis, and parallel translation. A transverse section of a cylinder is not a vertical section because it is not free to rotate. An axial section of a cylinder (i.e. a plane that must pass through the central axis of the cylinder) is not a vertical section because it is not free to shift away from the axis.

8.1.2 Random vertical planes

Adopt Cartesian coordinates in which the z axis is parallel to the vertical direction and the (x,y) plane is horizontal. Then we define a vertical plane V in $\mathbf{R}^3$ as any plane of the form

$$V = V_{\theta,p} = \{(x,y,z) : x\cos\theta + y\sin\theta = p\},$$

where $\theta \in [0,\pi)$ and $p \in \mathbf{R}$. This also specifies a system of polar coordinates (θ, p) for vertical planes. See Figure 8.3.

In geometrical terms, the position of any vertical plane V is determined by the position of its baseline, which is the line where V intersects the horizontal plane (and also its orthogonal projection onto the horizontal plane). The coordinates (θ, p) specify the position of the baseline in the (x,y) plane, exactly as in section 4.1.2.

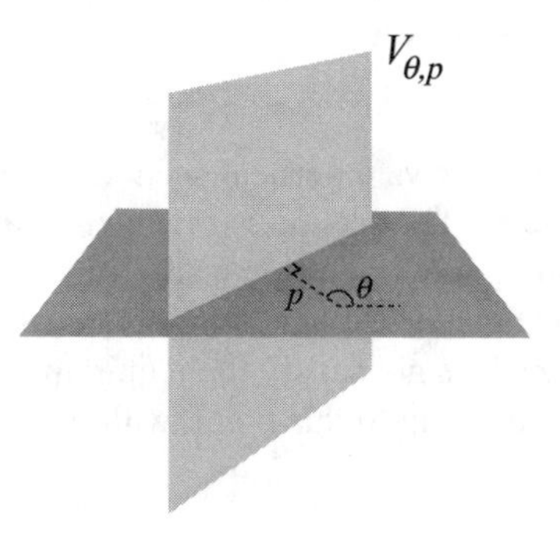

Figure 8.3 *A vertical plane can be specified by its polar coordinates* θ *and* p.

Until now we have used the letter T to denote all types of probes. In this chapter we manipulate several types of probes simultaneously. We shall use the symbols V, E and T to denote vertical planes, planes with general orientations, and lines, respectively.

The natural uniform measure for vertical planes in $\mathbf{R}^3$ is

$$dV = dp\,d\theta \tag{8.1}$$

corresponding to the uniform measure for lines in the plane (section 4.1.2).

The theory of random lines in the plane (Sections 4.1.2 and 5.2) carries over directly to a theory for random vertical planes in $\mathbf{R}^3$. We can define a *vertical uniform random* **(VUR)** plane hitting a domain X in $\mathbf{R}^3$ as a random vertical plane $V_{\Theta,P}$ whose coordinates Θ, P are jointly uniformly distributed over the set of all (θ, p) pairs such that $X \cap V_{\theta,p} \neq \emptyset$. Equivalently, the baseline $T_{\Theta,P}$ is an IUR line hitting the horizontal projection of X. Similarly, a *VUR system of serial sections* is an array

$$\mathcal{V} = \{V_{\Theta,U+mt} : m \in \mathbf{Z}\}$$

of parallel vertical planes at constant spacing $t > 0$ such that Θ and U are independent with $\Theta \sim \mathsf{Unif}([0,\pi))$ and $U \sim \mathsf{Unif}([0,t])$. Equivalently, the baselines of these planes form an IUR system of parallel lines in the horizontal plane.

It should be understood that VUR planes constitute a biased sample of three-dimensional materials. For example, consider a line segment in three dimensions. The probability that the segment is hit by a VUR plane is proportional to the length of the projection of the line segment onto the horizontal plane. A segment which is nearly vertical has almost zero probability of being cut by a VUR plane. Thus, there is an orientation-dependent bias if we sample curves in space using vertical planes. A similar statement is true about the sampling of surfaces using vertical planes.

8.1.3 Geometrical identities for vertical planes

Clearly, the volume of an object Y in $\mathbf{R}^3$ can be determined from its cross-sectional areas on vertical planes:

$$\int_{\text{vertical planes}} A(Y \cap V)\,\mathrm{d}V = \int_0^{\pi} \int_{-\infty}^{\infty} A(Y \cap V_{\theta,p})\,\mathrm{d}p\,\mathrm{d}\theta = \pi V(Y) \qquad (8.2)$$

by Fubini's theorem or (4.18). Thus, it will be possible to estimate absolute volume and volume fraction from VUR planes or VUR serial sections. The estimation procedures are equivalent to those which use FUR planes or FUR serial sections.

A more challenging problem is to find geometric identities which would enable us to determine lower-dimensional quantities such as surface area and length. This is indeed possible for surface area, without specific assumptions about the orientation and other characteristics of the surfaces. A corresponding result for spatial curve length seems unobtainable.

Factorisation of line measure

In order to derive a geometric identity for surface area $S(Y)$ for vertical planes, recall that the area of a surface can be determined from its intersections with line probes:

$$\int_{\text{lines}} N(Y \cap T)\,\mathrm{d}T = \pi S(Y), \qquad (8.3)$$

from equation (4.17), where the integral is over all lines T in $\mathbf{R}^3$ with respect to the invariant measure $\mathrm{d}T$ defined in (4.15).

A key observation is that every test line T in $\mathbf{R}^3$ is contained in a vertical plane, which is unique if the line is not vertical (not parallel to the z-axis), cf. Figure 8.4. It is therefore possible to generate any straight line in $\mathbf{R}^3$ by taking a vertical plane and then taking a line within this plane.

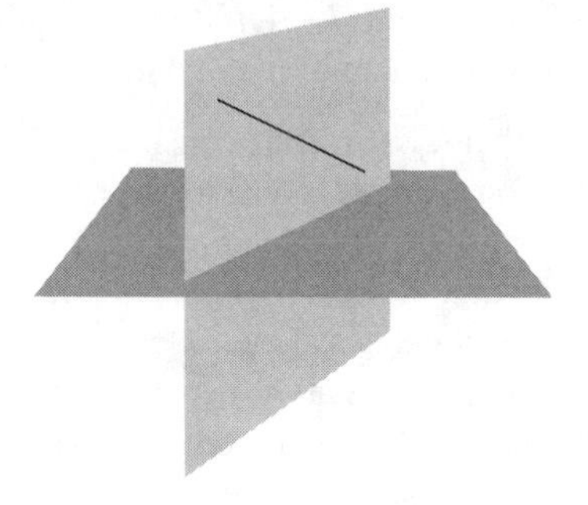

Figure 8.4 *Any non-vertical line is contained in a unique vertical plane.*

Let us try to rewrite (8.3) as a double integral over vertical planes V and test lines T inside V. We need to decompose the uniform density for lines in $\mathbf{R}^3$

$$\mathrm{d}T = \mathrm{d}t\,\mathrm{d}\omega$$

in terms of the uniform density on vertical planes. Recall that $\omega \in S^2_+$ is the direction of the line while $t \in \omega^\perp$ indicates the position of the line.

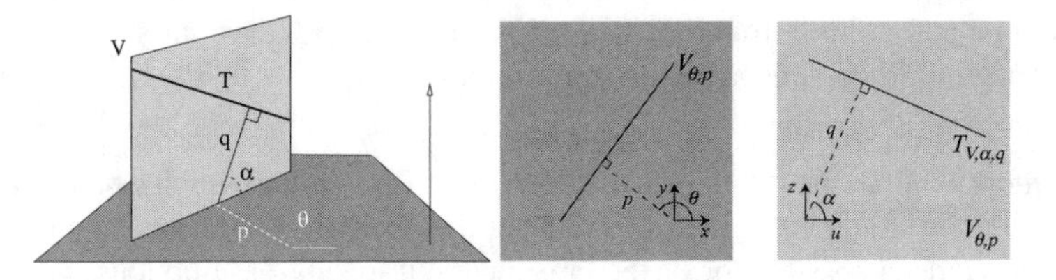

Figure 8.5 *Coordinates for specifying a test line T inside a vertical section plane V. Left: three-dimensional view. Middle: top view of vertical plane. Right: parametrisation of line inside the vertical plane.*

Consider an arbitrary test line T which is not parallel to the z-axis. Let $V = V_{\theta,p}$ be the unique vertical plane containing T. Introduce coordinates (α, q) specifying the position of the test line $T = T_{V,\alpha,q}$ within V as shown in Figure 8.5. For fixed V, these are just the polar coordinates of a line in the plane. Then the invariant measure $\mathrm{d}T$ for lines in space can be expressed as

$$\mathrm{d}T = \mathrm{d}t\,\mathrm{d}\omega = \sin\alpha\,\mathrm{d}\alpha\,\mathrm{d}q\,\mathrm{d}V. \tag{8.4}$$

Details of the proof of (8.4) are given at the end of the section.

We recognise $\mathrm{d}\alpha\,\mathrm{d}q$ as the invariant measure for lines $T_{V,\alpha,q}$ in the vertical plane V. We might write this as $\mathrm{d}\alpha\,\mathrm{d}q = \mathrm{d}T_V$ so that (8.4) reads

$$\mathrm{d}T = \sin\alpha\,\mathrm{d}T_V\,\mathrm{d}V.$$

Identity for surface area

Using the decomposition (8.4), we can now rewrite equation (8.3) using Fubini's theorem:

$$\begin{aligned}\pi S(Y) &= \int_{\text{lines}} N(Y\cap T)\,\mathrm{d}T \\ &= \int_{\text{vertical planes}} \int_{\text{lines} T\subset V} N(Y\cap T_{V,\alpha,q})\sin\alpha\,\mathrm{d}\alpha\,\mathrm{d}q\,\mathrm{d}V \\ &= \int_{\text{vertical planes}} \int_{\text{lines} T\subset V} N(Y\cap T_{V,\alpha,q})\sin\alpha\,\mathrm{d}T_V\,\mathrm{d}V \end{aligned} \quad (8.5)$$

holding for any two-dimensional curved surface Y in $\mathbf{R}^3$ of finite surface area $S(Y)$ satisfying weak regularity and rectifiability conditions. Here $N(C\cap T)$ is the number of intersection points between a curve C and the test line T.

It follows that we can determine surface area $S(Y)$ stereologically from vertical sections. Within each vertical section plane V, we take test lines T, count the number of intersections $N(Y\cap V\cap T)$ between each test line T and the surface profile $Y\cap V$, and *weight these counts by the sine of the angle between the test lines and the vertical direction.*

We may also rewrite (8.5) as

$$\pi S(Y) = \int_{\text{vertical planes}} W(Y\cap V)\,\mathrm{d}V \quad (8.6)$$

where $W(Y\cap V)$ denotes the inner integral

$$W(C) := \int_{\text{lines} T\subset V} N(C\cap T_{V,\alpha,q})\sin\alpha\,\mathrm{d}q\,\mathrm{d}\alpha \quad (8.7)$$

for any rectifiable curve C in V. If C is differentiable, and $\beta(s)$ denotes the angle between the vertical axis and the tangent to the curve at point s, then

$$\begin{aligned} W(C) &= \int_0^{\pi}\int_C |\sin(\beta(s)-\alpha)|\sin\alpha\,\mathrm{d}s\,\mathrm{d}\alpha \\ &= \int_C ((\frac{\pi}{2}-\beta)\cos\beta+\sin\beta)\,\mathrm{d}s. \end{aligned} \quad (8.8)$$

These results were obtained in the 1980's by Baddeley [25, 26, 27] and collaborators [33].

Proof of factorisation

In order to prove (8.4), let us use the 'Cartesian' coordinates (u,z) inside V, where $u = -x\sin\theta + y\cos\theta$. Then a line in V can be specified by the equation $u\cos\alpha + z\sin\alpha = q$. Since the coordinate axes in V are $(-\sin\theta, \cos\theta, 0)$ and $(0,0,1)$, we get

$$T_{V,\alpha,q} = \{(x,y,z) \;:\; x\cos\theta + y\sin\theta = p, \\ -x\sin\theta\cos\alpha + y\cos\theta\cos\alpha + z\sin\alpha = q\}.$$

Observe that the direction vector of the line is

$$\begin{aligned} \omega &= \cos(\alpha + \frac{\pi}{2})(-\sin\theta, \cos\theta, 0) + \sin(\alpha + \frac{\pi}{2})(0,0,1) \\ &= (\sin\theta\sin\alpha, -\cos\theta\sin\alpha, \cos\alpha). \end{aligned}$$

It follows that

$$d\omega = \sin\alpha\, d\alpha\, d\theta.$$

Furthermore, the plane through the origin orthogonal to $T_{V,\alpha,q}$ is spanned by the two orthonormal vectors $(\cos\theta, \sin\theta, 0)$ and $(-\sin\theta\cos\alpha, \cos\theta\cos\alpha, \sin\alpha)$. The projection of T onto this plane has coordinates (p,q) in this coordinate system. Therefore,

$$dt = dp\, dq.$$

Accordingly, the invariant measure for lines in $\mathbf{R}^3$ can be factorised as

$$dT = dt\, d\omega = (dp\, dq)(\sin\alpha\, d\alpha\, d\theta) = \sin\alpha\, d\alpha\, dq\, dV.$$

8.1.4 Estimating surface area from vertical sections

Surface area can be estimated from vertical sections using (8.5) and (8.6), in several ways, depending on the stochastic interpretation. We illustrate this with several designs based on systematic sampling.

Serial vertical sections

Consider again a stack of vertical uniform random (VUR) serial sections with fixed spacing $t > 0$,

$$\mathcal{V} = \{V_{\Theta, U+mt} : m \in \mathbf{Z}\},$$

where Θ and U are independent, $\Theta \sim \mathsf{Unif}([0,\pi))$ and $U \sim \mathsf{Unif}([0,t))$, respectively. The identity (8.6) for vertical planes yields

$$\mathbf{E}\left[\sum_m W(Y \cap V_{\Theta, U+mt})\right] = \frac{1}{t} S(Y). \tag{8.9}$$

Thus, surface area can be estimated from observation on a systematic set of vertical planes. The unbiased estimator takes the form

$$\widehat{S}(Y) = t \sum_m W(Y \cap V_{\Theta,U+mt}). \tag{8.10}$$

This estimator can sometimes be implemented in digital image analysis. If the surface profile $Y \cap V$ has been identified as a 1-pixel-thick region in the image, then we can compute the right hand side of (8.7) in software by numerical integration, evaluating $N(Y \cap V \cap T)$ for all lines T using the Hough transform. Alternatively if the profile $Y \cap V$ is available as a smooth curve, perhaps obtained by curve-following techniques, then we may compute (8.8) by direct integration.

Parallel test lines

It will often be too laborious or inefficient to measure $W(Y \cap V)$ exactly. Instead we may estimate $W(Y \cap V)$ using a grid of parallel test lines in the vertical plane.

In the representation (8.7), a weighting factor of $\sin\alpha$ is applied to intersection counts $N(Y \cap T)$ with test lines at orientation α. Note that $\sin\alpha$ is the sine of the angle between the test lines and the vertical direction (since $\min(\alpha, \pi - \alpha)$ is the angle between the common direction of the lines and the z-axis).

One way to incorporate the weighting factor $\sin\alpha$ which does not require extensive calculation is to generate a grid of parallel test lines with a weighted orientation distribution, say

$$\mathcal{T} = \{T_{V,\alpha,M+kd} : k \in \mathbf{Z}\} \tag{8.11}$$

where α is random with the *sin-weighted density*

$$f(\alpha) = \frac{1}{2}\sin\alpha, \quad \alpha \in [0, \pi), \tag{8.12}$$

α and M are independent, and $M \sim \mathsf{Unif}([0,d))$ where $d > 0$. Then an unbiased estimator of $W(C)$ is

$$\widehat{W}(C) = 2d \sum_k N(C \cap T_{V,\alpha,Q+kd}). \tag{8.13}$$

Estimating each summand in (8.10) this way gives an unbiased estimator of $S(Y)$,

$$\widehat{S}(Y) = 2td \sum_m \sum_k N(Y \cap V_m \cap T_{mk}) \tag{8.14}$$

where T_{mk} denotes the kth test line in the mth vertical section.

Notice that (8.14) has exactly the same form as the estimator (7.20) of surface area from IUR parallel test lines inside IUR serial sections (page 167). This occurs because in both designs *the test lines are IUR in three dimensions*.

Cycloid test curves

Another way to incorporate the weighting factor $\sin\alpha$ is to use a system of test *curves*.

Suppose that Q is a bounded curve whose orientation distribution* has the sin-weighted density (8.12). Consider a uniformly random test system (Section 7.1.3) consisting of translated copies $U + Q_j$ of Q where $U \sim UD_0$. Then by equation (4.33) we have, for any rectifiable plane curve C,

$$\mathbf{E}\left[\sum_j N(C \cap (U + Q_j))\right] = \frac{L(Q)}{A(D_0)} \mathbf{E}[|\sin(\Theta - \Psi)|] L(C)$$

where Θ and Ψ are independent random variables drawn from the orientation distributions of Q and C, respectively. Since Θ has the sin-weighted density (8.12) we get from (8.8) that

$$\mathbf{E}\left[\sum_j N(C \cap (U + Q_j))\right] = \frac{L(Q)}{2A(D_0)} W(C)$$

so that an unbiased estimator of $W(C)$ is

$$\widehat{W}(C) = 2\frac{A(D_0)}{L(Q)} \sum_j N(C \cap (U + Q_j)), \tag{8.15}$$

which requires simply counting the number of intersections between the test curves and the curve C.

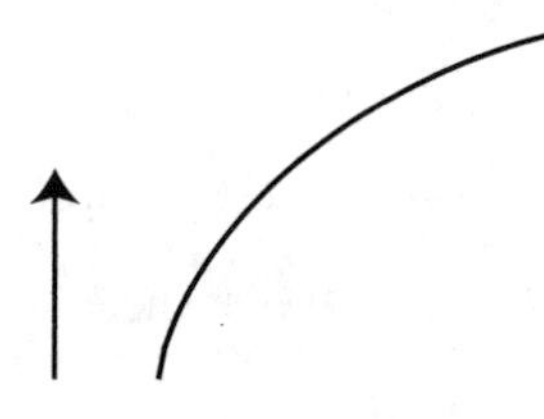

Figure 8.6 *The cycloid. Parametric equation* $x = t - \sin t$, $y = 1 - \cos t$. *The arc shown here corresponds to* $t \in [0, \pi]$ *and has height* 2, *width* π, *and curve length* 4. *The cycloid has orientation distribution proportional to* $\sin\alpha$, *where* α *is the angle to the vertical axis (indicated by the arrow).*

A curve which has sin-weighted orientation distribution (8.12) is shown in Figure 8.6. This is the *cycloid* studied in classical geometry: the curve traced by a point on the rim of a rolling wheel.

* The 'orientation distribution' of a curve was defined in Section 4.2.4 and used in Section 7.2.2.

Two examples of test systems composed of cycloids are shown in Figure 8.6. In applications, the arrow at the left must be aligned with the vertical direction in the vertical section. The test system should then be randomly translated. Note that the diagonal lines are not part of the test system but are drawn to facilitate counting.

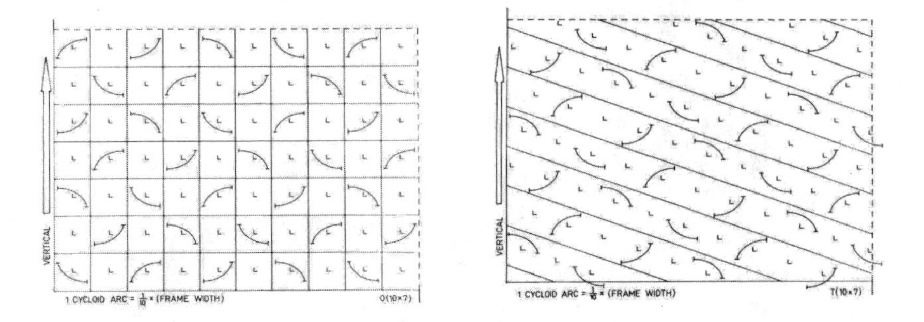

Figure 8.7 *Two test systems composed of cycloid arcs. The minor axis of each cycloid is parallel to the vertical arrow, which must be aligned with the vertical direction on the plane section. Reproduced from [33] by kind permission of Blackwell Publishing. Copyright © 1986 Blackwell Publishing.*

The final estimator of total surface area $S(Y)$ from VUR serial sections and cycloid test curves is

$$\widehat{S}(Y) = 2t\frac{A(D_0)}{L(Q_0)}\sum_m\sum_j N(Y \cap V_m \cap Q_{mj}) \tag{8.16}$$

where V_m denotes the mth vertical section and Q_{mj} denotes the jth cycloid test curve in V_m.

Note again that this estimator has exactly the same form as an analogous estimator of $S(Y)$ from IUR serial sections using isotropically oriented test curves. Uniformly-translated cycloid test curves in VUR sections are effectively equivalent to isotropic test lines in three dimensions.

Figure 8.8 shows a real application of vertical sections to estimate the macroscopic surface area of an object. Panel (a) shows the object whose surface area is to be estimated. In panel (b) the object has been embedded in plaster-of-Paris and cut into three slabs by a set of parallel but arbitrarily-oriented cutting planes. On each slab, a pair of roughly longitudinal marks is made on roughly opposite sides of the base. In panel (c) each slab has been placed roughly at the centre of a circle graduated evenly with marks labelled 1 to m ($m = 36$). A systematic sample of the integers from 1 to m was chosen, with period $k = m/2n$ where $n = 3$ is the number of slabs. The first slab was oriented so that the pencil mark was aligned with the mark numbered r, the second with $r + k$ and the

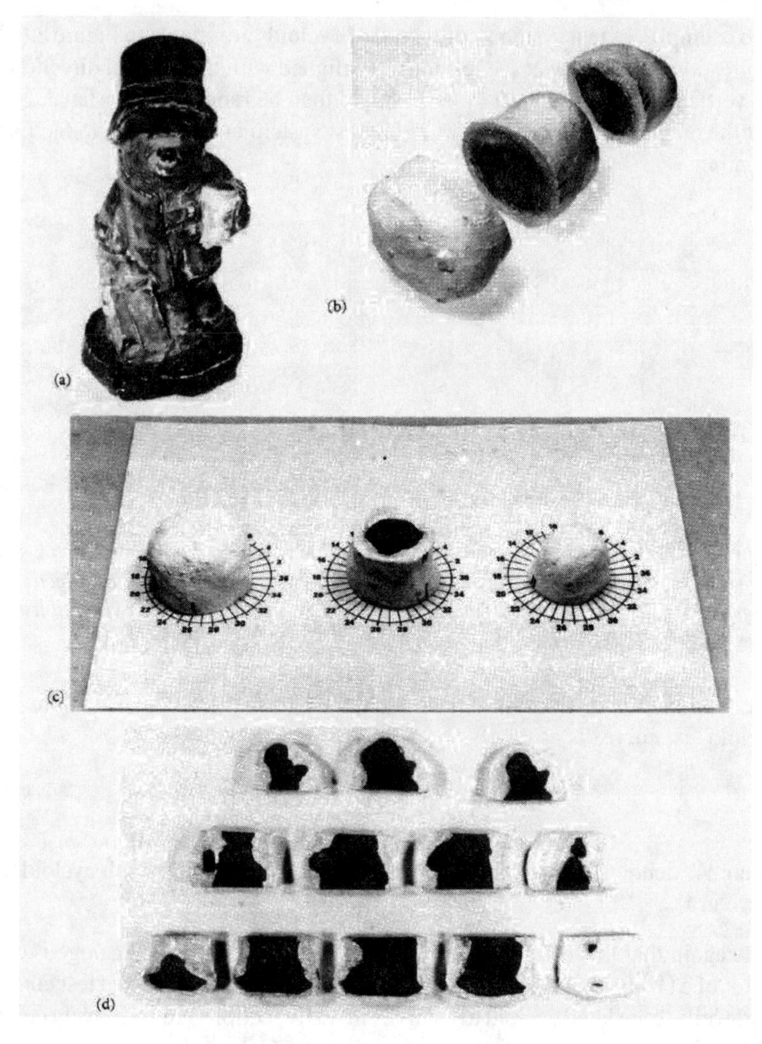

Figure 8.8 *Vertical section sampling design for estimating the surface area of a macroscopic object. See text for explanation. Reproduced from [33] by kind permission of Blackwell Publishing. Copyright © 1986 Blackwell Publishing.*

third with $r + 2k$. Then each slab was cut exhaustively into vertical sections (normal to the paper) with baseline parallel to the long edge of the paper (the direction marked by the numerals 18 and 36 on the circles). Panel (d) shows the complete set of vertical sections, with each row originating from one of the slabs. The vertical direction is identifiable in each section: it is the direction perpendicular to the straight-edge boundary of the section. A cycloid test system like those in Figure 8.7 is then superimposed on each of the vertical sections, and the estimate (8.16) is computed.

Other vertical section designs are sketched in Section 12.4.1 and in [33, 225].

8.1.5 Model-based estimation of S_V

A closely-related, model-based technique allows us to estimate the surface density S_V from a vertical section. Assume that the material is statistically homogeneous with respect to translations, but is isotropic only with respect to rotations about the vertical axis. Then it can be shown from (8.5) that

$$S_V = 2I_L \tag{8.17}$$

where I_L is the number of intersections per unit length of cycloid test curve in a vertical section.

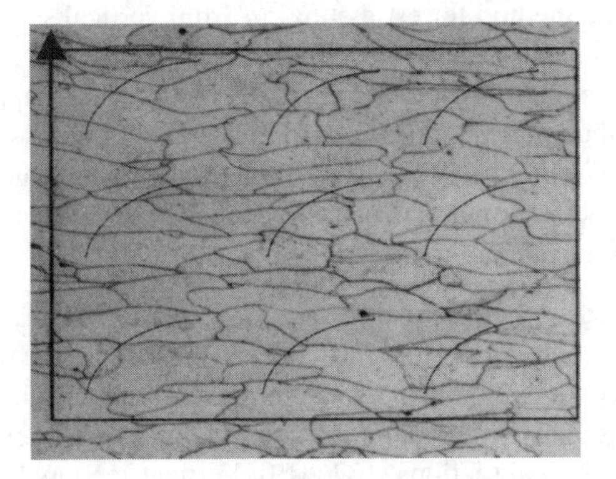

Figure 8.9 *Cycloid test system superimposed on a vertical metallographic section of cold rolled extra low carbon steel. The chosen vertical direction (arrow) is the direction perpendicular to the faces of the rolled plate. The number of intersections between the test curves and the anisotropic grain boundaries is $I = 36$. The true length of the minor axes of the cycloids was 138 μm. The length of a cycloid quarter-arc is two times its minor axis. Therefore the observed number of intersections per unit length of test curve is $I_L = 36/(9\times 2\times 138\ \mu m) = 0.0145\ \mu m^{-1}$. By courtesy of Prof. A. Gokhale, Materials Science and Engineering, Georgia Institute of Technology, Atlanta.*

Figure 8.9 shows an application to estimating S_V for the grain boundaries in rolled steel. The section has been taken perpendicular to the faces of the rolled plate. The isotropy of the steel microstructure is destroyed by the rolling process, violating the condition of isotropy which is essential for the classical method of estimating S_V described in Chapter 2. Instead, a cycloid test system is superimposed on the vertical section image. The vertical arrow on the test system must be aligned with the vertical direction in the section. Then S_V is estimated via (8.17) resulting in an estimate of $S_V = 0.029\,\mu\text{m}^{-1}$.

This model-based technique was developed in 1960 by Spektor as the 'method of the intersecting cycloid' [531] and later rediscovered by Hilliard [260]. Both authors obtained this as part of a theory of stereological estimation for orientation distributions of surfaces [259, 464, 530]. See also [495, p. 186 ff.].

This method applies only under the assumption that the material is isotropic with respect to rotation about the vertical axis — or at least that the orientation distribution of the surface of interest has this property. We emphasise that the design-based method for estimating surface area from vertical sections requires no assumptions about the surface orientation.

8.1.6 Applications of vertical sections

The model-based method for estimating S_V from vertical sections is now frequently applied in materials science [328, sect. 3.7, pp. 121–129]. The assumption of isotropy about the vertical axis is justified in many applications [6, p. 2601]. The method is commonly used to investigate fracture surfaces, following the work of Wojnar [610]. It is an effective way to estimate surface roughness [205] without assuming a stochastic model of the surface geometry. It is also used to investigate grain growth [276, 426].

The design-based method for estimating S or S_V from vertical sections is employed in many contexts in biological science and medicine. In neuroscience, vertical sections allow the microscopist to detect transitions between different layers [162]. Similarly in dermatology and in bone [569]. Membrane loss can be measured in vertical sections [137, 319]. Vertical section designs are useful in sampling flat or inaccessible tissues such as the gums and teeth [275, 611] and plants [565] and complex tissue such as the lung [393, 451, 605]. Vertical section sampling designs are discussed further in Chapter 12.

A modification of this technique is the *trisector*, a group of three vertical sections at equal angles, proposed in [204]. It yields more efficient estimates of S or S_V, when it can be applied.

8.2 Vertical slices and vertical projections

8.2.1 Motivation

Vertical plane sections do not allow us to estimate curve length. There is no known geometrical identity expressing the length $L(Y)$ of an arbitrary space curve Y in terms of its intersection counts $N(Y \cap V)$ with vertical planes V. This appears to be impossible: for example, a line segment that is aligned with the vertical axis has zero intersections with almost all vertical planes.

However, curve length may be estimated stereologically when the curve is *projected* onto a vertical plane. Figure 8.10 shows an example of an application in which the roots of grass plants, growing in a transparent medium, have been projected onto a plane which is parallel to the vertical direction (the direction of gravity). The aim is to determine the total length of roots.

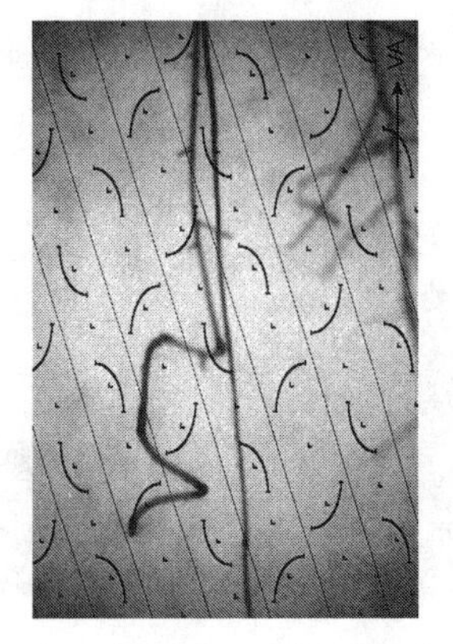

Figure 8.10 *Projection of vertical slice of root system of crested wheatgrass [612]. Seedlings growing in transparent gel were visualised through one eyepiece of a stereo macroscope. Video camera input, cycloid test system superimposed in vector graphics by CAST-Grid system, computer monitor output shown. True size of viewing frame* 12.9 × 18.5 *mm. Copyright © Kluwer Academic Publishers. Reproduced from [612] by kind permission of Kluwer Academic Publishers and Dr D. Wulfsohn (University of Saskatchewan, Canada and Royal Veterinary and Agricultural University (KVL) Denmark).*

Other examples arise when

- concrete is sawn into slices by cutting vertically (parallel to the direction of gravity or load stress), and the wire reinforcing filaments in the concrete are visualised using X-ray projected images;
- thick sections of brain cortex are sliced perpendicular to the outer surface of the cortex, and dendrites (thin filaments associated with each neuron) are seen in opaque projection;
- a bone biopsy, a core sample of ice, or a metal rod is sliced longitudinally (parallel to the central axis) and a thick section is viewed in opaque projection.

This type of sampling was first described by materials scientist and stereologist A. Gokhale [201].

8.2.2 *Derivation of basic identity*

Recall from Figure 2.23 that, when we project a space curve onto a plane, we may regard test lines in the plane of projection as equivalent to test *planes* in space. Now the length of a space curve is determined by its intersections with all planes through equation (4.14),

$$\int_{\text{planes}} N(Y \cap E)\,\mathrm{d}E = \pi L(Y) \tag{8.18}$$

where $\mathrm{d}E$ is the invariant measure for planes in $\mathbf{R}^3$, equation (4.8).

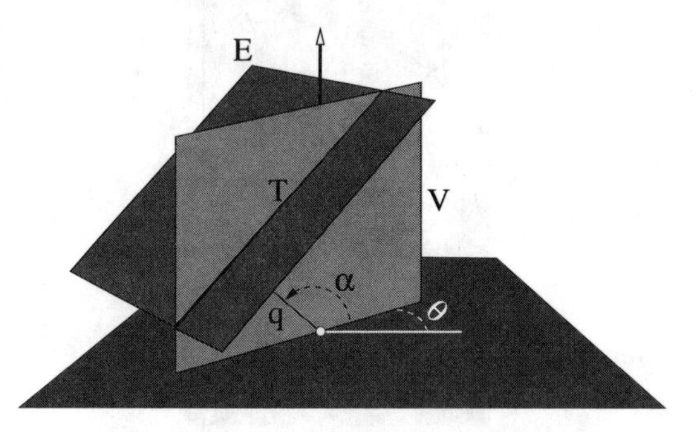

Figure 8.11 *A test line T in a vertical plane V corresponds to a test plane E in space.*

Let V be a vertical plane, and suppose that we project three-dimensional space orthogonally onto V. Any test line T inside V corresponds to a test plane E in space. Here E is the plane perpendicular to V such that $E \cap V = T$.

Without loss of generality we will suppose that V passes through the origin, say $V = V_{\theta,0}$. Using the coordinate system for test lines T in V that was adopted in Section 8.1, the test line is $T_{V,\alpha,q}$ and the corresponding plane E (shown in Figure 8.11) is the plane with equation

$$E_{\theta,\alpha,q} = \{(x,y,z) : \quad -x\sin\theta\cos\alpha + y\cos\theta\cos\alpha + z\sin\alpha = q\}$$

which has normal vector

$$\omega = (-\sin\theta\cos\alpha, \cos\theta\cos\alpha, \sin\alpha)$$

and distance q from the origin. Comparing with (4.8) and (4.9) in Section 4.1.3, we find that the invariant measure for planes in three dimensions is equivalent to

$$\mathrm{d}E = |\cos\alpha|\;\mathrm{d}q\,\mathrm{d}\alpha\;\mathrm{d}\theta.$$

This yields the geometrical identity

$$\int\!\!\int N(Y \cap E_{\theta,p,\alpha,q})|\cos\alpha|\;\mathrm{d}q\,\mathrm{d}\alpha\;\mathrm{d}\theta = \pi L(Y)$$

or equivalently

$$\int_0^\pi \int_{\text{lines } T \subset V} N(p_V(Y) \cap T) |\cos \alpha| \, dT_V \, d\theta = \pi L(Y) \tag{8.19}$$

where $p_V(Y)$ is the orthogonal projection of the curve Y onto the vertical plane $V = V_{\theta,0}$, and again dT_V is the invariant measure for lines within V. The outer integral involves simply rotating the plane of projection V about the vertical axis.

Thus, test planes perpendicular to the vertical plane V should be weighted by the *cosine* of their angle to the vertical. This also leads to estimators of curve length using cycloid test curves in the vertical plane, with the minor axis *perpendicular* to the vertical axis, instead of parallel to the vertical axis as before. See Figure 8.12.

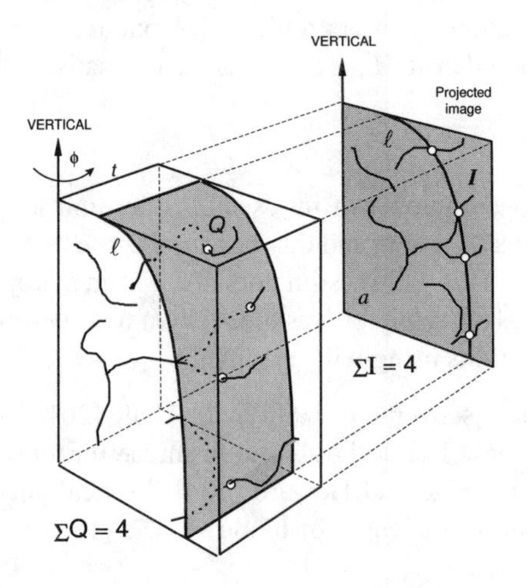

Figure 8.12 *A cycloid test curve in the projected image corresponds to a cycloidal test surface in space. After [201]. Copyright © Kluwer Academic Publishers. Reproduced from [612] by kind permission of Kluwer Academic Publishers and Dr D. Wulfsohn.*

8.2.3 Estimation of curve length from vertical projections

To estimate the length $L(Y)$ of a curve from vertical projections, we first project the curve onto vertical planes at several different angles of rotation $\theta_1, \ldots, \theta_m$. These angles should constitute a uniform sample from $[0, \pi)$, typically a systematic sample. Then we superimpose a test system of cycloids Q_{jk} onto the projected images (with uniformly random translation of the test system) and count the number of crossing points between cycloids and projected curves.

The estimator of $L(Y)$ is the numerical approximation to the integral (8.19),

$$\widehat{L}(Y) = \frac{1}{m} \sum_k \sum_j N(Q_{jk} \cap p_{V_{\theta_k,0}}(Y)). \tag{8.20}$$

8.2.4 *Estimation of curve length from vertical slices*

The stereology of projected thick sections was mentioned in Section 2.7. In this context, length density L_V can be estimated from *vertical* projected thick sections, known as *vertical slices*. A vertical slice of fixed thickness $t > 0$ is the region between two parallel vertical planes $V_{\theta,p}, V_{\theta,p+t}$. The contents of the slice is projected orthogonally onto one of these boundary vertical planes.

In a model-based setting, we assume that the experimental material is homogeneous under translation and is isotropic under rotations about the vertical axis. Then the true length density L_V of the space curve satisfies

$$L_V = \frac{2}{t} I'_L \tag{8.21}$$

where I'_L is the mean number of intersections per unit length of test system, between a cycloid test system and the *projected* curve resulting from a vertical slice of thickness t. This yields estimators of L_V from a projected vertical slice. Note that (8.21) has the same form as the relation for determining L_V from IUR projected thick sections using IUR test lines.

These results were discovered in 1990 by Gokhale [201, 202] initially for the case of thick sections. The design-based formulae for total curve length were developed by Cruz-Orive and Howard [134]. Vertical slice techniques have been used to estimate the length of blood vessels [18, 40, 484] and dendritic trees [271].

The 'vertical spatial grid' [135] is a virtual spatial probe allowing estimation of surface area and length.

8.3 Local lines

8.3.1 *Motivation*

In local stereology [293], geometrical quantities are estimated from lines and planes passing through reference points.

Figure 8.13 shows a plane section which is effectively a random plane through a fixed reference point in space. The fixed point is the tiny dot visible at the intersection of the cross-hairs.

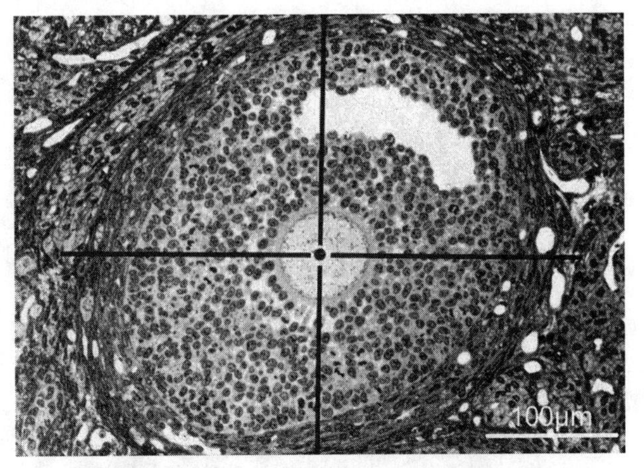

Figure 8.13 *Local section of mouse ovarian follicle. The section passes through a predetermined fixed point, the oocyte nucleolus, visible at the intersection of the cross-hairs. Reproduced from Bagger et al [34] by kind permission of Blackwell Publishing. Copyright © 1993 Blackwell Publishing.*

This is a section of the ovarian follicle of a mouse, and the fixed point is the nucleolus of the oocyte. The moderately sized circular region of white space around the nucleolus is (the profile of) the oocyte, while the large elliptical region is the profile of the ovarian follicle. Inside the follicle, around the oocyte, are granulosa cells, of which several hundred profiles are shown.

To obtain Figure 8.13, tissue from the ovarian follicle was cut in IUR serial sections. All sections were inspected and this section was selected because it contains a profile of the nucleolus. Hence, the sampling procedure effectively *conditions* on the event that this pointlike nucleolus is hit by the section plane.

This special position of the section plane is quite informative about the spatial organisation of the follicle. The question is what stereological inferences can be drawn from such preferential samples.

A different type of local sampling is the point-sampled intercepts technique demonstrated in Figure 8.14. The image is an IUR plane section of a particle population. A system of test points (+) has been superimposed on the image. For each test point which hits a particle profile (7 of the 9 test points in the Figure) we draw a line from the test point in a random direction, terminating at the boundary of the particle profile. This line is called a 'point-sampled intercept'. Sample moments of point-sampled intercept length are stereologically related to population weighted moments of particle volume.

A fundamental difference between Figures 8.13 and 8.14 is that the reference point is fixed (and dependent on the material) in Figure 8.13, while the test

Figure 8.14 *Point-sampled intercepts technique applied to plane section of ceramic-metal composite (from Figure 2.16).*

points are randomly located in Figure 8.14 (and contribute to the sample if they fall inside a particle). These samples have different probabilistic interpretations. In Figure 8.13 the population of nucleoli is sampled with uniform probability, while in Figure 8.14 the particles are sampled with probability proportional to their volume.

The local geometrical identities described here have been developed mainly for applications to particle analysis. In such applications, the relevant identity is applied to each particle from a sample of particles; see Chapter 11. Each sampled particle is analysed using a probe passing through a reference point in the particle.

In the remainder of this section, we consider a measurable subset Y of $\mathbf{R}^3$ with finite volume, and discuss how to estimate its volume using such local probes. Although the results are formulated without reference to particles, the prime example concerns a particle Y.

8.3.2 *Convex and star-shaped sets*

We will use the following geometrical terms; see e.g. [192].

A set Y in two or three dimensions is called *convex* if, for any two points $y, y' \in Y$, the line segment $[y, y']$ joining these points is included in Y. The set Y is called *star-shaped* with respect to a point x if, for any point $y \in Y$, the line segment $[x, y]$ is included in Y. See Figure 8.15. If Y is convex, then Y is star-shaped relative to any point in $\mathbf{R}^3$.

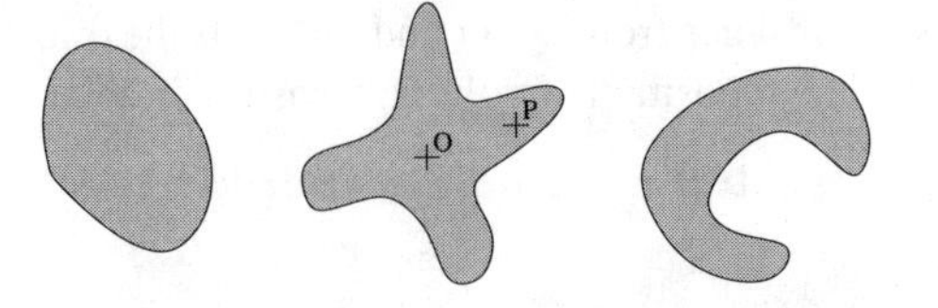

Figure 8.15 *Left: a convex set. Middle: a set which is star-shaped with respect to point O but not point P. Right: a set that is neither convex nor star-shaped.*

For sets Y of general shape, it is convenient to define the *star set* of Y with respect to a point x in Y as the set

$$\mathsf{star}(Y,x) = \{y \in Y : [x,y] \subset Y\}$$

of all points y in Y such that $[x,y]$ is included in Y. See Figure 8.16. If Y is a cave and x is the location of a strong light source, then $\mathsf{star}(Y,x)$ is the part of the cave that is illuminated.

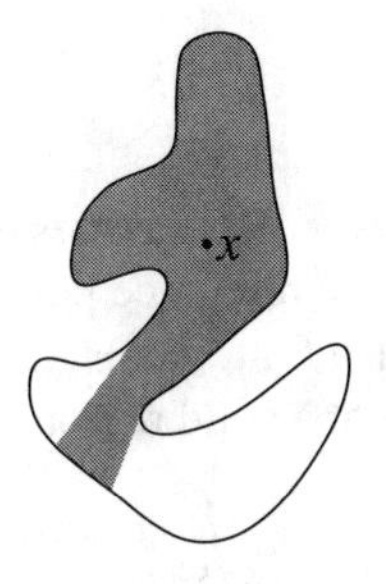

Figure 8.16 *The star set of an object Y from a point x is the set* $\mathsf{star}(Y,x)$ *of points in Y that can be seen directly from x.*

8.3.3 *Geometrical identities*

The volume of a solid can be expressed as an integral in Cartesian, polar or cylindrical coordinates. Each such expression leads to a different integral-geometric identity and stereological method for volume.

Here we deal with the case of polar coordinates, which leads us to the techniques of point-sampled intercepts and the 'nucleator' method.

Volume in polar coordinates

Any point $x \in \mathbf{R}^3$ can be expressed in polar coordinates as

$$x = r\omega = r(\sin\theta\cos\varphi, \sin\theta\sin\varphi, \cos\theta), \quad (\theta,\varphi) \in [-\frac{\pi}{2}, \frac{\pi}{2}] \times [0, 2\pi),\ r \geq 0$$

where $r = |x|$ is the distance from O to x and $\omega \in S^2$ is the direction of the vector from O to x. The decomposition of volume measure in polar coordinates is

$$\mathrm{d}x = r^2\,\mathrm{d}r\,\mathrm{d}\omega = r^2 \sin\theta\,\mathrm{d}r\,\mathrm{d}\varphi\,\mathrm{d}\theta. \tag{8.22}$$

See Figure 8.17.

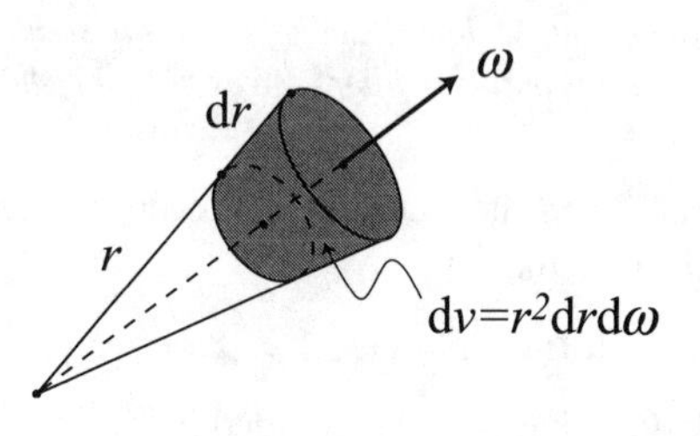

Figure 8.17 *Representation of volume in polar coordinates.*

Rays through a fixed point

This leads to some integral geometric identities involving lines that pass through a fixed reference point. Take the reference point to be the origin O. Let $R_\omega = \{r\omega : r \geq 0\}$ be the 'ray' (half-infinite line) extending from O in a given direction $\omega \in S^2$. Then the volume of a solid region Y in $\mathbf{R}^3$ can be written

$$\begin{aligned} V(Y) &= \int_{S^2} \int_{Y \cap R_\omega} r^2\,\mathrm{d}r\,\mathrm{d}\omega \\ &= \int_{S^2} \alpha(Y \cap R_\omega)\,\mathrm{d}\omega \end{aligned} \tag{8.23}$$

where (identifying R_ω with $[0,\infty)$) for $A \subset [0,\infty)$

$$\alpha(A) = \int_A r^2\,\mathrm{d}r. \tag{8.24}$$

This takes a simple form if Y is convex or star-shaped. Suppose O lies inside Y, and that Y is convex, or more generally that Y is star-shaped with respect to O. Then $Y \cap R_\omega$ is a line segment $[0,y]$. In this case

$$\alpha(Y \cap R_\omega) = \int_0^{|y|} r^2\,\mathrm{d}r = \frac{1}{3}|y|^3$$

is a multiple of the cubed length of the intercept, and (8.23) becomes

$$V(Y) = \frac{1}{3} \int_{S^2} L(Y \cap R_\omega)^3\,\mathrm{d}\omega. \tag{8.25}$$

Since the total measure of directions is $\int_{S^2} \mathrm{d}\omega = 4\pi$, we obtain the fact that, if Ω is an isotropic random direction (in S^2) and Y is star-shaped with respect to

O, then

$$\mathbf{E}\left[\frac{4\pi}{3}L(Y\cap R_\omega)^3\right]=V(Y) \tag{8.26}$$

so that an unbiased estimator of the volume of a convex or star-shaped set is $4\pi/3$ times the cubed length of its intercept with a ray generated from O in a random direction. For example, if Y is a sphere of radius R and centre O, then this estimator is constant and equal to the true volume $(4\pi/3)R^3$.

Lines through a fixed point

Polar coordinates also yield a geometric identity for local lines, i.e. doubly-infinite lines passing through the reference point O. A line in $\mathbf{R}^3$ through O with unit direction vector $\omega\in S^2_+$ will be denoted by

$$T_\omega=\{x\in\mathbf{R}^3 : x=r\omega \text{ for some } r\in\mathbf{R}\}.$$

Then the volume $V(Y)$ of a measurable set Y in $\mathbf{R}^3$ can be written

$$V(Y)=\int_{S^2_+}\int_{Y\cap T_\omega} r^2\,\mathrm{d}r\,\mathrm{d}\omega=\int_0^{\frac{\pi}{2}}\int_0^{2\pi}\int_{Y\cap T_{\theta,\varphi}} r^2\sin\theta\,\mathrm{d}r\,\mathrm{d}\varphi\,\mathrm{d}\theta,$$

or, in more condensed form, writing $\mathrm{d}T=\mathrm{d}\omega$,

$$\int_{\text{local lines}}\alpha'(Y\cap T)\,\mathrm{d}T=V(Y), \tag{8.27}$$

where for $A\subset T$

$$\alpha'(A)=\int_A r^2\,\mathrm{d}r.$$

If A consists of a single line-segment, say $A=[y_1,y_2]$, then

$$\alpha'(A)=\begin{cases}\frac{1}{3}[d(y_1,O)^3+d(y_2,O)^3] & \text{if } O\in[y_1,y_2]\\ \frac{1}{3}|d(y_1,O)^3-d(y_2,O)^3| & \text{otherwise,}\end{cases} \tag{8.28}$$

where $d(x,y)$ denotes the distance between the two points x and y in $\mathbf{R}^3$. If A consists of a finite number of disjoint line-segments, say

$$A=[y_1,y_2]\cup[y_3,y_4]\cup\ldots\cup[y_{2m-1},y_{2m}],$$

then $\alpha'(A)$ is the sum of such contributions,

$$\alpha'(A)=\sum_k\alpha'([y_{2k-1},y_{2k}]). \tag{8.29}$$

8.3.4 *Estimation of volume using lines through a fixed point*

Let T be an isotropic line in $\mathbf{R}^3$ through the reference point, cf. Section 5.5. Without loss of generality we can use O as reference point. It follows from

(8.27) that

$$\widehat{V}_1(Y) = 2\pi \int_{Y \cap T} r^2 \, \mathrm{d}r \tag{8.30}$$

is an unbiased estimator of $V(Y)$. If $O \in Y$ and $Y \cap T = [y_1, y_2]$ is a single line-segment, $\widehat{V}_1(Y)$ reduces to

$$\widehat{V}_1(Y) = \frac{2\pi}{3}[d(y_1, O)^3 + d(y_2, O)^3]. \tag{8.31}$$

If $Y \cap T$ consists of a finite number of line-segments, the sum of cubed distances in (8.31) is replaced by an alternating sum of cubed distances.

If Y is star-shaped with respect to O, then $Y \cap T$ always consists of a single line-segment, and the volume estimator $\widehat{V}_1(Y)$ is always of the form (8.31).

Star volume

For a set Y of general shape, with $O \in Y$, the expression (8.31) has another use. Take y_1 and y_2 to be the boundary points of Y that are nearest to O in opposite directions. Then (8.31) is an estimator of volume of the star set $\mathsf{star}(Y, O)$ defined in Section 8.3.2:

$$\mathbf{E}\left[\frac{2\pi}{3}[d(y_1, O)^3 + d(y_2, O)^3]\right] = V(\mathsf{star}(Y, O)) \tag{8.32}$$

known as the *star volume* of Y with respect to O. See [231, 232], [513, pp. 98], [514, 325, 332, 347–350].

Nucleator

A further development of the estimator $\widehat{V}_1(Y)$ is the so-called *nucleator*, cf. [220],

$$\widetilde{V}_1(Y) = \frac{1}{2}[\widehat{V}_{1a}(Y) + \widehat{V}_{1b}(Y)], \tag{8.33}$$

where $\widehat{V}_{1a}(Y)$ and $\widehat{V}_{1b}(Y)$ are the volume estimators (8.30) from two perpendicular isotropic lines, cf. Figure 8.18. One way of generating two such lines is by first generating an isotropic plane E through O and then generate two perpendicular lines through O in E (one of the lines has a uniform angle in $[0, \pi/2)$ relative to a fixed axis through O in E), see also Section 5.5. More generally, $V(Y)$ may be estimated by using a systematic set of n isotropic lines through O, generated in an isotropic plane E through O. The variance of estimators based on such systematic designs is discussed in Chapter 13.

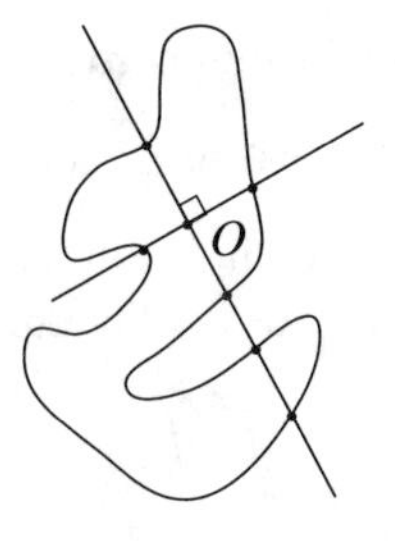

Figure 8.18 *The nucleator: volume estimation based on local lines.*

8.4 Local vertical sections

This section presents estimators of volume obtained from axial ('local vertical') sections, that is, sections that are constrained to contain a fixed axis. These estimators are obtained from the expression for volume in cylindrical coordinates.

8.4.1 Geometrical identities

We may express volume in cylindrical coordinates (z, u, ζ) where z is the height coordinate, $|u|$ is distance from the z-axis, and ζ is an angle of rotation about the z-axis. See Figure 8.19. Any point $x \in \mathbf{R}^3$ may be written

$$x = (u \cos\zeta, u \sin\zeta, z),$$

for $u \in \mathbf{R}$, $\zeta \in [0, \pi)$ and $z \in \mathbf{R}$. The corresponding decomposition of volume measure is

$$\mathrm{d}x = |u|\, \mathrm{d}u\, \mathrm{d}\zeta\, \mathrm{d}z.$$

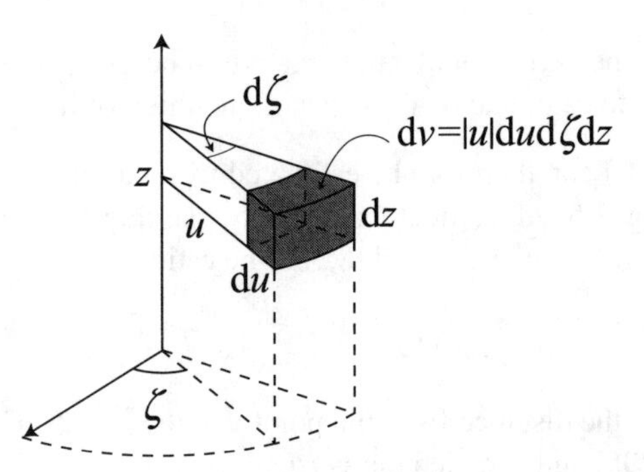

Figure 8.19 *Representation of volume in cylindrical coordinates.*

This gives a geometric identity involving *axial planes* ('local vertical planes')

$$V_{\theta,0} = \left\{ x \in \mathbf{R}^3 : x \cdot \omega_\theta = 0 \right\},$$

where $\omega_\theta = (\cos\theta, \sin\theta, 0)$. In cylindrical coordinates this is the plane $\zeta = \theta + \pi/2$. The identity states that, for a solid region Y in $\mathbf{R}^3$,

$$\begin{aligned} V(Y) &= \int_0^\pi \int_{Y \cap V_{\theta,0}} |u| \, du \, dz \, d\theta \\ &= \int_0^\pi \alpha^*(Y \cap V_{\theta,0}) \, d\theta \end{aligned} \tag{8.34}$$

where for $A \subset V = V_{\theta,0}$

$$\alpha^*(A) = \int_A |u| \, du \, dz = \int_A d(x,T) \, dx \tag{8.35}$$

where $d(x,T)$ denotes the distance from x to the vertical axis T (the z-axis in this case). The right-hand integral in (8.35) is with respect to two-dimensional Lebesgue measure.

8.4.2 *Estimation of volume from local vertical planes*

It follows that if V is a random axial plane (of the form $V = V_{\Theta,0}$ where $\Theta \in \mathsf{Unif}([0,\pi))$) then

$$\mathbf{E}\left[\pi\alpha^*(Y \cap V)\right] = V(Y) \tag{8.36}$$

so that an unbiased estimator of the volume of an arbitrary solid can be obtained by cutting Y with a random axial plane V and computing $\widehat{V}(Y) = \pi\alpha^*(Y \cap V)$.

Applying Fubini's theorem to the first integral in (8.35) we have

$$\int_A d(x,T) \, dx = \int_{-\infty}^{\infty} \int_{A \cap T_z} |u| \, du \, dz. \tag{8.37}$$

where T_z is the line perpendicular to the z-axis at height z. The inner integral of (8.37) is simple to calculate if $A \cap T_z$ is a finite union of line-segments.

The volume $V(Y)$ can therefore be estimated using a random axial plane $V = V_{\Theta,0}$, containing a fixed vertical axis T, say, passing through O. Here, Θ is uniform in $[0,\pi)$. Using (8.34) and (8.35), the estimator is

$$\widehat{V}_{2(1)}(Y) = \pi \int_{Y \cap V} d(y,T) \, dy, \tag{8.38}$$

where $d(y,T)$ is the distance from the point y to the line T. This estimator was developed in [295] and is called the *vertical rotator*.

In the special case where Y is a solid of revolution about the z-axis, with radius

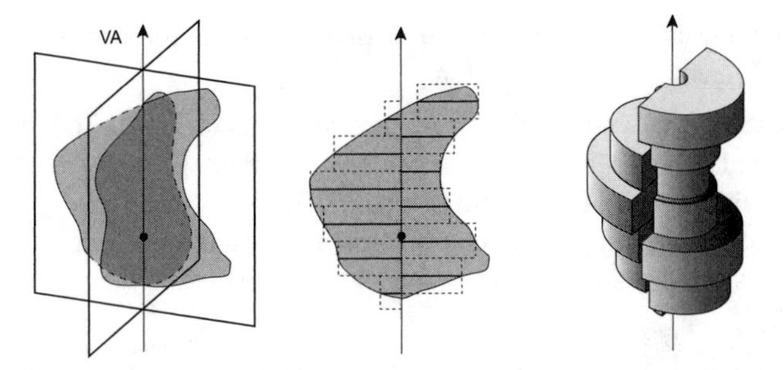

Figure 8.20 *The vertical rotator principle. The volume of an object is estimated from an axial plane section by computing the volume of the equivalent solid of revolution. Original drawing by kind permission of Prof. H.J.G. Gundersen.*

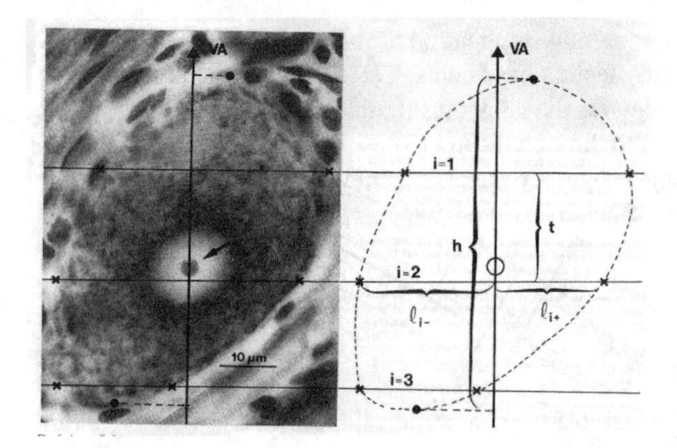

Figure 8.21 *Estimation of volume of a neuron using the vertical rotator principle. See text. Rat spinal ganglion, methacrylate embedded, 35 µm section. Reproduced from [295] by kind permission of Blackwell Publishing. Copyright © 1993 Blackwell Publishing.*

$f(z)$ at height z, all axial plane sections $Y \cap V$ are equivalent, and

$$\begin{aligned} \alpha^*(Y \cap V) &= \int \int_{-f(z)}^{f(z)} |u| \, du \, dz \\ &= \int f(z)^2 \, dz \end{aligned}$$

so that the estimator (8.38) is constant and equal to

$$\widehat{V}_{2(1)}(Y) = \pi \int f(z)^2 \, dz,$$

the exact volume of the solid of revolution. Thus, we may regard the vertical rotator method as estimating the volume of an object by computing the volume of the equivalent solid of revolution.

In applications that employ digital image analysis, the statistic $\alpha^*(Y \cap V)$ can be computed to high accuracy by tracing the boundary of $Y \cap V$ and performing numerical integration of the two-dimensional integral (8.35), or more typically by numerical integration of the one-dimensional inner integral in (8.37) which requires that for each quadrature point z_i we measure the horizontal transect $Y \cap V \cap T_{z_i}$.

In most practical applications, however, the one-dimensional integral (8.37) is estimated by systematic sampling of the z-axis with fairly small sample size. For each sampled height z_i we measure the horizontal transect $H = Y \cap V \cap T_{z_i}$ and compute $\int_H |u| \, du$. This involves measurements from boundary points on Y to a line in the section plane, see Figure 8.20.

Figure 8.21 shows the application of this principle to estimating the volume of a neuron. This section has been selected, from a stack of serial VUR sections, because it contains the nucleolus (indicated by the arrow). The focal plane has been positioned through the nucleolus. The vertical direction is determined by the sampling design: a vertical axis through the nucleolus is shown. A systematic sample of heights is then taken, with period t. At these heights, lines are drawn perpendicular to the vertical axis. The intersections of these lines with the cell boundary are marked by crosses.

8.5 Local isotropic sections

8.5.1 Geometrical identity

A geometrical identity of a slightly different kind exists for planes of arbitrary orientation passing through O. A plane in $\mathbf{R}^3$ through O with unit normal vector ω is of the form

$$E_\omega = \{x \in \mathbf{R}^3 : x \cdot \omega = 0\}.$$

The appropriate uniform measure here is given by $\mathrm{d}E = \mathrm{d}\omega$. Recall the decomposition (8.22) of the volume measure in terms of lines through the origin. In integral geometry, (8.22) is often written as

$$\mathrm{d}x^3 = d(x,O)^2 \, \mathrm{d}x^1 \, \mathrm{d}T,$$

where $\mathrm{d}x^3$ is the element of volume measure and $\mathrm{d}x^1$ is the element of length measure on T. This decomposition is a special case of a whole class of geometric decompositions of volume measure, called the Blaschke-Petkantschin formulae, cf. e.g. [293]. Among these formulae, we can also find a corresponding decomposition according to planes

$$\mathrm{d}x^3 = \frac{1}{\pi} d(x,O) \, \mathrm{d}x^2 \, \mathrm{d}E, \tag{8.39}$$

where dx^2 is the element of area measure on E. Using (8.39), we find for a measurable subset Y of $\mathbf{R}^3$

$$\int_{\text{local planes}} \alpha^\dagger(Y \cap E)\,\mathrm{d}E = V(Y), \tag{8.40}$$

where for $A \subset E$

$$\alpha^\dagger(A) = \frac{1}{\pi}\int_A d(x,O)\,\mathrm{d}x^2.$$

8.5.2 *Estimation of volume from local planes*

The local unbiased estimator of $V(Y)$ based on an isotropic plane E through O is given by

$$\widehat{V}_2(Y) = 2\int_{Y\cap E} d(y,O)\,\mathrm{d}y, \tag{8.41}$$

cf. (8.40).

8.5.3 *Estimation of surface area from local planes*

Under mild regularity conditions, it is also possible to express the area of a surface Y in $\mathbf{R}^3$ in terms of an integral with respect to local planes. This relationship is not trivial to derive but has a rather simple, explicit form

$$\int_{\text{local planes}} \alpha(Y \cap E)\,\mathrm{d}E = S(Y), \tag{8.42}$$

where for a rectifiable curve $C \subset E$

$$\alpha(C) = \frac{1}{\pi}\int_0^\pi \sum_{x\in C\cap T_1(\theta)} d(x,O)^2(1+\cot\beta(x;\theta)[\frac{\pi}{2}-\beta(x;\theta)])\,\mathrm{d}\theta. \tag{8.43}$$

Here, $T_1(\theta)$ is the line in E through O, making an angle θ with a fixed axis in E and $\beta(x;\theta)$ is the angle between $T_1(\theta)$ and the tangent to C at x, cf. Figure 8.22.

Surface area can be estimated using information on a local plane, more specifically an isotropic plane E through O. Using (8.42), an unbiased estimator of surface area can be constructed, based on information in a local plane, more specifically an isotropic plane E through O. The estimator takes the form

$$\widehat{S}(Y) = 2\int_0^\pi \sum_{x\in Y\cap T_1(\theta)} d(x,O)^2(1+\cot\beta(x;\theta)[\frac{\pi}{2}-\beta(x;\theta)])\,\mathrm{d}\theta. \tag{8.44}$$

The notation involved in this formula is explained just below (8.43). A discretised version of (8.44) is in stereological slang called the *surfactor*, see also [286]. Note that the variance of $\widehat{S}(Y)$ is 0 if Y is a ball, centred at the origin O.

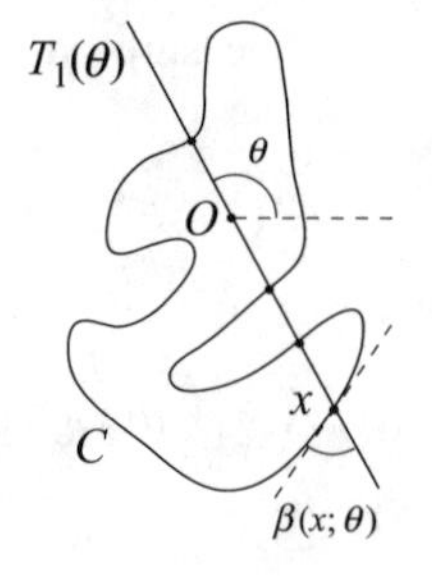

Figure 8.22 *Geometry for local estimation of surface area.*

8.6 Variance comparisons of local estimators

The volume estimators $\widehat{V}_1(Y)$, $\widetilde{V}_1(Y)$ and $\widehat{V}_2(Y)$ defined in (8.30), (8.33) and (8.41) respectively, all have variance 0 if Y is a ball centred at the origin O. Likewise, $\widehat{V}_{2(1)}(Y)$ has variance 0 if Y is rotation symmetric around the vertical axis T.

There is no closed form expression for the variances, but a nice relation can be derived between the variances of $\widehat{V}_1(Y)$ and $\widehat{V}_2(Y)$. To see this, recall that the isotropic line used in $\widehat{V}_1(Y)$ can be generated via the isotropic plane E used in $\widehat{V}_2(Y)$, see Section 5.5. It can be shown that

$$\mathbf{E}[\widehat{V}_1(Y)|E] = \widehat{V}_2(Y).$$

The estimators are thereby related by the Rao-Blackwell procedure, see also [32] and Section 13.3. It follows that

$$\begin{aligned}\mathsf{var}[\widehat{V}_1(Y)] &= \mathsf{var}[\mathbf{E}[\widehat{V}_1(Y)|E]] + \mathbf{E}\mathsf{var}[\widehat{V}_1(Y)|E] \\ &= \mathsf{var}[\widehat{V}_2(Y)] + \mathbf{E}\mathsf{var}[\widehat{V}_1(Y)|E] \\ &\geq \mathsf{var}[\widehat{V}_2(Y)].\end{aligned}$$

In [294], the coefficients of variation of $\widehat{V}_1(Y)$, $\widetilde{V}_1(Y)$ and $\widehat{V}_2(Y)$ have been determined by simulation for the case where Y is an ellipsoid centred at O with semi-axes of lengths $\beta_1 \geq \beta_2 \geq \beta_3$ where $\beta_1/\beta_2, \beta_2/\beta_3 \in \{1,2,4\}$. It was found that the coefficient of variation of $\widetilde{V}_1(Y)$ was about 2/3 of the coefficient of variation of $\widehat{V}_1(Y)$. For $\beta_1/\beta_2 \leq 2$ and $\beta_2/\beta_3 \leq 2$, the coefficients of variation of $\widetilde{V}_1(Y)$ and $\widehat{V}_2(Y)$ were both smaller than 1.

8.7 Exercises

Practical exercises

Exercise 8.1 Estimate the exterior surface area of a banana, and the surface area of the banana flesh, using vertical sections as follows. First, draw parallel lines, equally spaced 2 cm apart, on a sheet of paper. Place a banana on the sheet (no randomisation is necessary) and cut it into segments by following these guide lines. Then follow the method shown in Figure 8.8 panels (c)–(d) to obtain vertical systematic uniform sections of the entire banana with known section thickness. Estimate the surface areas with the help of a cycloid test system using equations (8.10) and (8.15).

Exercise 8.2 Take a slice of Emmenthaler cheese and apply the point sampled intercepts technique (Figure 8.14) to estimate the mean star volume of the air space (holes) by

$$\widehat{V}_{\text{star}} = \frac{4\pi}{3}\overline{\ell_0^3}$$

where $\overline{\ell_0^3}$ is the sample mean of the cubed lengths of the intercepts. See Exercise 8.4 for interpretation.

Theoretical exercises

Exercise 8.3 Derive the polar decomposition of Lebesgue measure given in (8.22), by considering the transformation

$$(r,\theta,\varphi) \to (r\sin\theta\cos\varphi, r\sin\theta\sin\varphi, r\cos\theta)$$

from $\mathbf{R}\backslash\{O\} \times [0,\frac{\pi}{2}) \times [0,2\pi)$ onto $\mathbf{R}^3\backslash\{O\}$.

Exercise 8.4 In Exercise 8.2 the estimated quantity V_{star} is the *average* star volume. In the *Random* setting, for example,

$$V_{\text{star}} = \mathbf{E}[V(\mathsf{star}(Y,O)) \mid O \in Y]$$

where Y is the stationary random set formed by the union of all holes. Suppose now that the holes are disjoint spheres, with radius distribution F. Show that V_{star} is the volume-weighted mean sphere volume of holes.

Exercise 8.5 Let C be a planar circle with centre O and radius R. Show that for such a C, (8.43) takes the form

$$\alpha(C) = 2\pi R^2.$$

Show then (8.42) for a sphere $Y \subset \mathbf{R}^3$ with centre O and radius R.

Exercise 8.6 This exercise concerns a local design for estimating the area of a bounded planar set Y.

1) Show that
$$A(Y) = \int_0^{\pi} \int_{Y \cap T_\theta} |r| \, \mathrm{d}r \, \mathrm{d}\theta,$$
where T_θ is the line through O with angle $\theta \in [0, \pi)$ relative to the x-axis.
2) Using 1), it is easily shown that
$$\widehat{A}(Y) = \pi \int_{Y \cap T_\Theta} |r| \, \mathrm{d}r$$
is an unbiased estimator of $A(Y)$ if $\Theta \sim \mathsf{Unif}([0, \pi))$. Find more explicit expressions for $\widehat{A}(Y)$ under various shape assumptions for Y.
3) Let the estimator in 2) be denoted by $\widehat{A}(Y; \Theta)$. In order to reduce the variability, systematic sampling may be used. The estimator based on n systematic lines takes the form
$$\widehat{A}_n(Y) = \frac{1}{n} \sum_{i=0}^{n-1} \widehat{A}(Y; \Theta + \frac{i}{n}\pi),$$
where $\Theta \sim \mathsf{Unif}([0, \pi/n))$. Show that $\widehat{A}_n(Y)$ is an unbiased estimator of $A(Y)$.

Methods for estimating the variance of $\widehat{A}_n(Y)$ are discussed in Chapter 13.

8.8 Bibliographic notes

Vertical sections

The cycloid was mentioned by Nicholas of Cusa, named by Galileo, and studied by Mersenne and Pascal. Sir Christopher Wren found the arc length of the cycloid.

Spektor [530] first developed techniques for directional analysis of surfaces (estimating the distribution of orientations of surface normals, and estimating the mean cosine of the normal). This led to an estimator of S_V from vertical sections which he called the ‘method of the intersecting cycloid’ [531]. It is applicable to “systems with a spatial axis of symmetry”. These results were rediscovered by J.E. Hilliard [260, 259, 464]. In a completely different context some of the results were found by Philip [463]. See [495, p. 186 ff.] for a summary of the model-based technique.

Baddeley [25, 26, 27] found the geometrical identities (8.5) and (8.6) relating absolute area S to vertical sections. These were developed into design-based techniques for estimating S in [33].

Vertical projections and vertical slices

Gokhale [201, 202] had the unprecedented insight that length can be determined from vertical projections. He developed the method for estimating L_V from vertical slices. Cruz-Orive and Howard [134] elaborated the principle for estimating absolute length L from vertical projections.

Star volume

Star volume [231, 232], [513, p. 98] is an important concept in mathematical morphology [514, pp. 325, 332, 347–350]. See also [280]. Estimation of star volume has been applied in the study of many spatial structures, including placenta [383, 378], synthetic porous media [485], trabecular bone [567, 568] and alveoli of the lung [613].

Local stereology

In 1979, R.E. Miles published some interesting integral geometric formulae valid for not necessarily convex sets [403]. These results triggered the development of local stereology. At the Second International Workshop on Stereology and Stochastic Geometry, held in Aarhus in 1983, Miles [405] and Jensen & Gundersen [283] both reported on new stereological estimators for moments of particle volume, valid without specific shape assumptions about the particles. Further developments were published in [284, 287, 407]. A comprehensive set of local geometric identities can be found in the monograph [293]. There exist many more identities than those presented in this chapter. Some of them require measurement of spatial angles which is difficult in optical microscopy. The local estimators called the nucleator, rotator and surfactor are described in detail in [220], [295] and [286], respectively.

CHAPTER 9

Ratio estimation

In 'ratio estimation', a ratio of two population parameters is estimated from sample data, usually by taking the ratio of the corresponding sample quantities. Our purpose may be to estimate the ratio for its own sake (estimation *of* ratios) or to use the ratio as an intermediate quantity in the estimation of another parameter (estimation *via* ratios) [83, chap. 6], [555, chap. 7].

Estimation of ratios is an essential technique in stereology. Many of the parameters of interest are ratios. It is traditional to derive new estimators by taking a ratio of two existing estimators, as if the 'fundamental formulae of stereology' held exactly. The statistical properties of these methods need to be understood in the context of ratio estimation.

Estimation via ratios is important in stereological sampling design. It enables us to deal efficiently with widely differing scales of organisation, especially in biological structures. It is also robust against tissue shrinkage and similar artefacts.

9.1 Ratio estimation in finite populations

This section summarises the theory of ratio estimation in finite population survey sampling. We use here standard notation from sampling theory.

9.1.1 Estimators

Consider a finite population of items numbered $1,\dots,N$ in which two quantities x_i and y_i are measurable for each item i. Define the population totals

$$X = \sum_{i=1}^{N} x_i, \qquad Y = \sum_{i=1}^{N} y_i$$

and their ratio

$$R = \frac{Y}{X} \tag{9.1}$$

which is, of course, also equal to the ratio of the population means. The usual estimator of R from a sample $\mathcal{S}$ is

$$\widehat{R} = \frac{\sum_{i \in \mathcal{S}} y_i}{\sum_{i \in \mathcal{S}} x_i}, \tag{9.2}$$

the ratio of the sample totals (or ratio of the sample means).

If additionally the value of X is *known*, then the identity

$$Y = RX \tag{9.3}$$

suggests that Y could be estimated indirectly by

$$\widehat{Y} = \widehat{R}X. \tag{9.4}$$

Estimators (9.2) and (9.4) are known as the *ratio estimators* of R and of Y, respectively.

The intuitive rationale for the ratio estimator $\widehat{Y}$ is that, if y_i and x_i are approximately proportional, then $\widehat{R}$ is an accurate estimator of R, and since X is assumed to be known exactly, $\widehat{Y}$ is an accurate estimator of Y.

A typical application in surveys would be to estimate the total wheat yield Y from all farms in a region, using the (known) total area X of wheat fields planted at the start of the season, and sampling a number of farms to estimate the average yield per unit area, R.

9.1.2 *Bias under random sampling*

The ratio estimators $\widehat{R}$ and $\widehat{Y}$ are biased in general.

Under any sampling design which has uniform sampling probability p for each item, the adjusted sample totals $\widetilde{Y} = \frac{1}{p}\sum_{i \in \mathcal{S}} y_i$ and $\widetilde{X} = \frac{1}{p}\sum_{i \in \mathcal{S}} x_i$ are unbiased estimators of Y and X, respectively. Thus $\widehat{R} = \widetilde{Y}/\widetilde{X}$ is a ratio of two unbiased estimators. In general the expectation of a ratio of two random variables is not equal to the ratio of their expectations, so $\widehat{R}$ is in general biased for R. The bias is [248]

$$\mathbf{E}[\widehat{R}] - R = -\mathsf{cov}\left(\widehat{R}, \frac{\widetilde{X}}{X}\right) \tag{9.5}$$

It follows that

$$\left|\frac{\mathbf{E}[\widehat{R}] - R}{\mathsf{sd}\left(\widehat{R}\right)}\right| \leq \frac{\mathsf{sd}\left(\widetilde{X}\right)}{X} \tag{9.6}$$

where $\mathsf{sd}(A) = \sqrt{\mathsf{var}[A]}$. Hence the bias tends to be unimportant in comparison to the standard error, even for samples of moderate size [83, p. 160], [555, p. 66].

If $\mathcal{S}$ is a systematic sample, the bias and variance in $\widehat{R}$ depend on the structure of the population. They are generally smaller than in the case of simple random sampling.

Lahiri [332] showed that $\widehat{R}$ and $\widehat{Y}$ are unbiased if the sample is drawn with probability proportional to its total size $\sum_{i \in \mathcal{S}} x_i$. Hájek [237], Midzuno [395] and Sen [512] developed a sampling design with this property.

9.1.3 Variance and efficiency

In general, the variance of the ratio $C = A/B$ of two random variables A and B depends on the joint distribution of A and B. There is no simple analytic form for $\mathsf{var}[A/B]$. However it may be approximated using the delta method, that is, by expanding the function $f(a,b) = a/b$ in a Taylor series at the point $(a,b) = (\mathbf{E}[A], \mathbf{E}[B])$ and evaluating the expectations of terms in the series. To second order this yields the approximation

$$\mathsf{var}[\frac{A}{B}] \approx \frac{\mathbf{E}[B]^2}{\mathbf{E}[A]^2} \left\{ \frac{\mathsf{var}[B]}{\mathbf{E}[B]^2} + \frac{\mathsf{var}[A]}{\mathbf{E}[A]^2} - 2\frac{\mathsf{cov}(A,\, B)}{\mathbf{E}[A]\,\mathbf{E}[B]} \right\}. \tag{9.7}$$

If A and B are consistent and asymptotically bivariate normal (as the sample size increases in some appropriate sense), then A/B is asymptotically normal with mean $\mathbf{E}[A]/\mathbf{E}[B]$ and variance equal to the right side of (9.7). Positive correlation between the numerator and denominator reduces the variance of the ratio, as we would expect intuitively.

Returning to the context of ratio estimation, the variance of

$$\widehat{R} = \sum_{i \in \mathcal{S}} y_i / \sum_{i \in \mathcal{S}} x_i$$

can be approximated by the right-hand side of (9.7) taking $A = \sum_{i \in \mathcal{S}} y_i$ and $B = \sum_{i \in \mathcal{S}} x_i$. The term in braces on the right of (9.7) is the squared coefficient of variation of $\widehat{R}$ and of $\widehat{Y}$.

For a simple random sample of size n with replacement from a population of size N, the ratio estimators $\widehat{R}$ and $\widehat{Y}$ have approximate variances

$$\begin{aligned} \mathsf{var}[\widehat{R}] &\approx \left(1 - \frac{n}{N}\right) \frac{N^2}{X^2} \frac{\sigma_R^2}{n} \\ \mathsf{var}[\widehat{Y}] &\approx \left(1 - \frac{n}{N}\right) N^2 \frac{\sigma_R^2}{n} \end{aligned}$$

for large N, where

$$\sigma_R^2 = \frac{1}{N-1} \sum_{i=1}^{N} (y_i - Rx_i)^2 \tag{9.8}$$

is the population variance of the 'errors' $e_i = y_i - Rx_i$ from a proportional relationship.

Thus, the ratio estimator $\widehat{Y}$ will be more accurate than the direct estimator $\frac{1}{p}\sum_{i \in \mathcal{S}} y_i$ in situations where σ_R^2 is less than the population variance of the values y_i. This is the case when the x and y values are positively correlated, especially when they are roughly proportional [555, p. 61].

The variances of $\widehat{R}$ and $\widehat{Y}$ can be estimated from a simple random sample:

$$\widetilde{\mathrm{var}[\widehat{R}]} = \frac{1}{\bar{x}^2}\frac{s_R^2}{n} \tag{9.9}$$

$$\widetilde{\mathrm{var}[\widehat{Y}]} = N^2\frac{s_R^2}{n} \tag{9.10}$$

where

$$s_R^2 = \frac{1}{n-1}\sum_{i \in \mathcal{S}} (y_i - \widehat{R}x_i)^2 \tag{9.11}$$

is the sample estimate of σ_R^2. Various alternative estimators are used in survey sampling [83, p. 155], [555, p. 61]. Bootstrap estimators may also be used [80, 81].

9.1.4 Best linear unbiased estimation

Conditions under which the ratio estimators are 'optimal' were first formulated by Brewer [61] and Royall [488].

A statistical model is assumed: the observations y_i are now treated as realisations of random variables Y_i, while the x_i are regarded as fixed. Suppose the random variables Y_i are uncorrelated, with means and variances

$$\mathbf{E}[Y_i] = Rx_i \tag{9.12}$$

$$\mathrm{var}[Y_i] = ax_i \tag{9.13}$$

where a is an unknown constant. In other words, there is a proportional regression of Y_i on x_i, with variance proportional to x_i.

Under this model, the ratio estimator $\widehat{R}$ is unbiased ('model-unbiased'), in the sense that

$$\mathbf{E}\left[\widehat{R}\right] = \frac{\sum_{i \in \mathcal{S}} \mathbf{E}[Y_i]}{\sum_{i \in \mathcal{S}} x_i} = R$$

for *every* choice of sample $\mathcal{S}$, where $\mathbf{E}$ denotes expectation with respect to the distribution of the random variables Y_i.

Standard theory of linear models tells us that $\widehat{R}$ is the **best linear unbiased**

estimator of R. That is, amongst all model-unbiased estimators of R of the form

$$R^* = \sum_{i \in \mathcal{S}} c_i y_i$$

the minimum variance is achieved by the ratio estimator $\widehat{R}$. Since $\widehat{Y} = X\widehat{R}$ and X is a known constant, similar comments apply to $\widehat{Y}$.

The variances of $\widehat{R}$ and $\widehat{Y}$ under the model, conditional on the sample $\mathcal{S}$, are

$$\text{var}[\widehat{R} \mid \mathcal{S}] = \frac{a}{\sum_{i \in \mathcal{S}} x_i} \tag{9.14}$$

$$\text{var}[\widehat{Y} \mid \mathcal{S}] = \frac{aX^2}{\sum_{i \in \mathcal{S}} x_i}. \tag{9.15}$$

Cruz-Orive [109] extended this theory for use in stereology, as we describe in Section 9.4.1.

9.2 Stereological estimation *via* ratios

This section concerns the stereological estimation of an absolute quantity using ratio estimation techniques.

We revert to our standard notation for stereological experiments (Section 6.1) in which Y denotes the feature of interest, X the specimen, and Z the reference space or containing space.

9.2.1 *One-stage ratio estimation*

As an example, suppose we wish to estimate the absolute volume of the cortex of a kidney. The cortex is the outer region occupying about one-third of the kidney's volume. Denote the cortex by Y and the entire kidney by Z. Writing the volume $V(Y)$ of the cortex as

$$V(Y) = \frac{V(Y)}{V(Z)} \times V(Z)$$

suggests a ratio estimator. First we can estimate the kidney volume $V(Z)$ directly, for example, by measuring the volume of fluid it displaces. Then taking plane sections of the kidney we can estimate $V_V = V(Y)/V(Z)$ by stereological methods, for example by the ratio of total areas

$$\widehat{V_V} = \frac{\sum_i A(Y \cap T_i)}{\sum_i A(Z \cap T_i)} \tag{9.16}$$

where T_i denotes the ith section plane. The ratio estimate of cortex volume is

$$\widehat{V}(Y) = \widehat{V_V} \times V^*(Z) \tag{9.17}$$

where $V^*(Z)$ is the estimate of kidney volume by fluid displacement.

Of course, $\widehat{V_V}$ is just the ratio of the Cavalieri estimators (Section 7.1.1) of the cortex and kidney. The cortex volume could have been estimated directly using the Cavalieri estimator

$$\widetilde{V}(Y) = t \sum_i A(Y \cap T_i) \tag{9.18}$$

where t is the section separation. This tends to be slightly less accurate than the ratio estimator, for reasons outlined in Section 9.1.2.

The ratio estimator $\widehat{V}(Y)$ has one important advantage: we can estimate $V_V = V(Y)/V(Z)$ under *different sampling conditions* from those applying when we measure $V(Z)$. For example, the volume fraction $V(Y)/V(Z)$ may be estimated on sections under the microscope at any convenient magnification, while the absolute volume $V(Z)$ is measured macroscopically. Indeed, we do not even need to know the magnification used, in order to estimate volume fraction.

Importantly, the ratio estimator is robust against tissue shrinkage. During 'fixation' (chemical treatment in preparation for sectioning), biological tissues may shrink as much as 25% in one dimension. In this context the Cavalieri estimator (9.18) is a good estimate of the volume of the cortex **after shrinkage**. The corresponding estimate of V_V from (9.16) is, therefore, a good estimate of the volume fraction after shrinkage. But if shrinkage is assumed to affect all compartments of an organ equally (to a first approximation) then the volume fractions before and after shrinkage are equal, so that (9.16) is a reliable estimate of the volume fraction *before* shrinkage. Since the fluid displacement volume $V^*(Z)$ was measured before fixation, the ratio estimator (9.17) is unaffected by tissue shrinkage, to a first approximation.

A related advantage is that the section separation t does not need to be known in order to compute the ratio estimator, whereas t must be known accurately for the Cavalieri estimator.

9.2.2 Multi-stage ratio estimation

Biological organs contain features at widely different scales. Figure 9.1 shows the example of the human lung: the natural scale for observing capillaries in the alveolar septa is 1000 times smaller than the scale of the gross anatomy of the lung.

Weibel and collaborators [575, 588, 591] proposed that the absolute size of a very small feature could be estimated using a cascade of sampling experiments at different magnifications. Effectively, we apply ratio estimation at each level: the feature of interest at one level becomes the reference space for the next

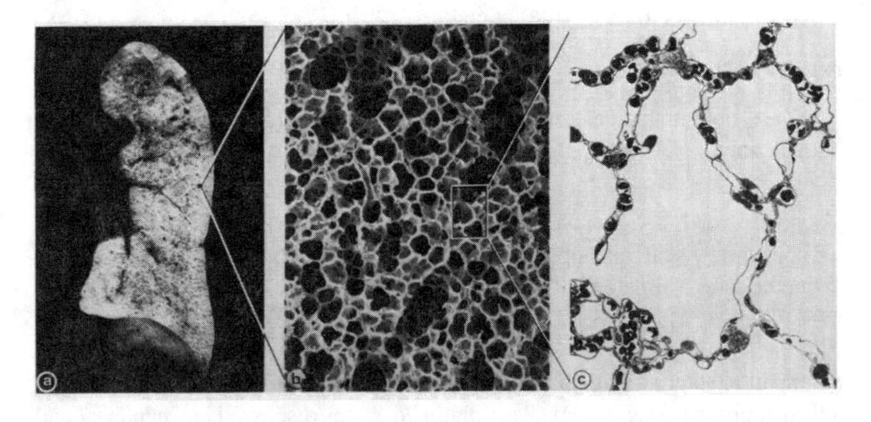

Figure 9.1 *The lung at different scales: (a) entire human lung, macroscopic view (magnification* 0.35×*) showing parenchyma and non-parenchyma (airways, blood vessels etc); (b) small slab of rat lung parenchyma, showing the foam-like structure of septa (scanning electron microscopy,* 95×*); (c) ultrathin section of interalveolar septa in gazelle lung, showing capillaries (transmission electron microscopy,* 700×*). Reproduced from [140] by kind permission of Professors L.M. Cruz-Orive and E.R. Weibel and Blackwell Publishing. Copyright © 1981 Blackwell Publishing.*

smallest level. This was formalised as multistage ratio estimation by Cruz-Orive and Weibel [140].

Let Z be the containing space and Y the feature of interest. Our aim is to estimate an absolute quantity $\beta(Y)$, where typically β is volume V, surface area S, length L or number N. The scale of Y is much smaller than that of Z. Suppose that there are other features $Y_1, \ldots, Y_m$ at intervening scales, and assume the features are nested so that Y_i contains Y_{i+1}:

$$Z \supset Y_1 \supset \cdots \supset Y_m \supset Y. \tag{9.19}$$

Examples of this scheme are [140]

Liver ⊃ Parenchyma ⊃ Hepatocytes ⊃ Hepatocytic mitochondria

Lung ⊃ Parenchyma ⊃ Interalveolar septa ⊃ Capillaries

Then we have

$$\beta(Y) = V(Z) \times \frac{V(Y_1)}{V(Z)} \times \frac{V(Y_2)}{V(Y_1)} \times \cdots \times \frac{V(Y_m)}{V(Y_{m-1})} \times \frac{\beta(Y)}{V(Y_m)}. \tag{9.20}$$

Writing $V_k = V(Y_k)$ and $R_k = V_k/V_{k-1}$ for $k = 1, \ldots, m$, this becomes

$$\beta(Y) = V_0 R_1 R_2 \cdots R_m R_{m+1} \tag{9.21}$$

where we also take $Y_0 = Z$ and write $R_{m+1} = \beta(Y)/V(Y_m)$.

We may estimate $\beta(Y)$ by applying ratio estimation at each successive stage. At stage 0, we estimate $V_0 = V(Z)$ macroscopically using an appropriate method, yielding an estimate $\widehat{V}_0$. At stage 1, we sample the material at an appropriate scale m_1, obtaining an estimate $\widehat{R}_1$ of the ratio $R_1 = V(Y_1)/V(Z)$, and form the ratio estimate of $V(Y_1)$

$$\widehat{V}_1 = \widehat{V}_0 \times \widehat{R}_1.$$

At stage k for $1 < k < m$ we *subsample the material from the previous stage* at an appropriate scale m_k, obtaining an estimate $\widehat{R}_k$ of $R_k = V(Y_k)/V(Y_{k-1})$, and form the ratio estimate of $V(Y_k)$

$$\widehat{V}_k = \widehat{V}_{k-1} \times \widehat{R}_k.$$

At the final stage, we again subsample, estimate the ratio $R_{m+1} = \beta(Y)/V(Y_m)$ by an appropriate stereological estimator $\widehat{R}_{m+1}$ (as discussed in Chapters 7 and 8), then finally form the multistage ratio estimate

$$\begin{aligned} \widehat{\beta}(Y) &= \widehat{V}_m \times \widehat{R}_{m+1} & (9.22) \\ &= \widehat{V}_0 \widehat{R}_1 \cdots \widehat{R}_m \widehat{R}_{m+1}. & (9.23) \end{aligned}$$

The variance of the estimator (9.23) is, to a first approximation,

$$\mathsf{var}[\widehat{\beta}(Y)] \approx \beta(Y)^2 \left(\frac{\mathsf{var}[\widehat{V}_0]}{V_0^2} + \sum_{k=1}^{m+1} \frac{\mathsf{var}[\widehat{R}_k]}{R_k^2} \right). \qquad (9.24)$$

by the delta method, assuming the estimators $\widehat{R}_1, \ldots, \widehat{R}_{m+1}$ are uncorrelated.

Cruz-Orive and Weibel [140, p. 239] discuss efficiency and optimal design of multistage ratio estimation. They show that equation (9.24) implies that multistage estimation will often be more efficient than single-stage ratio estimation. For an efficient design, the coefficient of variation of each ratio estimator $\widehat{R}_k$ should be as small as possible. To achieve this, they recommend choosing the intervening features Y_k recursively so that Y_k is the smallest convenient feature that contains Y_{k+1} and in which the numerator of the ratio is observable. After choosing the features Y_k one should choose m_k to be the smallest magnification which allows observation and measurement of Y_{k+1}. In this way a maximum amount of the structure Y_{k+1} is visible in the sample at stage k.

9.3 Estimation of stereological ratios

This section concerns the estimation of stereological ratio and density parameters.

Densities such as the volume fraction V_V are the parameters of primary interest in classical stereology and in modern model-based stereology. Ratios such as the volume fraction $V(Y)/V(Z)$ of a feature of interest Y in a reference space

Z are of interest in their own right, and serve as intermediate quantities in the ratio estimation of an absolute quantity, as explained in the previous section.

Ratio estimation is implicit in classical stereological methods, as we sketched in Section 2.2.4. Densities such as V_V are estimated from plane sections by the area fraction, the ratio of areas observed in the section. In turn this ratio may be estimated by subsampling the plane section using test systems (point counting and intercept counting).

9.3.1 General

In the *Restricted* or *Extended* case the parameter we want to estimate is the ratio $\beta(Y)/\nu(Z)$ of the size of the feature of interest Y to the size of the reference space Z, where β, ν are appropriate geometrical quantities. In the *Random* case the parameter is a stereological density β_V, defined as $\beta_V = \mathbf{E}[\beta(Y \cap X)]/\nu(X)$ for any specimen X. For convenience we denote the parameter by β_V in either case.

The ratio or density parameter β_V will be estimated by a ratio

$$\widehat{\beta_V} = \frac{\alpha(Y \cap T)}{\mu(Z \cap T)} \tag{9.25}$$

of the observed 'sizes' of Y and Z on a probe T. The probe may be a single plane or a stack of planes, an array of sampling rectangles, etc. The quantities α, μ are stereologically related to β, ν, respectively, through geometrical identities (Chapter 4) which imply that

$$\frac{\mathbf{E}[\alpha(Y \cap T)]}{\mathbf{E}[\mu(Z \cap T)]} = \beta_V, \tag{9.26}$$

where $\mathbf{E}$ is with respect to the randomisation of the sample *and/or* with respect to the postulated random contents of the specimen.

Recall that in the *Restricted* case, $X = Z$ and (9.25) reduces to

$$\widehat{\beta_V} = \frac{\alpha(Y \cap T)}{\mu(X \cap T)}.$$

In the *Extended* case, T may be a random bounded plane probe (for instance, an IUR 2-quadrat in $\mathbf{R}^3$). In the *Random* case, both Z and T are regarded as non-random, implying that the denominator of $\widehat{\beta_V}$ is non-random.

The basic principles of ratio estimation for finite populations, stated in Section 9.1, remain valid in this context. Assuming (9.26) holds, our estimator $\widehat{\beta_V}$ is the ratio of two unbiased estimators. In general $\widehat{\beta_V}$ is biased for β_V. Expressions for the bias and variance of $\widehat{\beta_V}$ can be obtained from the general results in equations (9.5) and (9.7), respectively, although these may be difficult to evaluate.

Three strategies may be employed to ensure that $\widehat{\beta_V}$ is a good estimator (with small bias and small variance):

1. Choose a design in which the sample size $\mu(Z \cap T)$ is *constant*. Then $\widehat{\beta_V}$ is an unbiased estimator of β_V.
2. Ensure that the estimators $\alpha(Y \cap T)$ and $\mu(Z \cap T)$ have small variance. In this context the bias in $\widehat{\beta_V}$ will generally be unimportant in comparison to its standard error.
3. Modify the sampling design to one in which samples are selected with probability proportional to sample size $\mu(Z \cap T)$. Then the same estimator $\widehat{\beta_V}$ becomes exactly unbiased under this design.

We consider these three strategies in turn.

9.3.2 *Fixed sample size*

Fixed sample size is the case where the denominator $\mu(Z \cap T)$ of (9.25) is constant. Stereological estimation is less complicated in this case, since equation (9.26) implies that $\widehat{\beta_V}$ is an unbiased estimator of β_V.

A microscope image or field-of-view is a sample of fixed size, assuming that the magnification and image size are predetermined, and that the reference space is defined to include the entire field-of-view. Figure 1.1 is an example, if the reference space is the entire rock. Within such a field-of-view, a fixed grid of test points or test lines also constitutes a sample of fixed size.

Recall Cruz-Orive's distinction between *Model I* and *Model II* illustrated in Figure 6.4 on page 143. In Model I, the entire field-of-view or probe T is included in the reference space Z with probability one:

$$\mathbf{P}[T \subseteq Z] = 1.$$

In Model II, this is not the case, i.e. there is nonzero probability that $Z \cap T$ is a proper subset of T.

Fixed sample size is achieved by taking a probe T of fixed size and shape, **and** ensuring that Model I holds. Ignoring some very unusual situations (cf. Exercise 9.1), these conditions can only be satisfied in the *Extended* or *Random* cases of stereological inference. In the *Extended* case, typically T would be a random quadrat (FUR or IUR) inside the reference space Z. In the *Random* case, typically T would be a fixed quadrat.

Some examples of stereological estimation with fixed sample size are the following.

Two-dimensional quadrats in sections

A two-dimensional microscope image of a plane section of material enables us to estimate the stereological ratio or density parameters V_V, S_V, L_V with fixed sample size. In either the *Extended* or *Random* cases, under Model I, we have

$$\mathbf{E}\left[\frac{A(Y\cap T)}{A(Z\cap T)}\right] = V_V \tag{9.27}$$

$$\mathbf{E}\left[\frac{4}{\pi}\frac{L(Y\cap T)}{A(Z\cap T)}\right] = S_V \tag{9.28}$$

$$\mathbf{E}\left[2\frac{N(Y\cap T)}{A(Z\cap T)}\right] = L_V \tag{9.29}$$

when the intrinsic dimension of Y is 3, 2 or 1, respectively. In particular, the classical stereological estimators based on A_A, L_A and N_A are unbiased under the classical assumptions of homogeneity and isotropy.

Test systems

A test system of points, lines or curves, which are placed at **fixed** positions in the microscope image, also has fixed sample size. In either the *Extended* or *Random* cases, under Model I, we obtain for a grid T of test points

$$\mathbf{E}\left[\frac{N(Y\cap T)}{N(Z\cap T)}\right] = V_V \tag{9.30}$$

(corresponding to the fundamental formula $P_P = V_V$) and for a grid T of test lines

$$\mathbf{E}\left[\frac{L(Y\cap T)}{L(Z\cap T)}\right] = V_V \tag{9.31}$$

$$\mathbf{E}\left[2\frac{N(Y\cap T)}{L(Z\cap T)}\right] = S_V \tag{9.32}$$

(corresponding to $L_L = V_V$ and $2I_L = S_V$, respectively). Again the assumptions of homogeneity and isotropy are required.

Note that these estimators are not necessarily less efficient than those in (9.27)–(9.29), see Section 13.3.

9.3.3 Systematic sampling

Systematic sampling can yield accurate estimates of absolute quantities, and can, therefore, be used to estimate the ratio of two absolute quantities. Two important applications are the estimation of volume fraction from serial sections using Cavalieri's principle for numerator and denominator, and the estimation of stereological ratios or densities using test systems for both the numerator and denominator. Ratio estimation theory (Section 9.1) is highly relevant here.

Serial plane sections

Using systematic serial section planes, the absolute volume of a solid can be estimated by Cavalieri's principle, as described in Section 7.1.1.

The context is the *Restricted* case: we assume the feature of interest Y and the specimen $X = Z$ are fixed sets with $Y \subset Z$. Take an FUR serial section stack $\{T_i\}$ with constant separation s (cf. Section 7.1.1). Estimating both $V(Y)$ and $V(Z)$ by Cavalieri's principle we get

$$\frac{\mathbf{E}\sum_i A(Y \cap T_i)}{\mathbf{E}\sum_i A(Z \cap T_i)} = \frac{V(Y)}{V(Z)}, \tag{9.33}$$

so that the ratio estimator of $V(Y)/V(Z)$ is

$$\frac{\sum_i A(Y \cap T_i)}{\sum_i A(Z \cap T_i)}. \tag{9.34}$$

The Cavalieri estimators of $V(Y)$ and $V(Z)$ are unbiased and (for sufficiently narrow section spacing s) have small variance, so that (9.34) is approximately unbiased and has small variance.

The estimator (9.34) is frequently used to estimate the volume fraction $V_V = V(Y)/V(X)$ of macroscopic objects (such as the fraction of kidney volume occupied by cortex, see Figure 3.3) and microscopic single objects (such as the fraction of chick embryo volume occupied by the brain) when the feature of interest is not directly accessible.

Note that (9.34) does not require the value of the section separation s. This is an advantage in practice, since section thickness or blade separation often cannot be measured directly, and may differ considerably from the nominal setting of the cutting device.

For the estimation of other quantities such as $S(Y)/V(Z)$ and $L(Y)/V(Z)$, isotropic random orientation of the section plane is required. Take an IUR serial section stack $\{T_i\}$, cf. Section 7.2.3. Then we have

$$\frac{\mathbf{E}\sum_i \alpha(Y \cap T_i)}{\mathbf{E}\sum_i A(Z \cap T_i)} = \beta_V, \tag{9.35}$$

where the quantities (α, β) may be (A, V) as above, or $(\frac{4}{\pi}L, S)$ or $(2N, L)$. See Exercise 9.3. The corresponding ratio estimator is

$$\frac{\sum_i \alpha(Y \cap T_i)}{\sum_i A(Z \cap T_i)}. \tag{9.36}$$

For sufficiently small s, since the denominator of (9.36) is unbiased and has small variance, the estimator (9.36) is approximately unbiased.

However, it should be noted that the expectation in (9.35) includes averaging over all spatial orientations of the stack of section planes. A single realisation of the section stack involves only a single orientation of the sections. Conditional on the section orientation, the expected value of the numerator of (9.36) may depend very strongly on the orientation, as we saw under the name of the "breadcrust theorem" in Section 4.1.5. Hence, the estimator (9.36) based on an IUR section stack is generally too variable to be useful for stereology of a single object.

If it is desired to estimate $S(Y)/V(Z)$ and $L(Y)/V(Z)$ for a single object, the usual remedy would be cut the specimen $X = Z$ arbitrarily into several lumps or blocks of material, and take IUR serial sections of each block, with independently randomised orientation in each block. For biological tissues this can be achieved by independently rotating the blocks, then embedding them in a medium which hardens (paraffin wax, Epon plastic, etc), then serially sectioning in a single direction. See Chapter 12. The resulting estimator is again a ratio of sums

$$\frac{\sum_b \sum_i \alpha(Y_b \cap T_i)}{\sum_b \sum_i A(Z_b \cap T_i)} \tag{9.37}$$

where b denotes the block. The numerator and denominator are still unbiased estimators of $V(Y)/s$ and $V(Z)/s$, respectively, and they now have small variance, so that this estimator of $\beta(Y)/V(Z)$ is approximately unbiased and has small variance.

Test systems

As noted in Section 2.2.4 it is traditional in stereology to substitute a stereological estimate for the exact value of any geometrical quantity. Thus, it would be considered unremarkable to replace the area A of a plane region by its point-counting estimate $a \times P$ where a is the area of one tile in the test point grid and P is the point count, cf. Section 7.1.3. Replacing both the numerator and denominator of a ratio by such 'surrogate' estimates yields a ratio estimator.

Figure 9.2 illustrates the standard techniques for estimating area fraction A_A by point counting. In the left panel, the area of both the feature of interest Y and the reference space Z are estimated using the same point grid T. The estimate of A_A is the ratio

$$\widehat{A_A} = \frac{N(Y \cap T)}{N(Z \cap T)} \tag{9.38}$$

(often denoted by P_P). In the right panel, the feature and reference space are counted using different grids T_1 and T_2, respectively, where T_2 is coarser and

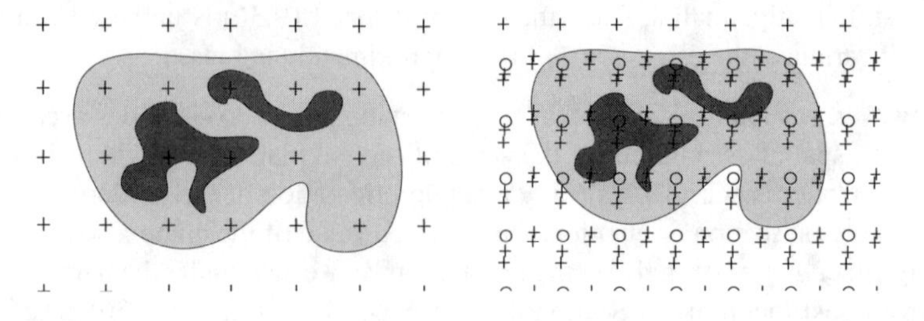

Figure 9.2 *Estimation of A_A from point counts using one (left) or two point grids with different density (right).*

is typically a subset of T_1. The main practical motivation is economy of effort. The estimate of A_A is the ratio

$$\widehat{A_A} = \frac{p_2}{p_1} \frac{N(Y \cap T_1)}{N(Z \cap T_2)} \tag{9.39}$$

where p_2/p_1 is the proportion of points of T_2 to points of T_1. In the Figure, T_2 consists of every second test point on every second row, so that $p_2/p_1 = 1/4$.

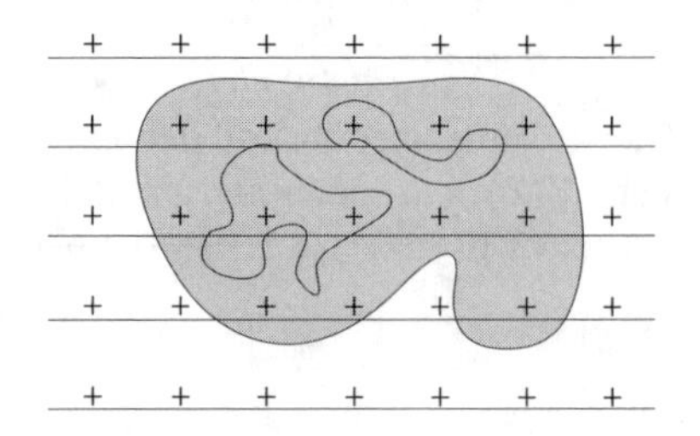

Figure 9.3 *Estimation of L_A from intersection and point counts.*

Figure 9.3 illustrates a method for estimating length fraction L_A when the feature of interest Y is a planar curve. A grid T_1 of test lines is used to count intersections with the feature Y (bold curves) and a grid T_0 of test points is used to point-count the reference space Z (shaded). The estimate of L_A is the ratio estimate

$$\widehat{L_A} = \frac{\pi}{2} \frac{p}{\ell} \frac{N(Y \cap T_1)}{N(Z \cap T_0)} \tag{9.40}$$

where ℓ/p is the length of test line per test point. Isotropy is required.

For these three estimators, the theory of ratio estimation applies under several different sampling designs, whose requirements are subtly different. Consider the point count estimator (9.39). The appropriate form of (9.26) is

$$\frac{p_2}{p_1} \frac{\mathbf{E}N(Y \cap T_1)}{\mathbf{E}N(Z \cap T_2)} = A_A. \tag{9.41}$$

This holds under each of the following designs:

(a) the *Restricted* case, where T_1, T_2 are infinite periodic grids of points, randomly translated as indicated in Section 7.1.2. Here p_2/p_1 is the grid constant relating the two grids; see below.

(b) the *Extended* case, where T_1, T_2 are finite bounded point grids, randomly translated so that T_1 meets the reference space. That is, T_2 is a 0-dimensional quadrat of fixed orientation (Section 4.2). Here $p_2/p_1 = N(T_2)/N(T_1)$ is the ratio of the actual number of points in each grid.

(c) the *Random* case, where T_1, T_2 are fixed bounded point grids, and $p_2/p_1 = N(T_2)/N(T_1)$ is the ratio of the actual number of points.

(d) the *Random* or *Extended* case, in which T_1, T_2 are infinite periodic grids of points, randomly translated as indicated in Section 7.1.2, and intersected with a bounded sampling window. Here p_2/p_1 is the grid constant.

The abovementioned 'grid constant' relates two infinite periodic grids of points. It may be defined by dividing the plane into congruent tiles, each tile having similar contents of test points. Then p_2/p_1 is the ratio of the number of points of T_2 to those of T_1 in any single tile. A more formal treatment of infinite periodic grids has been presented in Section 7.1.3.

Note the subtlety that in some sampling situations we use a virtually infinite periodic grid of test points, while in others we use a fixed bounded set of test points. These two designs may have different values of the constant p_2/p_1, so the estimation procedure is subtly influenced by the sampling context.

In all the cases listed above, it should be remembered that the basis of estimation is (9.41). The ratio estimator (9.39) is biased in general. Our aim should be to obtain accurate estimates of the numerator and denominator of (9.41). For example, in cases (c) and (d) above, if there are multiple sampling windows W_i, the natural estimator of A_A is the **ratio of sums**

$$\widehat{A_A} = \frac{p_2}{p_1} \frac{\sum_i N(Y \cap W_i \cap T_1)}{\sum_i N(Y \cap W_i \cap T_2)}. \tag{9.42}$$

Intuitively, one expects a ratio estimator based on counts to have higher variance than the corresponding ratio estimator based on measuring the areas exactly. However, this is not necessarily true. Notice that the numerator of (9.41) is not proportional to the observed area $A(Y \cap T_1)$, but to its *expectation* $\mathbf{E}A(Y \cap T_1)$. Intuitively, one expects $A(Y \cap T_1)$ to be the optimal estimator of $\mathbf{E}A(Y \cap T_1)$, but this may not be true, and indeed the point-counting estimator may be more accurate. For further details, see Chapter 13.

In the stereological literature, a composite grid with elements of dimensions 0, 1 and 2 is called an *integral test system* [282]. Yet another example is shown in Figure 9.4.

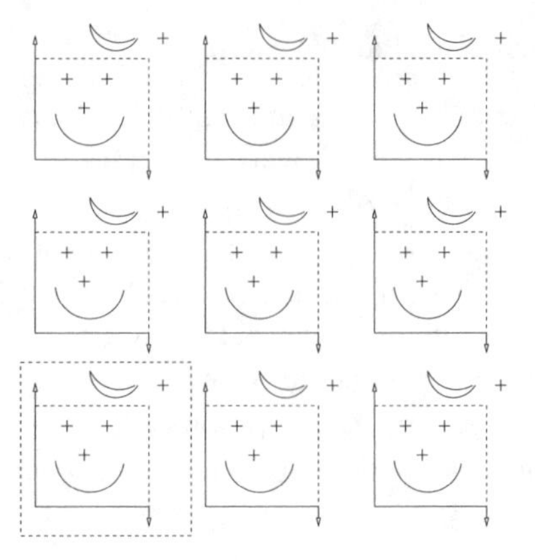

Figure 9.4 *An integral test system.*

9.3.4 Weighted sampling

Suppose (9.26) holds for a given sampling design. Consider the modified sampling design in which probes T are taken with probability proportional to sample size $\mu(Z \cap T)$ relative to the original design. Then under the new 'weighted' sampling design, the ratio estimator is unbiased:

$$\mathbf{E}\left[\frac{\alpha(Y\cap T)}{\mu(Z\cap T)}\right] = \beta_V. \tag{9.43}$$

In sampling theory this general principle was first described by Hájek [237]; see [238, pp. 126–128]. Weighted sampling designs were introduced in stereology by Miles and Davy [146, 147, 408].

As an example, in the *Restricted* case one may take a single planar section T with an area-weighted distribution, see Exercise 9.4. Then $\gamma(Y\cap T)/A(Z\cap T)$ is an unbiased estimator of $\beta_V = \beta(Y)/V(Z)$ for each of the pairs of quantities

$$(\gamma,\beta) = (A,V),(\frac{4}{\pi}L,S),(2N,L). \tag{9.44}$$

Techniques for weighted sampling are described in detail in Weibel's treatise [580, 581].

9.4 Data modelling

More efficient estimation of the ratio parameter is possible if we have variance-covariance information about the terms in the numerator and denominator of the ratio estimator. This calls for a statistical model of the observations.

Reverting to the notation of Section 9.1, assume we have bivariate observations of the numerator and denominator $(X_1,Y_1),\ldots,(X_n,Y_n)$ such that

$$\frac{\mathbf{E}\sum_i Y_i}{\mathbf{E}\sum_i X_i} = \theta, \tag{9.45}$$

where θ is the parameter of interest. The ratio estimator of θ is

$$\widehat{\theta} = \frac{\sum_i Y_i}{\sum_i X_i} \tag{9.46}$$

9.4.1 *Conditional BLUE*

Cruz-Orive [109] modified the theory of best linear unbiased estimation described in Section 9.1.4 for use in stereology. Suppose that x_i and y_i are both realisations of random variables, X_i and Y_i, respectively. Assume that the variables Y_i are conditionally uncorrelated given the sample values $X_i, i \in \mathcal{S}$, and satisfy a conditional proportional regression

$$\mathbf{E}[Y_i \mid X_i = x_i] = Rx_i \tag{9.47}$$

$$\mathsf{var}[Y_i \mid X_i = x_i] = ax_i^b \tag{9.48}$$

where a is unknown (but can be estimated from data) and b is known. These assumptions imply both $\mathbf{E}[Y_i]/\mathbf{E}[X_i] = R$ and $\mathbf{E}[Y_i/X_i] = R$. Then the best linear unbiased estimator of R is

$$\widehat{R}_b = \frac{\sum_{i \in c\mathcal{S}} X_i^{1-b} Y_i}{\sum_{i \in \mathcal{S}} X_i^{2-b}} = \frac{\sum_{i \in c\mathcal{S}} X_i^{2-b} R_i}{\sum_{i \in \mathcal{S}} X_i^{2-b}}, \tag{9.49}$$

the weighted mean of the simple estimates $R_i = Y_i/X_i$ with weights X_i^{2-b}. In the case $b = 1$ this reduces to the ratio estimator $\widehat{R}$.

The conditional variance of $\widehat{R}_b$ is

$$\mathsf{var}[\widehat{R}_b \mid X_i,\ i \in \mathcal{S}] = \frac{a}{\sum_{i \in \mathcal{S}} X_i^{2-b}}. \tag{9.50}$$

A model-unbiased estimator of a is

$$\hat{a} = \frac{1}{|\mathcal{S}|-1} \sum_{i \in \mathcal{S}} X_i^{2-b} (R_i - \widehat{R}_b)^2 \tag{9.51}$$

so that a model-unbiased estimator of the conditional variance (9.50) is

$$\frac{1}{|\mathcal{S}|-1} \sum_{i \in \mathcal{S}} Z_i (R_i - \widehat{R}_b)^2, \tag{9.52}$$

the weighted sample variance of the simple estimates R_i with weights $Z_i = X_i^{2-b} / \sum_{j \in \mathcal{S}} X_j^{2-b}$.

The detailed application of this theory depends on whether the sample sizes X_i are fixed or random, that is, on whether Cruz-Orive's Model I or Model II applies. Under Model I, the sample sizes X_i are fixed, and are controlled by the design. The classical theory of best linear unbiased estimation (Section 9.1.4) applies. The variance exponent b can be estimated from data (assuming at least two different sizes of probe are used) by grouping the observations Y_i according to probe size $X_i = x_i$, computing the sample variances $s^2(x_i)$ for each group, and fitting a power law $s^2 = ax^b$ to the sample variances [109].

In Model II, the sample sizes X_i are random. The exponent b may be estimated approximately by visual inspection of the scatter plot of (X_i, Y_i) or by fitting a parametric model [109].

Jensen and Sundberg [291] discussed the appropriateness of the model assumptions (9.47)–(9.48). Their conclusion was that in Model I these assumptions are reasonable, as they can be motivated by simple probabilistic approximations, but for Model II there is a difficulty with random effects and spatial inhomogeneity. If (9.47)–(9.48) hold at any location in space, but the coefficient a depends on location, then (9.47)–(9.48) do not hold for a uniform sample from the entire reference space. Jensen and Sundberg [291] suggest using a structural relationship instead of the conditional regression model for data collected in the framework of Model II.

In the same vein, Chia [80, 81] pointed out that one of the consequences of the conditional proportional regression model is that $X_i = 0$ implies $Y_i = 0$ a.s. since Y_i is a non-negative random variable. This is not necessarily true if an integral test system is used to collect the data, cf. Figures 9.2 and 9.3.

9.4.2 *Parametric modelling*

Chia [80, 81] considered parametric modelling of the bivariate data (X_i, Y_i). She focused on the case where (X_i, Y_i) are bivariate *counts*. Simple examples of parametric models are the Poisson and binomial proportional regressions

$$Y|X = x \sim \mathsf{Pois}(x\theta), \quad \theta > 0, \tag{9.53}$$

$$Y|X = x \sim \mathsf{Bin}(x, \theta), \quad 0 < \theta < 1. \tag{9.54}$$

are studied. Poisson proportional regression is a reasonable initial model for intercept counts with a test system, or particle counts from an unbiased sampling rule, under sparse sampling conditions. Binomial proportional regression is a reasonable model for point-counting data with a test system containing a fixed number of test points which are coarsely spaced. Both models (9.53) and (9.54) satisfy Cruz-Orive's conditional proportional regression assumptions (9.47) and (9.48) with exponent $b = 1$. For both the Poisson and binomial models, the maximum likelihood estimator of θ becomes the traditional ratio of sums estimator $\widehat{\theta} = \sum Y_i / \sum X_i$.

More generally, suppose the marginal distribution of X is governed by a parameter φ and that the parameter space for (θ, φ) is a Cartesian product. Then X is ancillary for θ, and inference for θ should be performed conditionally on X. This again reduces to Cruz-Orive's results.

Chia [80] presents several methods for testing or validating the model assumptions and for estimating the variance of $\widehat{\theta}$.

9.5 Exercises

Practical exercises

Exercise 9.1 Ice is manufactured in rectangular blocks of fixed dimensions. The volume fraction of air bubbles is of interest. A block is sliced end-to-end, parallel to one of its faces, into slabs of equal thickness s. Find conditions under which this is a suitable sampling design of **fixed** sample size in the *Restricted* case.

Exercise 9.2 Develop a technique for estimating the number of jellybeans in a large cylindrical jar.

Theoretical exercises

Exercise 9.3 Show (9.35).

Exercise 9.4 Let T be an area-weighted planar section hitting Z. Then the coordinates (ω, s) of the plane have joint density (with respect to $\mathrm{d}s\,\mathrm{d}\omega$) of the form

$$f(\omega, s) = \frac{A(Z \cap T_{\omega,s})}{2\pi V(Z)}. \tag{9.55}$$

1) Show that (9.55) defines a density, i.e.

$$\int_{S_+^2} \int_{\mathbf{R}} f(\omega, s)\,\mathrm{d}s\,\mathrm{d}\omega = 1.$$

2) Show, using the Crofton formula, that $\gamma(Y \cap T)/A(Z \cap T)$ is an unbiased estimator of $\beta(Y)/V(Z)$ where the possible pairs of quantities γ, β are given by (9.44).

Exercise 9.5 Show that the Poisson and binomial models given in (9.53) and (9.54), respectively, satisfy (9.47) and (9.48) with $b = 1$. Show also that $\widehat{\theta} = \sum Y_i / \sum X_i$ is the maximum likelihood estimator of θ in both models.

9.6 Bibliographic notes

Ratio estimation in survey sampling theory is amply described in textbooks, e.g. [83, chap. 6], [555, chap. 7] and [238].

Mayhew and Cruz-Orive [380] first drew attention to the fact that some stereological ratio estimators have non-constant denominator and are therefore biased. Miles and Davy [146, 147, 402, 408, 410] laid rigorous foundations for estimation theory in stereology, including for the unbiased estimation of ratios.

Unbiased estimation of a ratio using a weighted sampling design was developed by Hájek [237], Lahiri [332], Midzuno [395] and Sen [512] in survey sampling theory. Unbiased estimation of stereological ratios using weighted sampling was developed by Miles and Davy [146, 147, 408]. Practical techniques for weighted sampling are described in detail in Weibel's treatise [580, 581].

Stereological estimation of volume ratios by systematic sections was investigated in [138, 139, 431, 130]. Its efficiency was investigated in [368, 429, 117].

Weibel and collaborators [575, 588, 591] were apparently the first to propose ratio estimation using multiple sampling experiments at different magnifications. This was formalised as multistage ratio estimation by Cruz-Orive and Weibel [140].

Stereological ratio estimation based on counts from integral test systems was further developed by Jensen and Gundersen [282].

The theory of best linear unbiased estimation in sampling designs was initiated by Brewer [61] and Royall [488]. Cruz-Orive [109] extended it for stereological applications. See also Jensen and Sundberg [291] and Nagel [432]. Stereological ratio estimation was also studied by Davy [146], Dhani [158], Downie [165], and Chia [80, 81].

CHAPTER 10

Discrete sampling and counting

In a 'discrete' sampling design, the material under study is first cut into pieces; the pieces are then treated as individual units in a sampling population, and some of them are randomly selected. Inference about the material from this sample is justified by the random selection of the pieces, rather than by the way in which the pieces were cut.

Discrete sampling methods avoid many of the difficulties in classical stereology related to shape and size of sections, such as non-constant section thickness, error in measuring thickness, and deflection of the cutting blade by the contents of the material. These are not important to the validity of discrete sampling inference. Indeed, we are free to cut the material in any fashion, and we may use this freedom to maximise the convenience and efficiency of the sampling.

The 'particle problem' described in Sections 2.3–2.6 is still present in this context. When a solid material, containing a population of particles, is cut into pieces, an individual particle may be cut into several separate pieces, although we still wish to count it as a single unit. Also, particles may be over- or under-represented in the sample, according to the number of pieces they yield.

This chapter canvasses two main solutions to the particle problem using discrete sampling: *unbiased sampling rules* and *reweighting*. An unbiased sampling rule is simply a rule for ensuring that all particles have equal probability of being sampled. Reweighting is a technique for correcting the bias in a biased sample by associating unequal weights with the sampled pieces. These techniques can be used profitably in other contexts too.

10.1 Basic principles

10.1.1 Discrete sampling

The basic idea of discrete sampling is that we nominate a containing space that includes all the material of interest; divide the containing space into a finite number of pieces; treat these pieces as the sampling units in a finite population; and select a random sample of pieces.

Let Z be the containing space, an arbitrary set in any space (usually in two- or three-dimensional space). Assume Z is partitioned into disjoint pieces $Z_1,\dots,Z_K$ in arbitrary fashion. The number of pieces, and their ordering, is also arbitrary.

Note that the material may or may not be physically cut into pieces. Other practical methods for partitioning material include superimposing onto a micrograph a transparency marked with boundary lines; laying marker ropes across a meadow; and staining different anatomical regions of an organ with different coloured inks. It is required only that the pieces Z_k constitute a partition of Z, meaning that they are disjoint ($Z_k \cap Z_l = \emptyset$ for $k \neq l$) and comprise the whole containing space ($\bigcup_k Z_k = Z$). Thus, if the material is physically cut, the cutting process must not destroy or remove any material.

Next we treat the pieces $Z_1,\dots,Z_K$ as the sampling units in a finite population, and select a random sample of them. It is convenient to identify each piece Z_k with its index k, so that the sampling population is the index set $\{1,2,\dots,K\}$, and the sample is a random subset $\mathcal{K}$ of $\{1,2,\dots,K\}$.

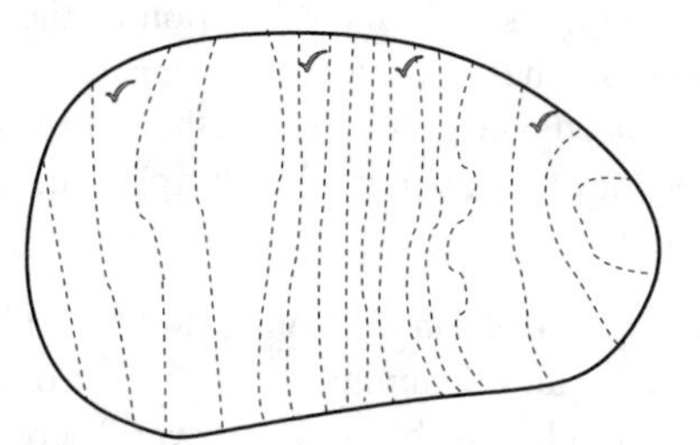

Figure 10.1 *A discrete sampling design. After the material is cut into pieces in an* arbitrary *fashion (dotted lines), the pieces are treated as sampling units, and a random sample of pieces is taken. The sketch shows a systematic random sample of the pieces with period 4.*

Figure 10.1 sketches an example, in which a solid object is cut into pieces in arbitrary fashion, and a systematic random sample of the pieces is taken with period 4.

It should be clear that discrete sampling is a form of cluster sampling (see Appendix A). Although cluster sampling is usually presented in the literature as a technique for sampling a finite population, it can be applied to any population whatsoever. By a basic principle of cluster sampling, a uniform random sample of pieces Z_k yields a uniform random sample of the containing space Z. That is, if the pieces Z_k are sampled with uniform probability p, then the sampled material constitutes a random sample of the containing space Z with uniform sampling probability p, in the sense that any point $x \in Z$ has probability p of being included in the sampled material. See Appendix A for details.

Thus, for example, the systematic design sketched in Figure 10.1 is also a sample *of the containing space* with uniform sampling probability $1/4$.

10.1.2 Unbiased estimation

Our objective is to estimate a stereological parameter φ defined for the entire material contained in Z. In order that we may estimate it from the sample using elementary sampling techniques, φ must be expressible as a sum of contributions from the individual pieces:

$$\varphi = \sum_{k=1}^{K} \varphi_k \tag{10.1}$$

where φ_k is a quantity depending only on the contents of Z_k. Assuming (10.1) holds and assuming each piece Z_k has the same probability p of being sampled, an unbiased estimator of the population total φ is the rescaled sample total,

$$\widehat{\varphi} = \frac{1}{p} \sum_{k \in \mathcal{K}} \varphi_k. \tag{10.2}$$

See Appendix A for details. Note this requires that the sampling probability p be known and positive.

Thus the general principle is that arbitrary division of the containing space into pieces, followed by random selection of a sample of pieces with equal sampling probability for each piece, allows unbiased estimation of total quantities, without any further information or assumptions about the geometry of the pieces.

10.1.3 Additive parameters

This method is suited to estimating any parameter φ which can be expressed as a population *total* of contributions from each piece of the containing space, as in (10.1).

Such parameters include the 'absolute' geometrical quantities such as volume, surface area and length. A practically useful example is area A in the plane, which we often need to estimate in micrographs and on maps. Let Y be an arbitrary feature of interest in a two-dimensional containing space, $Y \subset Z \subset \mathbf{R}^2$. Define the target parameter to be the absolute area

$$\varphi = A(Y).$$

Any partition of the containing space Z into pieces $Z_1,\ldots,Z_K$ causes Y to be partitioned into pieces $Y_k = Y \cap Z_k$ (see Figure 10.2) whose total area is

$$\sum_{k=1}^{K} A(Y_k) = A(Y)$$

so that (10.1) holds with

$$\varphi_k = A(Y_k) = A(Y \cap Z_k),$$

the area of the feature of interest within the kth piece.

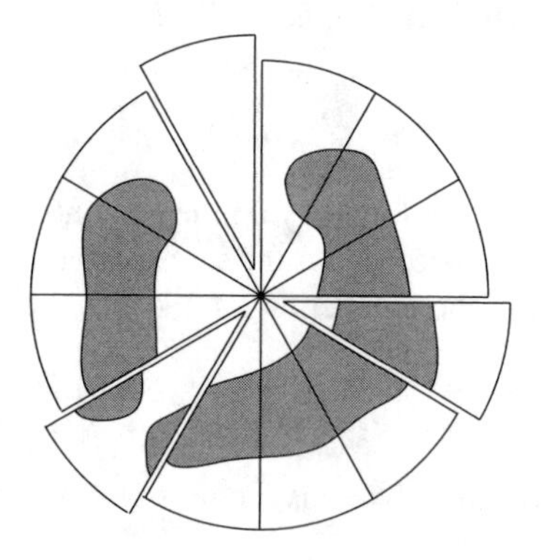

Figure 10.2 *Estimation of total area of cheese on a pizza by systematic sampling of slices. Every 4th slice has been selected. The total area of cheese on the sampled slices, multiplied by 4, is an unbiased estimate of the total area of cheese.*

Thus, the absolute area of a feature of interest can be estimated by cutting the containing space into pieces in an arbitrary fashion, taking a random sample of these pieces with equal sampling probability p, computing the sample total of the areas of the resulting pieces of the feature of interest, and estimating

$$\widehat{A}(Y) = \frac{1}{p} \sum_{k \in \mathcal{K}} A(Y \cap Z_k).$$

A version of this technique was presented in Section 7.1.3 on systematic subsampling of the plane. However, we now see that the division of space into tiles does not have to be geometrically regular. For example in cartography, the total area of lakes in a country which is divided into regions (the Départements in France or the States of Malaysia) could be estimated by randomly selecting a sample of these regions and measuring the lake area in each sampled region — even if some lakes lie across region boundaries.

The same argument applies for estimating any set-additive* quantity such as volume V, surface area S and length L.

A notable exception is N, the number of particles or components, which is not set-additive. Estimation of N requires special methods, as discussed below.

10.2 Unbiased sampling rules for particles

Suppose now that the containing space Z includes a finite population of particles $Y_1, \ldots, Y_N$. Particles are assumed to be disjoint, and each particle is a connected set, but no further geometrical assumptions are imposed. The containing space is again divided into pieces $Z_1, \ldots, Z_K$. A random sample of the pieces is taken, with uniform sampling probability for each piece Z_k.

The problem is that this does not directly define a random sample of the particle population. An individual particle Y_j may extend across several pieces Z_k. See Figure 10.3. Given a sample of pieces Z_k, how should this be interpreted as a sample of particles Y_j? We need a *sampling rule* to decide which of the particles Y_j that are partially represented in a sampled piece Z_k should be deemed to have been selected by this piece.

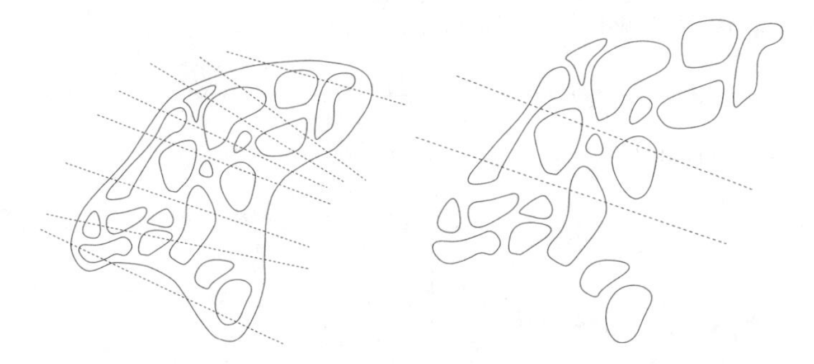

Figure 10.3 *The particle problem in a discrete setting.* Left: *When the containing space is partitioned into pieces (by the dotted lines), an individual particle may lie across several pieces.* Right: *Given a sampled piece (between dotted lines), we need a sampling rule to determine which particles are sampled.*

This section explains how to design the sampling rule so that particles are sampled with uniform probability. First we explain why this is desirable.

* A quantity μ is (finitely) set-additive if it satisfies $\mu(A \cup B) = \mu(A) + \mu(B) - \mu(A \cap B)$ for any sets A and B, and $\mu(\emptyset) = 0$ where $\emptyset$ is the empty set.

10.2.1 Sampling rules and sampling bias

A *particle sampling rule* is any criterion for deciding whether or not a given particle Y_j is 'sampled' by a given piece Z_k. A sampling rule is characterised by the indicator variables

$$I_{jk} = \begin{cases} 1 & \text{if } Y_j \text{ is selected by } Z_k \\ 0 & \text{if not.} \end{cases}$$

One example of a sampling rule (called 'plus-sampling' after [399]) is to select every particle that is visible wholly or partly in the sampled material. A particle Y_j is sampled by a selected piece Z_k if they overlap:

$$I^+_{jk} = \begin{cases} 1 & \text{if } Y_j \cap Z_k \neq \emptyset \\ 0 & \text{otherwise.} \end{cases}$$

However, this rule has an inherent sampling bias, as we shall see presently.

The effect of any sampling rule I_{jk} can be calculated as follows. Suppose we have selected a random sample of pieces Z_k with uniform probability. For each sampled piece Z_k, our sampling rule is applied to determine which particles are selected by that piece (namely the particles Y_j satisfying $I_{jk} = 1$). The particles selected by all sampled pieces Z_k, $k \in \mathcal{K}$ are combined to form a sample $\mathcal{S}$ from the particle population. In this sample, some particles may appear with multiplicity greater than 1, since a given particle Y_j might be selected by several different pieces Z_k according to the sampling rule in force. The sample $\mathcal{S}$ contains particle Y_j with multiplicity

$$M_j = \sum_{k \in \mathcal{K}} I_{jk} \, .$$

The expected multiplicity of particle Y_j in the sample $\mathcal{S}$ is

$$\begin{aligned} \mathbf{E}[M_j] &= \mathbf{E}\left[\sum_{k \in \mathcal{K}} I_{jk}\right] \\ &= \mathbf{E}\left[\sum_{k=1}^{K} \mathbf{1}\{k \in \mathcal{K}\} I_{jk}\right] \\ &= p \sum_{k=1}^{K} I_{jk}. \end{aligned}$$

In cases where a particle cannot be sampled in more than one piece ($M_j \leq 1$), this expectation $\mathbf{E}[M_j]$ is the probability that Y_j will be sampled. The value

$$m_j = \sum_{k=1}^{K} I_{jk}$$

is the number of pieces Z_k in the partition which, if selected, would sample the particle Y_j according to the sampling rule in force.

For the case of plus-sampling, this number is

$$m_j^+ = \sum_{k=1}^{K} I_{jk}^+ = \sum_{k=1}^{K} \mathbf{1}\{Y_j \cap Z_k \neq \emptyset\},$$

the number of nonempty fragments $Y_j \cap Z_k$ into which the particle Y_j is cut by the partition. This number is not constant for all particles. Thus, the sample of particles obtained by taking a uniform sample of pieces and applying the plus-sampling rule is a *biased* sample of the particle population.

The pitfalls of plus-sampling were discussed in Chapter 2 albeit in a different sampling context. When applied to sections of a three-dimensional particle population, this rule says that we should sample or count all profiles seen in the section. We saw in Section 2.3.3 that this overcounts particles and yields a biased sample of the particle population. When this rule is applied in two dimensions, it says that we should count all profiles that are visible wholly or partially. This overcounts profiles and yields a biased sample of profiles (Section 2.6).

10.2.2 Unbiased sampling rules and unbiased estimation

The term 'bias' is used in two senses: sampling bias, and estimation bias. A biased sample of a population is one in which the sampling probabilities of different items are not equal. A biased estimator of a parameter is one whose expected value does not equal the true value of the parameter.

In the present context, a sampling bias leads to an estimation bias. Suppose we wish to estimate the population total

$$\Phi = \sum_{j=1}^{N} \varphi(Y_j)$$

of some geometrical quantity φ for the particle population. For example, if $\varphi(Y_j) \equiv 1$ then Φ is the total number of particles, and if $\varphi(Y_j) \equiv V(Y_j)$ then Φ is the total volume of all particles. The *sample total* of this quantity,

$$S_\varphi = \sum_{k \in \mathcal{K}} \sum_j I_{jk}\, \varphi(Y_j)$$

has expected value

$$\begin{aligned} \mathbf{E}\left[S_\varphi\right] &= \mathbf{E}\left[\sum_j M_j \varphi(Y_j)\right] \\ &= p \sum_j m_j\, \varphi(Y_j). \end{aligned}$$

If the values m_j are not equal, then we cannot form an unbiased estimator of the population total Φ from the sample total S_φ.

We will say that the sampling rule is **unbiased** *if the values m_j are all equal, $m_j \equiv m$ for all particles Y_j.* If the sampling rule is unbiased, and we selected pieces Z_k with uniform sampling probability p, then we obtain a sample of the particle population with constant expected multiplicity $\mathbf{E}[M_j] = pm$ for all j.

For an unbiased sampling rule, the sample total yields an unbiased estimator of the particle population total. An unbiased estimator of total particle number N is

$$\widehat{N} = \frac{1}{mp} \sum_{k \in \mathcal{K}} N_k$$

where $N_k = \sum_j I_{jk}$ is the number of particles sampled in Z_k. More generally, an unbiased estimator of the population total $\Phi = \sum_j \varphi(Y_j)$ is

$$\widehat{\Phi} = \frac{1}{mp} \sum_{k \in \mathcal{K}} \sum_j I_{jk} \varphi(Y_j).$$

One way to obtain an unbiased sampling rule is to ensure that $m_j = 1$ for all j, that is, that **each particle is sampled by exactly one of the pieces** Z_k in the partition. We pursue this objective below.

10.2.3 Nondestructive partitions

A simple solution is to cut the containing space without cutting any particles, so that each particle Y_j is contained entirely within one piece Z_k. See Figure 10.4.

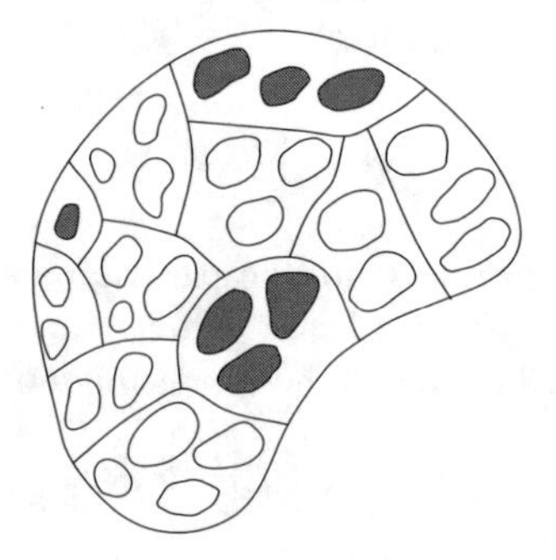

Figure 10.4 *Cluster sample of particles using a nondestructive partition. The containing space has been partitioned without cutting any particles into fragments. A cluster sample is taken. The hatched particles are sampled.*

The sampling rule is simply to take every particle Y_j which lies inside one of the sampled pieces Z_k. A uniform sample of the pieces Z_k with sampling probability p will then yield a uniform sample of the particle population with sampling probability p. This is a very direct form of cluster sampling.

This design has limited use in stereology. Usually we cannot avoid cutting particles into multiple pieces when physically slicing a solid material. However, it may be applied to two-dimensional microscope images when our background knowledge tells us how to partition the plane without splitting the units of interest. For example, if it is required on a microscope section image to count a large number of profiles of cells that are only found inside the tubules of the reproductive system, then we may identify all sections of tubules on the microscope image, select some of the tubule sections with equal probability, and count cell profiles inside the sampled tubule sections.

This design is also relevant when the particles are pointlike objects that are never split by the section plane.

10.2.4 'First-time' rule and the disector

One unbiased sampling rule with $m_j = 1$ for all j is to select a particle if it intersects the selected piece Z_k and does not intersect any of the previous pieces $Z_1,\ldots,Z_{k-1}$. Thus, a given particle Y_j will be sampled by the *first* piece (in the entire sequence $Z_1,\ldots,Z_K$) which intersects it, and not by any subsequent pieces. Every particle Y_j is thus sampled by exactly one of the pieces Z_k. See Figure 10.5.

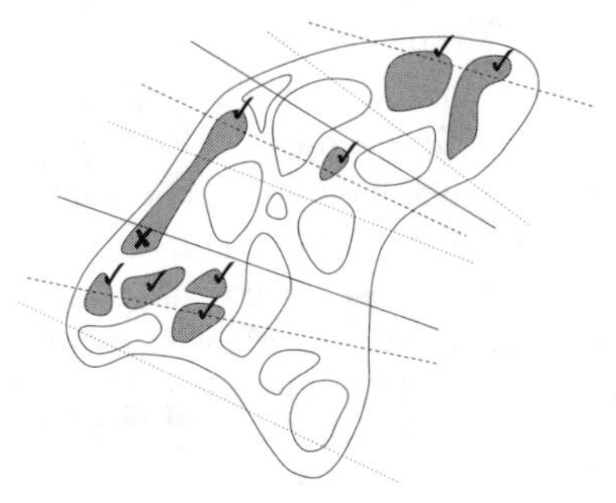

Figure 10.5 *First-time rule. The material has been cut into irregular strips Z_k which are ordered from top to bottom. Strips 1, 4 and 7 have been selected by a uniform sampling design. A particle is sampled by a strip (indicated by ticks) if it overlaps the strip in question and does not overlap any previous strip (i.e. does not overlap any higher strip).*

Again this rule may be interpreted as dividing the particle population into disjoint clusters. The cluster associated with the piece Z_k consists of all particles that meet Z_k and do not meet any previous piece $Z_1, \dots, Z_{k-1}$. A uniform sample of these clusters yields a uniform sample of particles.

In theory the pieces Z_k may be enumerated in any order, but in practice this rule is easier to implement if the pieces are enumerated in a natural order in which Z_{k-1} and Z_k are adjacent.

One important application of this rule in microscopy is the *disector* technique [538]. A three-dimensional material is cut into slabs (from left to right, say) by a series of section planes, and some of the slabs are selected. A particle is sampled if it is visible in one of the sampled slabs and does not cross the left boundary plane of the slab. This is the first-time rule when the slabs are ordered from left to right. See Figure 10.6.

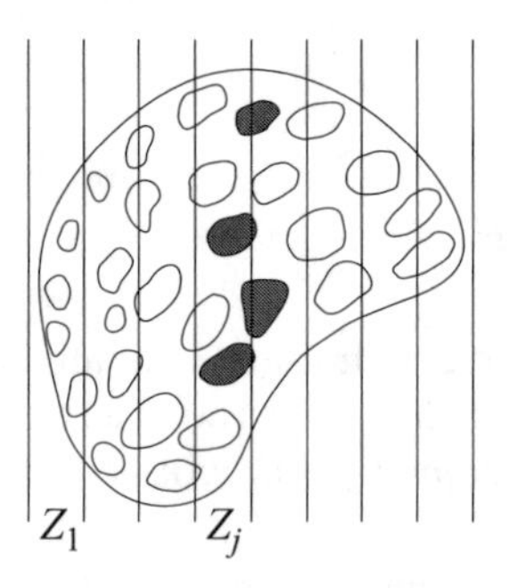

Figure 10.6 *2D illustration of the disector rule. The particles sampled in Z_k are shown hatched.*

A further practical simplification is possible when it is known that no particle is small enough to lie entirely inside one slab Z_k. Then we may simply inspect the two plane sections which delimit the slab, and sample any particle which is present on the right section plane but absent on the left plane. See also the 2D illustration in Figure 10.7. It may easily be verified that any given particle will be counted as belonging to exactly one of the slabs. Further details are given in Chapter 11. See also the practical illustration in Figure 3.7.

The disector rule is often used to estimate the total number N of particles in the particle population. In standard stereological notation, Q_k^- denotes the number of particles sampled by the disector rule in one sampled slab Z_k. Since this sampling rule partitions the particle population into disjoint subsets, we have

$$N = \sum_{k=1}^{K} Q_k^- . \tag{10.3}$$

Since we have expressed N as a sum of contributions from individual slabs Z_k, we may estimate N from a sample of slabs as indicated in Sections 10.1.2 and

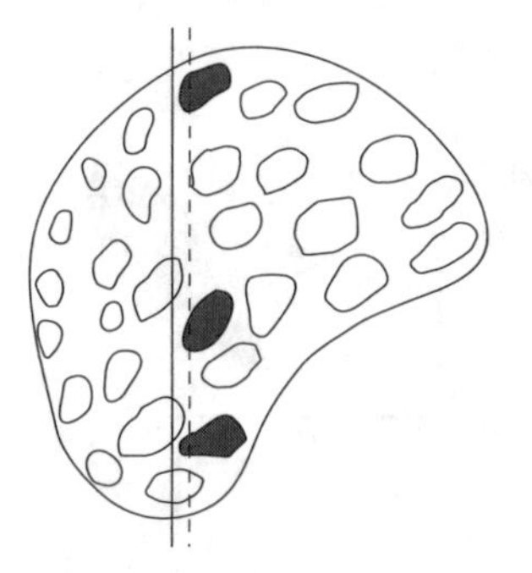

Figure 10.7 *'Physical disector' sampling rule in the plane with strips Z_k narrower than the horizontal extent of any of the particles. The stippled line is called a counting line while the fulldrawn line a lookup line.*

10.2.2. Using a random sample of slabs with uniform sampling probability p,

$$\widehat{N} = \frac{1}{p} \sum_{k \in \mathcal{K}} Q_k^-$$

is an unbiased estimator of N.

10.2.5 Tiling rule

A convenient way to sample a large region in the plane is to divide it into rectangular tiles and select some of the tiles. Suppose the plane is partitioned into rectangular sampling windows $\{Z_{j_1 j_2} : j_1, j_2 \in \mathbf{Z}\}$ where

$$Z_{j_1 j_2} = Z_0 + (j_1 h_1, j_2 h_2),$$

and h_1 and h_2 are the side lengths of the rectangle Z_0. Apply the 'first-time' sampling rule using the lexicographic ordering

$$Z_{j_1 j_2} \prec Z_{j'_1 j'_2} \Leftrightarrow (j_1 < j'_1) \text{ or } (j_1 = j'_1 \text{ and } j_2 < j'_2).$$

This important sampling rule was described in [216].

The solid boundary line drawn in Figure 10.8 separates the tile $Z_{j_1 j_2}$ from all the preceding tiles in the lexicographic ordering. Hence, a particle Y_j will be sampled by the tile $Z_{j_1 j_2}$ if and only if Y_j intersects the tile and does not cross the indicated boundary line. This gives us a simple practical test for applying the tiling rule.

The tiling rule is a simple, general solution to the problem of edge effects in counting planar profiles that was raised in Section 2.6. Applications are shown in Figure 2.8 on page 21, Figure 3.7 on page 70, and in Chapter 11.

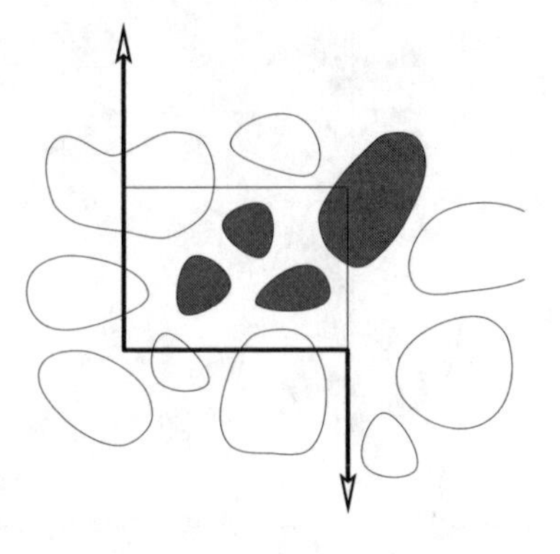

Figure 10.8 *Tiling rule. A particle will be sampled in the tile Z_k if it overlaps Z_k and does not cross the infinite boundary indicated by the solid lines. For details, see the text.*

It is convenient to combine the tiling rule with a 'double' systematic subsample of tiles. We select every m_1th tile in every m_2th column starting from a random initial tile. The selected tiles are

$$Z_{M_1+i_1m_1,M_2+i_2m_2}, \quad i_1,i_2 \in \mathbf{Z}$$

where M_1,M_2 are independent and M_l is a uniform random integer between 1 and m_l, $l=1,2$. Each tile has probability $p=1/(m_1m_2)$ of being sampled. When combined with the tiling rule, this yields a sample of the particle population with uniform sampling probability $p=1/(m_1m_2)$.

Figure 10.9 sketches an example with $m_1=m_2=3$. The sampled objects are shown hatched.

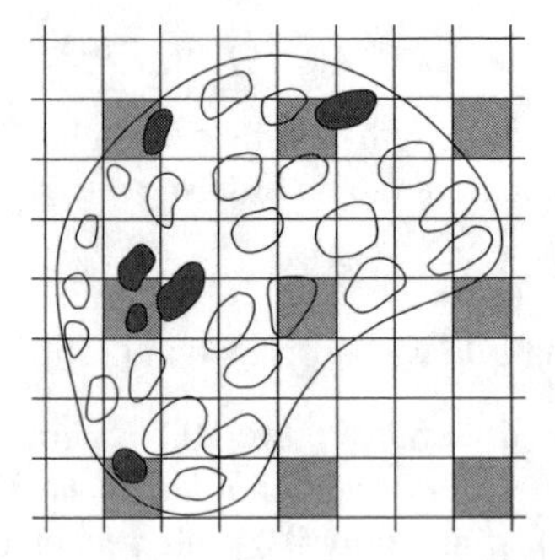

Figure 10.9 *Tiling rule applied to a systematic subsample of tiles.*

Similar kinds of sampling rule can be applied to non-rectangular tessellations of the plane, and in higher dimensions. The first-time rule always translates into a rule involving a forbidden boundary, as follows. Let $U_k=\bigcup_{\ell\leq k} Z_\ell$ be the union of the first k pieces, for $1 \leq k \leq K$. Assume U_k is a regular closed set and that particles are connected sets. Then the first-time sampling rule for Z_k is equivalent to sampling any particle Y_j that intersects Z_k but does not intersect the boundary ∂U_{k-1}. A three-dimensional example is described in the next chapter.

10.2.6 Associated point rule

The basic problem in constructing an unbiased sampling rule is to find a way of 'eliminating' the extent of the particles. One natural idea is to associate to each particle a point, and to sample these Associated Points (APs) instead of the particles.

There are many ways of defining an AP for each particle. For planar particles, options include the centre of gravity, the circumcentre (centre of the smallest disc containing the particle), and the bounding box midpoint (centre of the smallest rectangle containing the particle and with sides parallel to the coordinate axes).

Additionally, in microscopy there are some materials and tissues in which the particles contain a unique, point-like object that may be used as an associated point. A biological cell may be sampled using its nucleus or nucleolus as an associated point (assuming all cells have exactly one nucleus or nucleolus).

Let $a(Y_j)$ denote the associated point for a particle Y_j. The sampling rule is simply that Y_j is sampled by Z_k if $a(Y_j) \in Z_k$. Since $a(Y_j)$ is a point, it must belong to exactly one of the pieces Z_k, so this yields an unbiased sampling rule.

In the present context, the only constraints on the mapping a are that $a(Y_j) \in Z$ for all j, and that there must be a 1–1 correspondence between APs and particles. In other contexts, it is also required that the mapping be invariant under translation.

Further possibilities for APs are tangent points: extreme θ-points and (θ_1, θ_2)-points, cf. [402]. These latter types of APs are based on the concept of a $\theta-$tangent of a planar particle which is the unique orientated tangent of orientation $\theta \in [0, 2\pi)$ to the particle, having no part of the particle on its right side. The particle boundaries must be continuously differentiable closed curves. The extreme $\theta-$point of a particle is the point at which the $\theta-$ tangent is tangent, provided this point is unique. Otherwise, the extreme $\theta-$point is undefined. The definition of an extreme $\theta-$point is illustrated in Figure 10.10 for $\theta = \pi/2$. The $(\theta_1, \theta_2)-$point of a particle is the intersection point of the θ_1- and θ_2-tangents. The AP rule combined with 'double systematic subsampling is illustrated in Figure 10.11.

The same ideas can in principle be used to define APs for spatial particles.

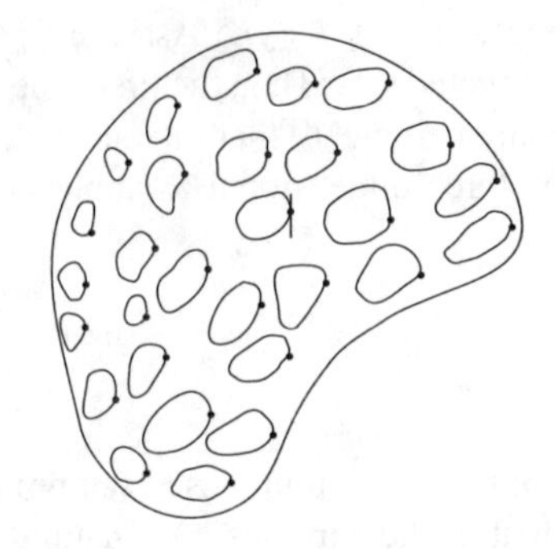

Figure 10.10 *Extreme θ−points for $\theta = \pi/2$.*

Their practicability depends on the representation available for the particles. In microscopical analyses of cell populations the nucleus or an identifiable subunit like a nucleolus has been used as AP in cases where a 'thick' tissue block can be studied using optical sections.

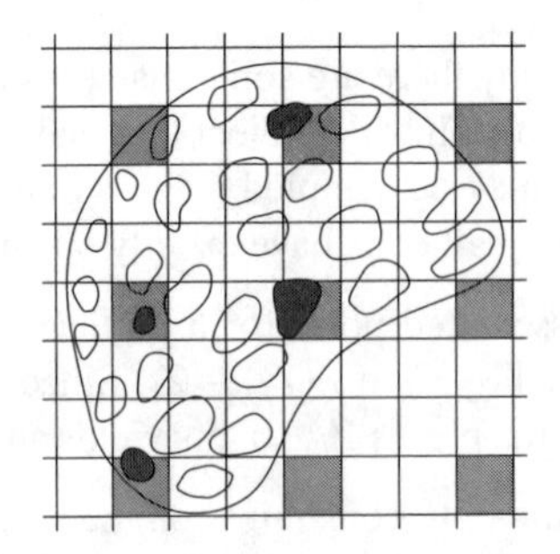

Figure 10.11 *Profiles sampled by the associated point rule, taking the AP of a profile to be its rightmost point, for a systematic subsample of tiles.*

10.3 Decomposition

An old joke says that the best way to count sheep is to count their legs and divide by four. Similar techniques are useful in stereology.

10.3.1 Multiple associated points

An extension of the foregoing strategy is to allow particles or profiles Y_j to be counted potentially more than once, but to compensate by assigning each object a fractional weight [152, 239, 290, 402].

For example, in a spatial pattern of line segments, Hall [239], [240, pp. 216–217] observed that since every line segment has two endpoints, we could estimate the expected number of segments per unit area by counting the number of segment endpoints in a sampling window and dividing by 2 times the area of the window.

Independently Jensen and Sundberg [290] proposed associating $q > 1$ points to each particle. Let $\mathcal{P} = \{1,\dots,N\}$ and $\mathcal{P}' = \{1,\dots,Nq\}$ represent the population of particles and the population of all APs, respectively. We may then estimate particle number N or a particle population total Φ by constructing an unbiased sampling rule for $\mathcal{P}'$. If $\tau : \mathcal{P}' \to \mathcal{P}$ is the mapping associating an AP with its particle number, then a particle population total can be written as

$$\Phi = \sum_{j \in \mathcal{P}} \varphi(Y_j) = \sum_{l \in \mathcal{P}'} \frac{1}{q} \varphi(Y_{\tau(l)}).$$

From a random sample $\mathcal{S}' \subseteq \mathcal{P}'$ of APs we can then estimate Φ by

$$\widehat{\Phi} = p^{-1} \sum_{l \in \mathcal{S}'} \frac{1}{q} \varphi(Y_{\tau(l)}) = p^{-1} \sum_{j=1}^{N} \frac{q_j}{q} \varphi(Y_j),$$

where p is the sampling probability of an AP and q_j is the number of sampled APs associated to Y_j.

There are several reasons why this generalization is interesting. First of all, a number of sampling designs described in the literature are special cases of the generalized AP method: (i) the classical AP method ($q = 1$), (ii) the net tangent count of convex objects ($q = 2$), and (iii) total curvature estimation of convex objects. Secondly, for certain types of particles, it may be difficult to use the classical AP method but easy to use a generalized AP method with $q \geq 2$. This is true if the particles are connected curves of finite length, in which case it is natural to use the endpoints of the curves ($q = 2$) as APs, cf. Figure 10.12. Thirdly, if the particles are appreciable in size compared to the size of the sampling window, the variance of estimators will be reduced by using more than one AP.

10.3.2 More general weights

In general, suppose we attach a fractional weight W_{jk} to a particle Y_j when we sample piece Z_k. If we can arrange that

$$\sum_k W_{jk} = 1 \tag{10.4}$$

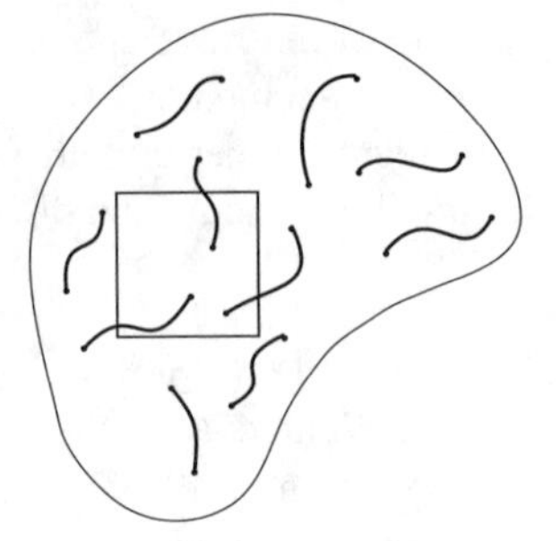

Figure 10.12 *The total number of fibres can be estimated using the total number of endpoints inside the sampling window.*

holds for all particles Y_j, then the bias is corrected and the weighted sample total gives an unbiased estimator of the population total:

$$\widehat{\Phi} = p^{-1} \sum_{k \in \mathcal{K}} \sum_{j=1}^{N} W_{jk} \varphi(Y_j).$$

Following are some mechanisms which guarantee (10.4):

- Equip particle Y_j with $q_j > 1$ associated points $a_1(Y_j), \ldots, a_{q_j}(Y_j)$ where q_j may not be constant, and weight Y_j by the fraction of its associated points which fall in the sampled pieces:

$$W_{jk} = \frac{1}{q_j} \sum_{\ell=1}^{q_j} \mathbf{1}\{a_\ell(Y_j) \in Z_k\};$$

- Weight each object Y_j proportional to its area of intersection with the sampled piece,

$$W_{jk} = \frac{A(Y_j \cap Z_k)}{A(Y_j)};$$

- Weight Y_j by the reciprocal of the number of pieces Z_k which intersect it:

$$W_{jk} = \frac{\mathbf{1}\{Y_j \cap Z_k \neq \emptyset\}}{\sum_{\ell=1}^{K} \mathbf{1}\{Y_j \cap Z_\ell \neq \emptyset\}}.$$

Other weight functions in $\mathbf{R}^2$ include the integral of curvature of $Z_k \cap \partial Y_j$, if the boundary ∂Y_j of each profile is a simple closed curve; and 4 minus the number of boundary intersections, $1 - \#(\partial Z_k \cap \partial Y_j)/4$, if both Y_j and Z_k are convex.

10.4 Reweighting

An alternative strategy, for achieving unbiased estimation of particle population parameters, is to use a biased sampling rule and to compensate for the bias.

A biased sampling rule is one which results in unequal sampling probabilities for different particles.

The Horvitz-Thompson device can be used to correct sampling bias in the estimation of total number N or, more generally, in the estimation of a total quantity $\Phi = \sum_{j=1}^{N} \varphi(Y_j)$ for the particle population.

Suppose our sampling rule leads to a sample $\mathcal{S}$ of particles with unequal sampling probabilities $p_j = \mathbf{P}[j \in \mathcal{S}]$. The Horvitz-Thompson estimator of φ is

$$\widehat{\Phi} = \sum_{j \in \mathcal{S}} \frac{\varphi(Y_j)}{p_j}. \tag{10.5}$$

This is an unbiased estimator of Φ, as shown in Appendix A. The only requirements are that all the sampling probabilities must be positive, and the sampling probabilities must be known exactly for *sampled particles* in order to compute the estimator.

10.4.1 Local serial sections — 'empirical method'

American pathologist W.R. Thompson, in a far-sighted article in the statistical journal *Biometrika* [557, 558] proposed the following technique for estimating the total number N of particles in a three-dimensional material.

The material is cut into slabs by a series of section planes. A random sample of slabs is selected, with uniform sampling probability p. We sample any particle which intersects a sampled slab.

For each sampled particle Y_j, we inspect the *neighbouring* slabs (which are not part of the sample) to determine how many slabs contain profiles of the same particle. Call this number m_j. See Figure 10.13.

Then the estimate of total particle number is

$$\widehat{N} = \frac{1}{p} \sum_{j \in \mathcal{S}} \frac{1}{m_j} \tag{10.6}$$

where the sum is over all sampled particles in all sampled slabs, and again m_j denotes the number of slabs that contain the particle Y_j.

To see that (10.6) is an unbiased estimator of N, we simply notice that it is an application of the Horvitz-Thompson estimator (10.5) to plus-sampling for serial sections. The probability that a given particle Y_j will be sampled by the plus-sampling rule is

$$p_j = p\, m_j = p\, \#\{k : Z_k \cap Y_j \neq \emptyset\}.$$

Hence (10.6) is exactly the Horvitz-Thompson estimator (10.5) for total particle number. The stack of slabs associated with each sampled particle would

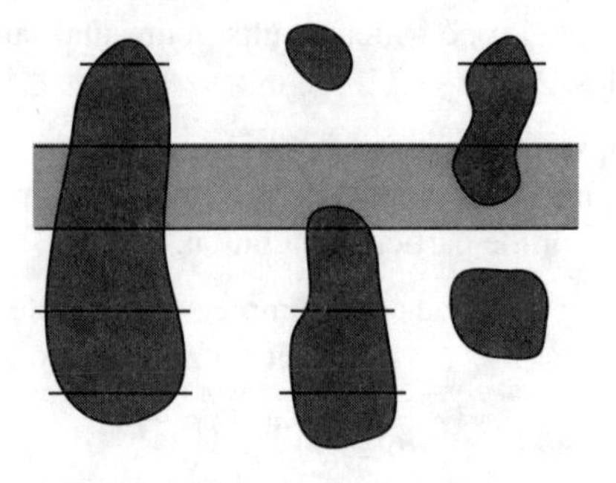

Figure 10.13 *The 'empirical method'. The sampled slab (light shading) is intersected by three particles (dark shading) which are therefore sampled. Inspecting the nearby slabs, we find that our three sampled particles intersect 6, 4 and 3 slabs, respectively. They contribute to the number estimate with weights* $1/6$, $1/4$ *and* $1/3$, *respectively.*

now be called an *adaptive cluster sample* [555, 556], and in that context the estimator (10.6) is the standard Horvitz-Thompson estimator of population size.

This technique was proposed briefly in 1896 by Gaule and Lewin. Thompson [557] reinvented it and gave a clear statistical argument for its validity and unbiasedness, based however on homogeneity assumptions. It was rediscovered and developed thoroughly by R. Coggeshall and collaborators [86, 87, 467] who called it the 'empirical method', because it is based on empirical observation of the relationship between the number of particle profiles in thick sections and the true number of particles in three dimensions. Cruz-Orive [116] developed closely related techniques.

Some important points about this estimator should be noted. The unbiasedness of (10.6) depends only on the fact that slabs are sampled with uniform sampling probability. The slabs do not have to be of equal thickness, or to have parallel planar surfaces. The tissue does not have to be homogeneous.

We can rewrite (10.6) as

$$\widehat{N} = \frac{1}{p}\frac{Q}{\bar{\bar{m}}} \tag{10.7}$$

where Q is the total count of particle profiles in the sampled slabs, and $\bar{\bar{m}}$ is the sample harmonic mean of the counts m_j,

$$\bar{\bar{m}} = \left(\frac{1}{n}\sum \frac{1}{m_j}\right)^{-1}.$$

This resembles the Abercrombie method (Section 2.5) in that $\bar{\bar{m}}$ plays a role analogous to the mean particle height. However the role of m_j is not to serve as a surrogate for particle height. Rather the numbers pm_j are the exact sampling probabilities. The main statistical difference between Abercrombie's method and the empirical method is that the numbers m_j are the sampling bias factors for the particles that are actually sampled, so that the Horvitz-Thompson principle applies.

The foregoing discussion assumes that it is impossible for a particle to be present in two of the sampled slabs. Typically we ensure this by taking a systematic sample of slabs with a spacing much larger than the diameter of any particle. In the general case where particles can be sampled with multiplicity greater than 1, the statements above still hold true when the sum in (10.6) is interpreted as a sum-with-multiplicity (so that a particle sampled with multiplicity M_j yields a total contribution of M_j/m_j to the sum).

Cruz-Orive [116] developed related techniques to estimate any parameter of the particle population, such as the mean particle volume. Intuitively we can regard the sample of particles Y_j with weights $1/m_j$ as equivalent to a uniform sample from the particle population. For a population total quantity $\Phi = \sum_{j=1}^{N} \varphi(Y_j)$, the section stack estimate is

$$\widehat{\Phi} = \frac{1}{p} \sum_{j \in \mathcal{S}} \frac{\varphi(Y_j)}{m_j}. \tag{10.8}$$

This is very useful in the case of particle volume, $\varphi(Y_j) = V(Y_j)$, since we can estimate the volume of each sampled particle accurately from its slab sections using the Cavalieri estimator [116].

10.4.2 Plus-sampling in rectangular tiles

Let us suppose that the aim is to estimate the number N of planar particles from observation in rectangular sampling windows. Let the situation be as in Section 10.2.5. Let $\mathcal{P}_{j_1 j_2}$ be the set of particles which are partly or fully contained in $Z_{j_1 j_2}$. Using 'double' systematic sampling, a sample of clusters are collected

$$\mathcal{S} = \cup_{(j_1,j_2) \in \mathcal{K}_1 \times \mathcal{K}_2} \mathcal{P}_{j_1 j_2}$$

from which the Horvitz-Thompson estimator of N may be constructed

$$\widehat{N} = \sum_{j \in \mathcal{S}} p_j^{-1}.$$

The problem is here to find the sampling probabilities since the clusters need not be disjoint. Suppose that m_1 and m_2 have been chosen so large that *sampled* clusters are disjoint, i.e.

$$\mathcal{P}_{j_1 j_2} \cap \mathcal{P}_{j_1' j_2'} = \emptyset \text{ for } |j_1 - j_1'| \geq m_1 \text{ or } |j_2 - j_2'| \geq m_2. \tag{10.9}$$

The sampling probability p_i is equal to the fraction of the $m_1 m_2$ systematic samples in which the ith particle appears. Under the condition (10.9), the number of systematic samples in which the ith particle appears is equal to the number n_i of sampling windows in which the ith particle appears and, accordingly,

$$p_j = \frac{n_j}{m_1 m_2}, \quad j = 1, \ldots, N.$$

10.5 Variance reduction

It is not possible to say much about the variance of the estimators presented in this chapter, because they are based on an arbitrary partition of the containing space Z. However, some strategies for reducing variance are available.

When estimating an additive parameter φ, we should choose the partition of Z so as to make the sample contributions φ_k as nearly equal as possible.

For example, Figure 10.14 sketches a population of dots which is to be counted. Two possible partitions of the material are shown. In the partition on the left, the pieces Z_k are of equal size, but the numbers of dots in each piece vary widely. In the partition on the right, each piece has roughly the same number of dots. Either partition gives us an unbiased estimator of the total number of dots, but the partition on the right yields an unbiased estimator with a lower variance.

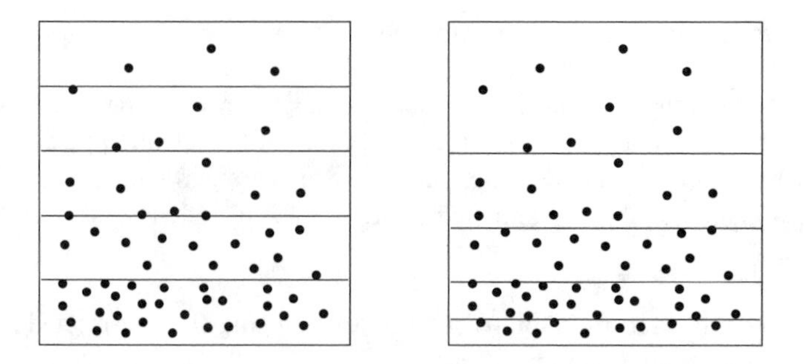

Figure 10.14 *Estimation of the number of dots by inefficient (left) and efficient (right) division of material into five pieces.*

This strategy is useful in applications where we have prior information about the microscopic structure of the material. In an organ where the local density of cells of interest is known to be higher at the distal end than at the proximal end, it may be sensible to cut tissue blocks of varying size, with smaller blocks at the distal end.

10.6 Fractionator

Sampling efficiency can be improved by sub-sampling techniques such as stratification and nested cluster sampling (see Appendix A). Multi-stage sampling designs, in which sampled material is subjected to further sub-sampling and sub-sub-sampling, etc, play a very important role in practical stereology. We discuss them extensively in Chapter 12.

In the discrete setting there is one important and popular multi-stage design, known as *'fractionator'* sampling [219, 224].

- At **stage 1**, the containing space Z is divided into pieces $Z_1, \ldots, Z_K$ arbitrarily. We select a uniform sample of pieces with sampling probability p_1. Let the sampled pieces Z_k be those indexed by $k \in \mathcal{K}^{[1]}$ where $\mathcal{K}^{[1]} \subseteq \{1, 2, \ldots, K\}$. The sampled material is

$$\mathcal{Z}^{[1]} = \bigcup_{k \in \mathcal{K}^{[1]}} Z_k.$$

- At **stage 2**, the sampled material $\mathcal{Z}^{[1]}$ is further subdivided into pieces $Z_{2,1}$, $\ldots, Z_{2,K_2}$ arbitrarily. We select a uniform sample of these pieces with sampling probability p_2. The sampled pieces are $Z_{2,k}$ for $k \in \mathcal{K}^{[2]} \subseteq \{1, 2, \ldots, K_2\}$. The sampled material is

$$\mathcal{Z}^{[2]} = \bigcup_{k \in \mathcal{K}^{[2]}} Z_{2,k}.$$

- At **each stage** ℓ for $1 < \ell \leq L$, the sampled material $Z^{[\ell-1]}$ from the previous stage is further subdivided into pieces $Z_{\ell,1}, \ldots, Z_{\ell,K_\ell}$ arbitrarily. We select a uniform sample of these pieces with sampling probability p_ℓ. The sampled pieces are $Z_{\ell,k}$ for $k \in \mathcal{K}^{[\ell]} \subseteq \{1, 2, \ldots, K_\ell\}$. The sampled material is

$$\mathcal{Z}^{[\ell]} = \bigcup_{k \in \mathcal{K}^{[\ell]}} Z_{\ell,k}.$$

By the basic principle of cluster sampling, the sampled material $\mathcal{Z}^{[\ell]}$ at each stage ℓ is a uniform sample of the material at stage $\ell - 1$, with sampling probability p_ℓ. Hence the final sample $\mathcal{Z}^{[L]}$ is a uniform sample of the containing space Z with sampling probability

$$p = p_1 p_2 \cdots p_L.$$

It follows that any additive quantity (10.1) can be estimated from a fractionator sample, by rescaling the sample total by $1/p$.

10.7 Exercises

Practical exercises

Exercise 10.1 Devise a method for estimating the size of a crowd of people from an aerial photograph.

Exercise 10.2 In this exercise you will estimate the number of typed characters on a page of newspaper, using the fractionator.

- Remove a page from the newspaper. If desired, you may trim off any white space or pictures that do not contain typed characters.
- **Stage 1**: Tear or cut the page into about 20–30 pieces in any fashion you wish. Arrange them in a sequence. Take a systematic random sample of period 3. Retain the sampled pieces and discard the others.
- **Stage 2**: Tear or cut the sampled material from stage 1 into about 20 pieces altogether, by tearing the larger ones into several pieces. Arrange them in a sequence. Take a systematic random sample of period 2. Retain only the sampled pieces.
- **Stage 3**: Again tear or cut the sampled material from stage 2, to yield at least 15 pieces, and take a systematic sample of period 3. (Further sampling stages may be conducted if desired.) The result is a uniform sample of the newspaper with sampling probability $1/18$.
- **Counting:** choose an unbiased sampling rule for the typed characters that are wholly or partially visible on the sampled pieces of newspaper. (For example, use the associated point rule that the lower left corner of the character must be visible on the sample.) Count the sample total of characters according to this sampling rule.
- **Estimation:** multiply by 18 to obtain an estimate of the total number of characters.

Exercise 10.3 Bake a cake, containing a known number of glacé cherries, in a cylindrical tin. Cut the cake in the usual way. Use the first-seen rule to estimate the total number of cherries from a few slices.

Exercise 10.4 Take a uniform sample of particle profiles from Figure 2.16 with sampling fraction $1/9$ using the two-dimensional disector, the tiling rule, and the associated point method.

Theoretical exercises

Exercise 10.5 Find the exact distribution of the 2D disector estimator $\widehat{N}$ for the example population shown in Figure 10.6 when every second cluster is taken. What is the distribution of the 2D disector estimator if the ordering is reversed and still every second cluster is taken? The two estimators $\widehat{N}_1$ and $\widehat{N}_2$ can be combined into a single estimator

$$\frac{1}{2}(\widehat{N}_1 + \widehat{N}_2),$$

where $\widehat{N}_1$ and $\widehat{N}_2$ are identically distributed but generally dependent. Show that this estimator is unbiased in general. Show that in this case

$$\text{Var}\frac{1}{2}(\widehat{N}_1 + \widehat{N}_2) \leq \text{Var}\widehat{N}_1.$$

Exercise 10.6 Find the distribution of the estimator $\widehat{N}$, based on the tiling rule, for the population shown in Figure 10.9. The sampling is 'double' systematic sampling with $m_1 = m_2 = 3$. Show also directly that $\mathbf{E}\widehat{N} = 29$.

Exercise 10.7 Find the distribution of the estimator $\widehat{N}$, based on the associated point rule, for the population in Figure 10.11. The sampling is 'double' systematic sampling with $m_1 = m_2 = 3$. Show also directly that $\mathbf{E}\widehat{N} = 29$.

Exercise 10.8 Let $\widehat{\Phi}$ be the Horvitz-Thompson estimator of $\Phi = \sum_{j=1}^{N} \varphi_j$, where $\varphi_j = \varphi(Y_j)$, cf. (10.5).

1) Show that $\widehat{\Phi}$ is an unbiased estimator of Φ.

2) Show that

$$\text{var}[\widehat{\Phi}] = \sum_{i=1}^{N}\sum_{j=1}^{N} \frac{p_{ij}}{p_i p_j}\varphi_i\varphi_j - (\sum_{i=1}^{N}\varphi_i)^2,$$

where

$$p_{ij} = \mathbf{P}[i \in S,\ j \in S],$$

$i, j = 1, \ldots, N$, are the so-called sampling probabilities of second-order. Note that $p_{ii} = p_i$.

Exercise 10.9 Verify that the tiling rule described in Section 10.2.5 can be regarded as 2-stage fractionator sampling where at each stage 2D disector sampling is used.

10.8 Bibliographic notes

Cluster sampling is a classical technique in survey sampling [83]. Cluster sampling and stratified sampling have long been used in cartography [350, Chap. 8]. They have also been applied to spatial sampling, for ecological and environmental surveys, in recent decades [555, 66].

In stereology, the first clear descriptions of discrete uniform sampling — that do not presuppose any kind of uniformity in location and/or orientation — appear to be those of Miles and Davy [410, Section 4] and Gundersen [219]. Almost simultaneously, Cruz-Orive [116] noted that the first-seen rule can be used without knowledge of the size and shape of the pieces. See also [140].

Counting and sampling rules have a long history, indeed stretching back to the beginnings of microscopy. History of particle counting is recounted in [44, 254]. The disector is usually attributed to the *nom de plume* D.C. Sterio [538] but was first described in 1895 by Miller & Carlton [413] and has been rediscovered numerous times [59, 317, 557, 558, 477]. A closely related sampling technique was introduced in textile sampling by British statisticians H.E. Daniels [143] and D.R. Cox [99, 100].

The disector was originally defined [538] as "a pair of parallel planes a fixed, known distance apart" (hence "di" for a pair of planes). The justification required uniform random placement of the planes in three dimensions. Cruz-Orive [116] pointed out that the section spacing need not be known or fixed, and established that sections need only be selected with equal probability, calling such a section pair a "selector". Later usage has blurred this distinction and indeed introduced several different techniques which go under the name of 'disector': a pair of section planes ('physical disector') or a slab bounded by two planes ('optical disector').

The papers by Thompson [557, 558] are rich in ideas. He describes the Cavalieri estimator [557, p. 21], mentions random sampling of an organ to estimate its absolute and relative characteristics [557, p. 22], argues the desirability of sampling particles ("islets") with equal probability [557, p. 22], states [557, pp. 22, 24] that it may be possible to work with samples which have nonuniform probability where the probabilities of selection are positive and are known for the sampled items, and proposes [557, p. 26, eq. (12)] the Horvitz-Thompson estimator of particle number. The H-T estimator of a general quantity is effectively proposed in equation (3) in [557, p. 23]. He uses this to estimate mean particle volume, using Cavalieri for the individual estimates, in the accompanying paper [558, p. 32]. Thompson [557, p. 26] also effectively proposes the disector rule. However the arguments are based on uniform positioning of the sections in space.

The 'empirical method' first appeared in 1896 and has been rediscovered several times [51, 193, 344]. See the historical note [254]. It was rediscovered and perfected by Coggeshall and collaborators from 1984 onward [86, 87, 467]. It became clear (apparently) that the empirical method was inefficient compared to the disector, so it was abandoned in favour of the disector [467, 85].

Miles' paper on edge effects [399] stimulated stereological research on sampling rules and introduced the first technique of Horvitz-Thompson type. The tiling rule was introduced by Gundersen [216, 217]. See also [513, p. 100]. Miles [402] introduced the associated point rule and proved its validity in a model-based setting. Lantuéjoul [335] proposed a Horvitz-Thompson technique. Fractional weights are used in [152, 239, 290, 402] and [240, pp. 216–217].

The 'unbiased brick' was described by Gundersen [217, 218] and others [273, 606, 607]. The selector technique was proposed by Cruz-Orive [116]. Practical discussion of particle sampling rules is offered in [219, 224]. A general theory of sampling rules is developed in [31].

In applications, Coggeshall [85] reviews biased and unbiased counting methods for cell numbers, and cites proof that biases in the classical methods are severe. Hedreen [253] also reviews different methods and discusses [252] the effect of lost caps. West [597] gives a detailed explanation of the principles of unbiased sampling rules, and a detailed discussion of bias and variance, using worked examples of synthetic particle populations. Unbiased counting rules have been adopted widely [164, 203, 426, 594]. Bertram [50] gives a very clear description of unbiased counting rules in the kidney.

There is still much controversy about particle counting in biology. Schmitz et al [507] point to wide differences in the results for design-based methods and attribute this to 'observer bias' and/or high variance. West [596] responds that this is more likely to be due to incorrect application of unbiased counting rules and other essential principles.

CHAPTER 11

Inference for particle populations

This chapter describes stereological methods for inference about a population of particles of very general shape.

Classical methods usually required that all particles in the population have a common, simple, known shape such as a sphere or cube. While this was perfectly appropriate for some applications (e.g. in materials science), it was a severe limitation in other applications, such as biological science. Cells in a cell population have different, complex shapes, which are not directly observable in a plane section.

It is, therefore, an important advance in flexibility that stereological methods of estimating particle number and size have been developed for arbitrarily shaped particles [293, 406, 407, 538]. These methods require only the condition of *identifiability*, namely that the different profiles belonging to the same particle can be identified.

We begin in Section 11.1 with a brief overview of different strategies for particle stereology. Section 11.2 presents the modern methods in the design-based context (*Restricted* and *Extended* cases) and Section 11.3 the model-based case (*Random* case).

11.1 Strategies for particle stereology

11.1.1 The problem

Classical stereological literature identified the fundamental problems involved in inference for particles. Most important of these is the sampling bias which is inherent in observing a population of three-dimensional particles using a two-dimensional section plane.

In Chapter 2, it was shown heuristically that particle number cannot be estimated from observations in a single planar section, without further assumptions about particle shape. The basic problem is that a section plane samples particles with probabilities proportional to their heights. The key, showing that particle size and particle number are confounded, is the Rhines-DeHoff relation

$$N_A = \mathbf{E}_N[H] \cdot N_V, \tag{11.1}$$

where N_A is the number of particle traces per unit section area, $\mathbf{E}_N[H]$ is the mean particle height and N_V is the number of particles per unit volume. This relation can be proved in a rigorous manner, both in the design-based and model-based case, for particles of arbitrary shape. In Exercises 11.1 and 11.2, rigorous design-based justifications are developed for the *Restricted* and *Extended* cases. A rigorous model-based justification can be found in [543].

In statistical terms, the main obstacle is that the sampling bias is not observable. The sampling probability for an individual particle is proportional to its height normal to the section plane. This height is not directly observable, at least not in thin sections of opaque materials.

11.1.2 Geometrical models

Most classical techniques avoided this obstacle by assuming the particles all have a common, simple, geometrical shape, such as a sphere. This enables us to relate the observed sizes of particle profiles in the plane section to the unobserved heights of particles. For example, as we saw in Section 2.4.2, the mean diameter of a population of spheres can be estimated from the harmonic mean diameter of their plane sections. This gives us an estimate of $\mathbf{E}_N[H]$ which we combine with the observed numerical density of profiles N_A in (11.1) to obtain an estimate of N_V.

However, a relationship between profile size and particle height can only exist if the geometrical model is quite restrictive. Hence the classical methods, developed for spheres, cubes, cylinders and certain other shapes, cannot be generalised to arbitrarily complex shape models.

Alternatively we may try to estimate mean particle height in a separate experiment. Abercrombie [3] proposed measuring particle height in sections normal to the original sections in which profiles were counted ('Method 1'). However such methods also depend sensitively on the geometrical assumptions [3].

It was widely assumed that methods developed for spheres would be approximately correct for biological cells which are (asserted to be) approximately spherical. Very few published studies to date have attempted to test this claim empirically. One sobering study [391] applied several classical methods for estimating N_V and N to the same material, and obtained widely different results. The methods of Abercrombie and Floderus yielded results that were biologically implausible.

11.1.3 Sampling modifications

Two approaches to sampling bias were discussed in Chapter 10 and we add a third one here.

The first approach is to eliminate the sampling bias by modifying the sampling design so that it has uniform sampling probability. This approach is followed by the *disector* technique and its relatives.

The second approach is to reweight the sampled particle profiles so that estimators are unbiased. This is the rationale of the 'empirical method' described in Section 10.4.1.

A third approach, not mentioned earlier, is to modify the sampling design so that the sampling bias is proportional to a manageable characteristic of the particle. This approach is followed in the *point-sampled intercept* technique, in which the sampling bias is proportional to particle volume, leading to estimators of particle size moments in the volume-weighted distribution.

This chapter summarises and discusses these new developments in stereological particle analysis, see also the early developments in [111, 150]. We will discuss design-based inference as well as model-based inference for particle populations [280, 292].

11.2 The design-based case

In the design-based case, the particle aggregate is regarded as non-random. Randomness is introduced via the sampling design, e.g. the random positioning of a planar section. The statistical properties of estimators relate to the randomisation of the sample.

The framework is a finite particle population consisting of N particles $Y_1, \ldots, Y_N$ contained in a reference space Z in $\mathbf{R}^3$.

11.2.1 Particle number

As mentioned in Chapter 10, particle size and particle number are confounded if ordinary planar sampling is used. A solution to this problem is to change the sampling. In Chapter 10, a number of examples of unbiased sampling rules were given. Here, not only the particle aggregate was fixed but also the partition of the space containing the particles. The particle aggregate was partitioned into arbitrary clusters and a discrete unbiased sampling rule was constructed, typically by selecting a particle in the first cluster it appeared.

Below we present the corresponding continuous sampling designs. The continuous formulation is useful if we do not want to condition on the partition when evaluating the statistical properties of an estimator. Also, some sampling designs cannot be described, using a partition of space, for instance sampling in the plane with a systematic set of circular observation windows.

Significant applications in neuroscience of the methods of estimating particle number are described in [298, 449, 455, 474, 550]. Applications in materials science have been described in [303, 328]. An informal review has been given in [219].

Estimation of particle number in the Restricted *case*

In the *Restricted* case, the specimen available for observation is the whole reference space Z. Particle number can be estimated without specific assumptions about particle shape, using the discrete disector design described in Section 10.2.4. We will here present the continuous analogue, using planes with uniform random position.

We consider a systematic set of planes with a uniform start

$$\mathcal{T} = \{T_{U+mt} : m \in \mathbf{Z}\},$$

where T_h is a plane in $\mathbf{R}^3$ of fixed orientation and position $h \in \mathbf{R}$, and t is the distance between neighbour planes. The start U is uniform in $[0,t)$. Associate to $\mathcal{T}$ a systematic set of slabs of thickness d

$$\mathcal{J} = \{J_{U+mt} : m \in \mathbf{Z}\},$$

where J_{U+mt} is the slab bounded by the planes T_{U+mt} and T_{U+d+mt}, and $0 < d < t$, cf. Figure 11.1. A particle is sampled in a particular slab if it is first seen in this slab. That is, Y_i is sampled in J_{U+mt} if Y_i intersects J_{U+mt} but Y_i does not intersect $J_{U+\ell t}$ for any $\ell < m$.

If we let $u_i = \min\{p(y) : y \in Y_i\}$, where p is the projection onto a line perpendicular to the planes, the sampled particles are those with indices in $\mathcal{S} \subseteq \{1,\ldots,N\}$ where

$$\mathcal{S} = \{i : U + mt \leq u_i \leq U + d + mt \text{ for some } m\}.$$

The sampling probabilities are constant:

$$\mathbf{P}[Y_i \text{ sampled}] = \sum_m \mathbf{P}[U + mt \leq u_i \leq U + d + mt] = d/t.$$

An unbiased estimator of N is

$$\widehat{N} = \frac{t}{d} N(\mathcal{S}),$$

where $N(\mathcal{S})$ is the number of sampled particles. Note that if t is a multiple of

d, the conditional distribution of the sample $\mathcal{S}$, given U mod d, is the same as for the systematic discrete disector design, described in Section 10.2.4.

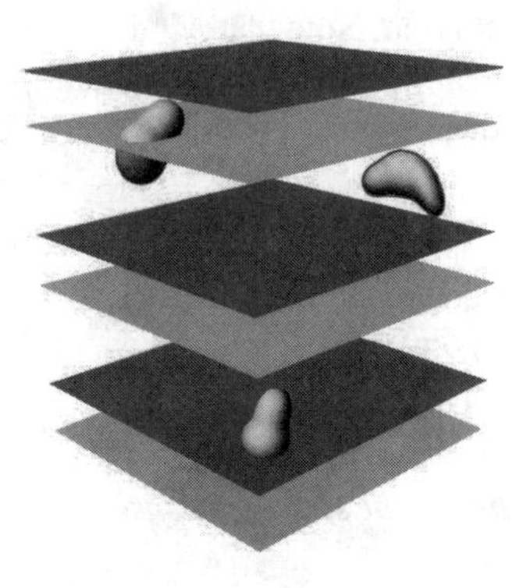

Figure 11.1 *Sampling particles with a systematic set of disectors. A particle is sampled if it first appears in one of the slabs bounded by neighbouring light grey and dark grey planes. If the planes are ordered from top to bottom, only the uppermost particle is sampled by this procedure.*

A few remarks about practical implementations are in order. To count the number $N(\mathcal{S})$ of disector-sampled particles, in principle we need to inspect the entire contents of each sampled slab (between the two bounding planes). The typical application where this is possible is a thick section of a translucent biological tissue observed in optical transmission microscopy (conventional or confocal). By adjusting the position of the focal plane, the microscopist is able to observe the entire contents of the thick section, and therefore to apply this counting rule directly. Hence this approach is sometimes called the *optical disector*.

In certain other applications, it is only possible to visualise a two-dimensional plane section. Examples are polished sections of opaque materials, thin sections of translucent materials, and ultrathin sections in transmission electron microscopy. The disector rule can still be applied if we are assured that no particle is small enough to lie between two bounding planes of a slab. If the projection $p(Y_i)$ of each particle Y_i is a line-segment of length at least d, then a particle Y_i is sampled if there exists an integer m such that Y_i hits T_{U+d+mt} (reference plane) but not T_{U+mt} (look-up plane). In this case, it is enough to collect information *on* the section planes and the design is called the *physical disector*. It is clearly still necessary to be able to identify particle traces on the two planes that belong to the same particle.

Using the results in Section 7.1.1, the ratio N_V can be estimated by

$$\widehat{N_V} = \frac{1}{d} \frac{N(\mathcal{S})}{A(Z \cap \mathcal{T})}. \tag{11.2}$$

The estimator $\widehat{N_V}$ is ratio-unbiased. Note that $\widehat{N_V}$ does not depend on the thickness t. The estimator (11.2) may be combined with an independent estimator of $V(Z)$ (obtained by fluid displacement, for instance) to obtain an alternative estimator of the absolute number N.

It might be needed to subsample in each slab. One possibility is to project the particles sampled in a slab onto one of the planes of the slab and apply the tiling rule or the associated point rule, see Sections 10.2.5 and 10.2.6, on the projected particles.

Estimation of particle number in the Extended *case*

The specimen available is here but a small portion of the much larger reference space Z. We suppose that the specimen is a 3-dimensional FUR box Q, hitting Z. Such a box is distributed as $u + Q_0$ where Q_0 is the reference position of Q and u has density

$$f(u) = \begin{cases} \frac{1}{V(Z \oplus \check{Q}_0)} & \text{if } Z \cap (u + Q_0) \neq \emptyset \\ 0 & \text{otherwise.} \end{cases}$$

There are different ways of sampling particles according to an unbiased sampling rule. One way is to consider Q as part of a 'hypothetical' tessellation of $\mathbf{R}^3$ with boxes congruent to Q. A particle Y_i is sampled in Q if Q is the first box where Y_i appears according to the lexicographic ordering in $\mathbf{R}^3$, cf. Figure 11.2. If a point is associated to each particle, an alternative sampling procedure is to sample all particles with associated point in Q, cf. Figure 11.3. Both procedures have the property

$$V(\{u : Y_i \text{ sampled in } u + Q_0\}) = V(Q_0).$$

Therefore,

$$\mathbf{P}[Y_i \text{ sampled}] = \frac{V(Q_0)}{V(Z \oplus \check{Q}_0)},$$

and an unbiased estimator of N is

$$\widehat{N} = \frac{V(Z \oplus \check{Q}_0)}{V(Q_0)} N(\mathcal{S}),$$

where $N(\mathcal{S})$ is the number of sampled particles. Using (4.22) with $Y = Z$ and $d = m = k = 3$, we find

$$\widehat{V}(Z) = \frac{V(Z \oplus \check{Q}_0)}{V(Q_0)} V(Z \cap Q)$$

is an unbiased estimator of $V(Z)$. Accordingly, a ratio-unbiased estimator of N_V is

$$\widehat{N_V} = \frac{N(\mathcal{S})}{V(Z \cap Q)}.$$

Note that $\widehat{N_V}$ does not involve the quantity $V(Z \oplus \check{Q}_0)$ which is often unknown. Instead of measuring $V(Z \cap Q)$ exactly, $V(Z \cap Q)$ may be replaced by a stereological estimator, for instance a point count from points rigidly attached to Q.

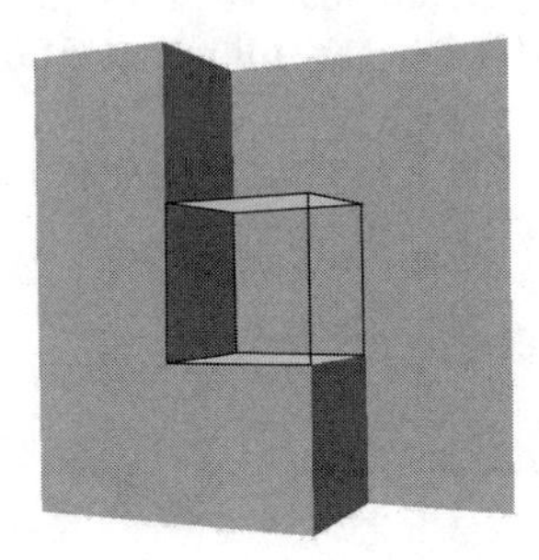

Figure 11.2 *Uniform sampling of particles using the lexicographic ordering in* $\mathbf{R}^3$. *A particle is sampled if it hits the 3D box but not the most distant vertical light grey plane* and *the layer to the left and below the box.*

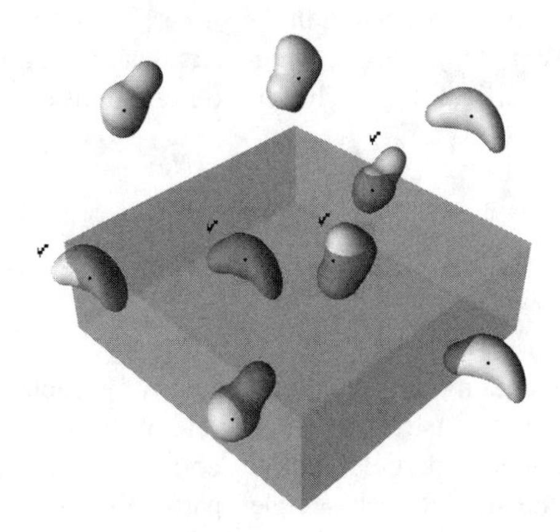

Figure 11.3 *Uniform sampling of particles using associated points. A particle is sampled if its associated point belongs to the 3D box. Sampled particles are indicated by ticks.*

An estimator of N can be obtained by combining $\widehat{N_V}$ with an estimator of $V(Z)$.

11.2.2 Mean particle size

An appropriate definition of particle size clearly depends on the particle population considered. The particles may be of full dimension in which case volume, surface area, diameter (maximal distance between pairs of points in the particle), and height (usually direction dependent) may be appropriate size parameters that do not require that the particles are of the same shape. For some purposes it may be convenient to consider collections of spatial curve segments and point clusters as particle populations, cf. Figure 11.4. For such particle populations, diameter and height are still meaningful size parameters but here it is also natural to consider curve length or the number of points in a cluster as size parameters.

Figure 11.4 *Particle populations with particles of dimension 1 and 0, respectively.*

Let us write $\mathbf{E}_N[V]$, $\mathbf{E}_N[S]$ and $\mathbf{E}_N[H]$ for the mean particle volume, surface area and height. Combining the Fundamental Formulae of Stereology with one of the estimators of N_V presented in the previous subsection, we can construct estimators of these three mean size parameters that are valid without specific assumptions about particle shape. Thus, for observations on a plane, we have

$$\widehat{\mathbf{E}_N[V]} = \widehat{A_A}/\widehat{N_V}$$
$$\widehat{\mathbf{E}_N[S]} = \frac{4}{\pi}\widehat{L_A}/\widehat{N_V}$$
$$\widehat{\mathbf{E}_N[H]} = \widehat{N_A}/\widehat{N_V}.$$

Mean particle size, and more generally moments of particle size, may also be estimated without first constructing an estimate of N_V. The basic idea is here first to collect a sample of particles with the correct sampling probabilities and then to estimate the size of each sampled particle by a local technique, see Sections 8.3–8.5 and [293]. If an ordinary (unweighted) mean particle size

$$\mathbf{E}_N[\varphi] = \frac{1}{N}\sum_{i=1}^{N}\varphi(Y_i)$$

is to be estimated, particles are sampled with equal probability. If instead a weighted mean is to be estimated, this weighting is implemented in the sampling. For instance, if weights are proportional to volume

$$w_i = V(Y_i)/\sum_i V(Y_i),$$

corresponding to a volume-weighted mean particle size

$$\mathbf{E}_V[\varphi] = \sum_i V(Y_i)\varphi(Y_i)/\sum_i V(Y_i),$$

then particles are sampled with probability proportional to volume. Volume-weighted mean particle volume has proven to be a useful parameter in cancer diagnostics, cf. [330, 331, 527].

Below, we consider estimation of unweighted and volume-weighted mean particle size in more detail.

Unweighted mean particle size

We suppose that a point y_i is associated to each particle Y_i. Using these associated points, we can construct an unbiased sampling rule such that

$$\mathbf{P}[i \in \mathcal{S}] = p, \quad i = 1, \ldots, N.$$

In the *Restricted* case, we may use a systematic set of disectors, as shown in Figure 11.1, and sample a particle if its associated point lies in one of the slabs bounded by neighbouring light grey and dark grey planes. In the *Extended* case, the sample may consist of those particles with associated points in a 3D box, cf. Figure 11.2.

The next step is to construct a conditionally unbiased estimator $\widehat{\varphi}(Y_i)$ of $\varphi(Y_i)$ for each sampled particle. Such an estimator satisfies

$$\mathbf{E}[\widehat{\varphi}(Y_i) \mid i \in \mathcal{S}] = \varphi(Y_i).$$

If the size parameter φ is either volume or surface area, an estimator of $\varphi(Y_i)$ may be constructed using a random plane T through y_i, cf. Figure 11.5, as discussed in Chapter 8. The plane may either be an isotropic or vertical random plane through y_i. An isotropic or sine-weighted line through y_i, contained in the plane T, may also be involved in the estimation. In the *Restricted* case, T may be taken to be parallel to the planes of the disectors if the orientation of these planes has been randomised in the appropriate way. In the *Extended* case, T may be parallel to one of the faces of the sampling box if the orientation of the box has been randomised appropriately.

It can be shown that the estimator

$$\sum_{i \in \mathcal{S}} \widehat{\varphi}(Y_i)/N(\mathcal{S}),$$

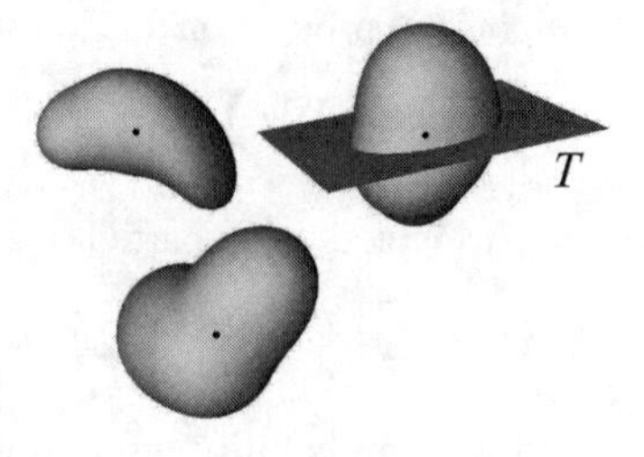

Figure 11.5 *Local sectioning.*

resulting from this two-step procedure is a ratio-unbiased estimator of the mean $\mathbf{E}_N[\varphi]$. Thus,

$$\begin{aligned}
\mathbf{E}\sum_{i\in\mathcal{S}}\widehat{\varphi}(Y_i) &= \mathbf{E}\sum_{i=1}^{N}\mathbf{1}\{i\in\mathcal{S}\}\widehat{\varphi}(Y_i) \\
&= \sum_{i=1}^{N}\mathbf{P}[i\in\mathcal{S}]\,\mathbf{E}[\widehat{\varphi}(Y_i)\mid i\in\mathcal{S}] \\
&= \sum_{i=1}^{N}p\varphi(Y_i).
\end{aligned}$$

In particular,

$$\mathbf{E}N(\mathcal{S}) = pN.$$

It follows that

$$\frac{\mathbf{E}\sum_{i\in\mathcal{S}}\widehat{\varphi}(Y_i)}{\mathbf{E}N(\mathcal{S})} = \frac{\sum_{i=1}^{N}\varphi(Y_i)}{N} = \mathbf{E}_N[\varphi].$$

Local geometric identities can be used to construct $\widehat{\phi}(Y_i)$. Local estimators of volume are given in (8.30), (8.41) and (8.38) while a local estimator of surface area is presented in (8.44), see also [220, 295].

It is also possible to develop estimators of mean particle size $\mathbf{E}_N[\varphi]$ in the case where the particles are either 1- or 0-dimensional, cf. Figure 11.4, and φ is either length or number. The initial sampling of the particles is as described above but the individual size estimators $\widehat{\varphi}(Y_i)$ are based on more comprehensive information from a local slice centred at the associated point y_i of the particle Y_i, cf. [293, Chapter 6].

Let us indicate how such an estimator can be constructed in the case where $\varphi = N$ and $Y_i = \{z_{i1},\ldots,z_{iN_i}\}$ is a finite set of points. The local estimate of $N(Y_i) = N_i$ is based on observation in a random slice $y_i + T$ of thickness t through y_i. Let us here concentrate on the case where the slice is isotropic, i.e. its mid-plane is isotropic. For a general exposition of stereological estimation based on local slices, see [313, 552]. If T is isotropic, the probability $p(y_i, z_{ij})$ that z_{ij} lies in $y_i + T$ is

$$p(y_i, z_{ij}) = \begin{cases} t/[2d(y_i, z_{ij})] & \text{if } d(y_i, z_{ij}) > \frac{t}{2} \\ 1 & \text{otherwise.} \end{cases}$$

Note that the different points z_{ij} in the particle Y_i are not sampled with equal probability. The Horvitz-Thompson estimator (see Appendix A) can be used in such cases. We get

$$\widehat{N}_i = \sum_{z_{ij} \in y_i + T} p(y_i, z_{ij})^{-1}.$$

This technique is in particular of interest if

$$Y_i = \{y_j \mid d(y_i, y_j) \leq r, j \neq i\},$$

say. Here, $\mathbf{E}_N[\varphi]$ becomes the average number of other associated points y_j at a distance at most r to a point y_i, see also the 2D illustration in Figure 11.6. This is a design-based analogue of the K−function from point process theory, see [543]. The K−function describes the second-order properties of the points $\{y_i\}$. Second-order stereology is described more generally in [288]. The first principles were laid down in [174].

It has earlier been investigated whether the K−function can be estimated from observations in a single section through the particle population. This is possible under specific shape assumptions. The case of spherical particles has been treated in [244, 245].

Higher order moments like the second-order moment in the particle volume distribution can also be estimated by local techniques, cf. e.g. [284, 289, 293].

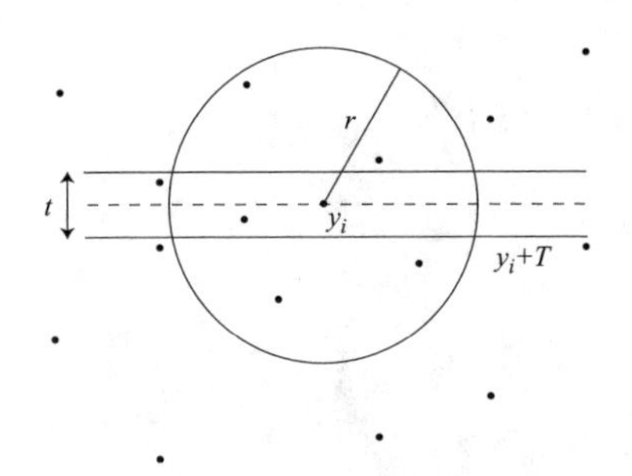

Figure 11.6 *Estimation of number of points in a circular disk of radius r centred at y_i from observation in an isotropic band of breadth t.*

Volume-weighted mean particle size

In this subsection we will concentrate on the case where particle size is volume. The aim is to estimate the volume-weighted mean particle volume

$$\mathbf{E}_V[V] = \sum_i \frac{V(Y_i)}{\sum_j V(Y_j)} V(Y_i) = \sum_i V(Y_i)^2 / \sum_i V(Y_i).$$

In order to estimate $\mathbf{E}_V[V]$, it is a good idea to sample particles by uniform random points because such points sample particles with a probability proportional to their volume. Thus, if $U \sim \mathsf{Unif}(Z)$, then

$$\mathbf{P}[U \in Y_i] = \frac{V(Y_i)}{V(Z)}.$$

Independent replication of the point-sampling procedure is usually not applied because it is inefficient to use only one point per linear or planar section. In the *Restricted* case, a uniform grid of points on a line or plane may be used, cf. Figure 11.7. In the *Extended* case, quadrats equipped with point grids are appropriate, cf. Figure 11.8.

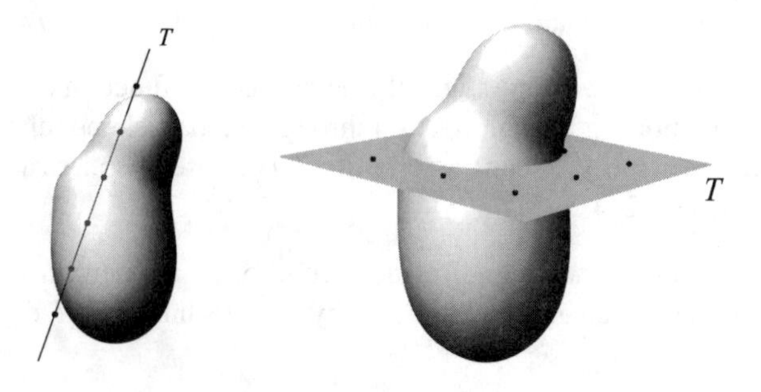

Figure 11.7 *Systematic point sampling in the restricted case.*

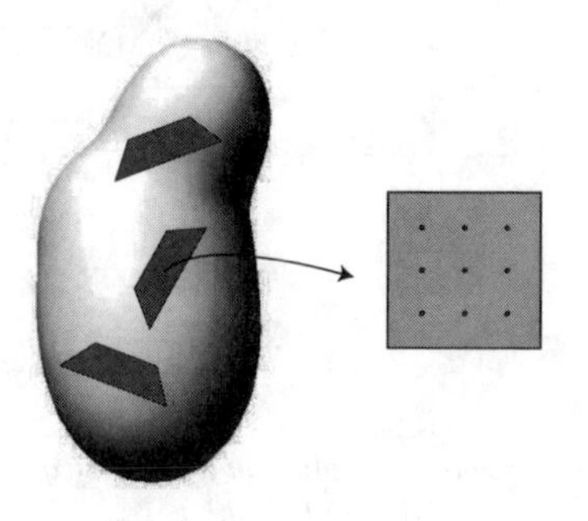

Figure 11.8 *Systematic point sampling in the extended case.*

If Y_i is sampled by a uniform random point U, then the volume of the particle can be estimated from observations in an isotropic section through U. For a linear section we can estimate $V(Y_i)$ by

$$\widehat{V}(Y_i) = 2\pi \int_{Y_i \cap (U+T)} d(x,U)^2 \, \mathrm{d}x, \tag{11.3}$$

where T is an isotropic line through O, cf. (8.30). In the literature, $Y_i \cap (U+T)$ is called a *point-sampled intercept*, cf. e.g. [224]. If $Y_i \cap (U+T)$ consists of a finite number of line-segments

$$I_j = [U_{j-}, U_{j+}], \quad j = 0, 1, \ldots, k,$$

numbered such that $U \in I_0$, then

$$\widehat{V}(Y_i) = \frac{2\pi}{3}[d(U_{0-},U)^3 + d(U_{0+},U)^3 + \sum_{j=1}^{k} \mid d(U_{j+},U)^3 - d(U_{j-},U)^3 \mid],$$

cf. Figure 11.9.

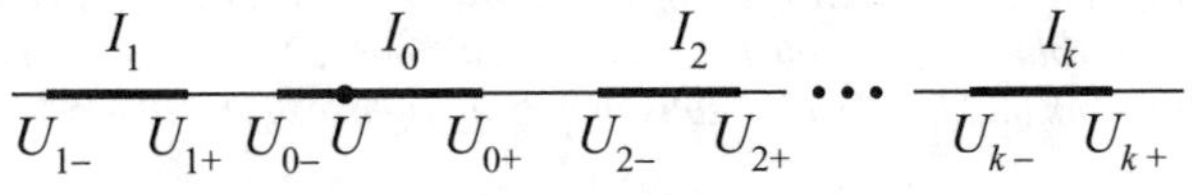

Figure 11.9 *Labelling of point-sampled intercepts.*

Using the fact that U is conditionally uniform in I_0, we can replace $d(U_{0-},U)^3 + d(U_{0+},U)^3$ by $d(U_{0-},U_{0+})^3/2$. If $Y_i \cap (U+T)$ consists of a single line-segment ($k=0$), the resulting estimator becomes

$$\frac{\pi}{3} d(U_{0-},U_{0+})^3.$$

Figure 11.10 shows an implementation of this estimator using vertical sections. The image is a vertical section of skin tissue. A sin-weighted random direction has been chosen (by a variant of the method described in Chapter 8) and a set of test lines is drawn parallel to this direction. The test lines carry evenly-spaced test points, which are used to sample profiles of cell nuclei (dark shapes). For each sampled nucleus, the ruler at bottom left is used to measure the cubed intercept length.

As suggested in [405, 407], it is possible to reduce the variance by replacing $\widehat{V}(Y_i)$ with the conditional mean value of $\widehat{V}(Y_i)$ given the line $U+T$. If $x_1, \ldots, x_{2(k+1)}$ are the end-points of $I_0, \ldots, I_k$ in order along the line, then the alternative estimator becomes

$$\widetilde{V}(Y_i) = \frac{\pi}{3L(Y_i \cap (U+T))} \sum_{u=1}^{2(k+1)} \sum_{v=u+1}^{2(k+1)} (-1)^{v-u+1} d(x_u,x_v)^4. \tag{11.4}$$

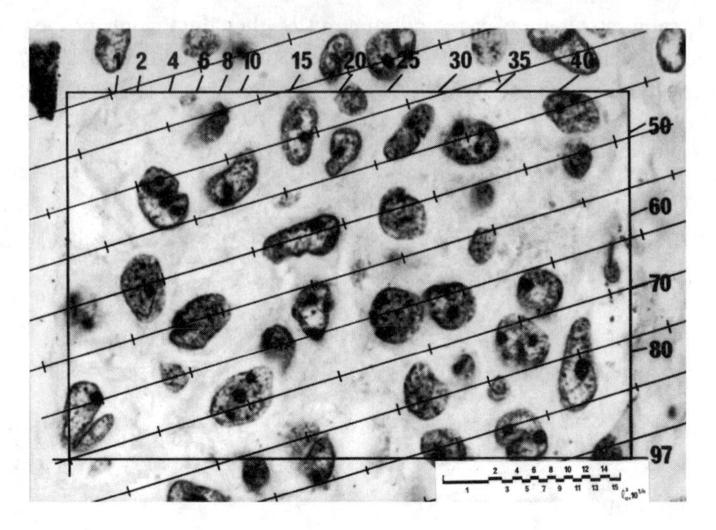

Figure 11.10 *Vertical section of skin tissue with test system for estimating volume-weighted mean volume of cell nuclei. Reproduced from [225, Fig. 6] with kind permission of the authors and Blackwell Publishing. Copyright © 1988 Blackwell Publishing.*

The double sum in (11.4) is called the 4-linc in [405, 407]. The general theory for particles on $\mathbf{R}^n$ was developed in [284, 403], see also [116, 228, 279].

In most applications, the line T is generated by first taking an IUR or VUR section plane E, then generating a random line T inside E. In this case, the sampling variance can be reduced further by taking the conditional expectation of $\widehat{V}(Y_i)$ given the plane E. The resulting estimator was described by Cabo and Baddeley [70] along with an evaluation of the gain in efficiency.

11.2.3 The particle size distribution

As mentioned earlier, it is generally recognised that particle size distributions can only be estimated from observations in a single section under specific shape assumptions for the particles. The case of spherical particles was treated as early as in 1925 [603, 604]. A design-based proof of Wicksell's integral equation can be found in [278]. Estimating sphere size distribution from observations of diameters of circular sections is an ill-posed problem that has occupied many mathematical statisticans, see e.g. the review [113]. The case of ellipsoidal particles has been treated in the two papers [107, 108]. As pointed out by Cruz-Orive, it is not possible to reconstruct from planar sections the shape and size distributions of the ellipsoids.

If the modern sampling methods described earlier in this section can be implemented, there is a possibility to get information about the particle size distribution, also in the case of arbitrarily shaped particles. The situation can be outlined as follows.

Let us suppose that we want to estimate the distribution of a size parameter φ. This distribution is characterised by the following distribution function

$$F(x) = \frac{1}{N}\sum_{i=1}^{N} \mathbf{1}\{\varphi(Y_i) \leq x\}, \quad x \in \mathbf{R}.$$

If each particle is sampled with probability p, the sample distribution coincides with the population distribution. Thus, the sample distribution has distribution function

$$\begin{aligned} G(x) &\propto \mathbf{E}\sum_{i \in S} \mathbf{1}\{\varphi(Y_i) \leq x\} \\ &= \sum_{i=1}^{N} \mathbf{E}\mathbf{1}\{i \in S\}\mathbf{1}\{\varphi(Y_i) \leq x\} \\ &= p\sum_{i=1}^{N} \mathbf{1}\{\varphi(Y_i) \leq x\} \\ &= NpF(x). \end{aligned}$$

Since $G(\infty) = F(\infty) = 1$, we must have that $G = F$.

If we are able to determine $\varphi(Y_i)$ with high precision, then we can regard

$$\varphi(Y_i), \quad i \in S,$$

as a sample from the distribution we are interested in. This is the case for optical sectioning of spherical particles where the size of a particle can be measured directly on a central section, see also [285]. However, the general situation is that the particles are of unknown varying shape. In such situations, it is important to be able to assess the variability of the estimator from sectional data. An empirical study can be found in [247].

11.3 Model-based case

In the model-based case, the particles are modelled by a marked point process. Number per unit volume is replaced by the intensity of the process and simple finite averages by mean values in the mark distribution.

The idea is to represent the ith particle by an associated point y_i and a mark $Z_i \subseteq \mathbf{R}^3$ such that the ith particle is $Y_i = y_i + Z_i$. The marks belong to the set M_d of bounded $d-$dimensional sets with finite $d-$dimensional volume.

The points $\Psi = \{y_i\}$ are assumed to constitute a point process in $\mathbf{R}^3$, i.e. a locally finite random set in $\mathbf{R}^3$. It is furthermore assumed that Ψ is stationary such that

$$\Psi + y = \{y_i + y\}$$

has the same distribution as Ψ for all $y \in \mathbf{R}^3$. If we let Λ denote the intensity measure of Ψ,

$$\Lambda(Q) = \mathbf{E}N(\Psi \cap Q), \quad Q \subseteq \mathbf{R}^3 \text{ bounded Borel},$$

then, because of the stationarity, Λ is proportional to volume measure

$$\Lambda(Q) = N_V V(Q),$$

the constant of proportionality N_V is the model-based definition of the number of particles per unit volume.

In order to describe not only the associated points but the whole particle system we need the concept of a marked point process. A marked point process is a random sequence

$$\Xi = \{[y_i; Z_i]\}$$

such that $\Psi = \{y_i\}$ is a point process in $\mathbf{R}^3$ (locally finite). The marks Z_i are subsets of $\mathbf{R}^3$ of dimension d, belonging to M_d. We assume that the marked point process Ξ is stationary, meaning that

$$\Xi + y = \{[y_i + y; Z_i]\}$$

has the same distribution as Ξ for all $y \in \mathbf{R}^3$.

In the special case where the marks are independent and identically distributed it is easy to define the particle size distribution, i.e. the distribution of $\varphi(Z_0)$ where Z_0 is a random mark with the common distribution of the marks. If the marks are dependent a particle size distribution may be defined via the so-called intensity measure of the marked point process. The intensity measure is for $Q \subseteq \mathbf{R}^3$ and $K \subseteq M_d$ given by

$$\Lambda_m(Q \times K) = \mathbf{E}N(\Xi \cap (Q \times K)) = \mathbf{E}\sum_i \mathbf{1}\{y_i \in Q, Z_i \in K\}.$$

Since Ξ is stationary, $\Lambda(\cdot \times K)$ is proportional to volume measure in $\mathbf{R}^3$

$$\Lambda_m(Q \times K) = N_V(K)V(Q), \tag{11.5}$$

the constant of proportionality being the mean number of particles per unit volume with marks in K. The mark distribution is defined as

$$\mathbf{P}_m(K) = N_V(K)/N_V. \tag{11.6}$$

The particle size distribution is simply the distribution of $\varphi(Z_0)$ where Z_0 is a random mark with distribution $\mathbf{P}_m$. Using (11.5) and (11.6) it can be shown for any measurable non-negative function g defined on $\mathbf{R}^3 \times M_d$,

$$\mathbf{E}g(y_i, Z_i) = N_V \int_{\mathbf{R}^3} \int_{M_d} g(y, Z) \mathbf{P}_m(\mathrm{d}Z)\, \mathrm{d}y. \tag{11.7}$$

The assumption of stationarity replaces the uniformity in the sampling in the design-based case. In a similar manner, isotropy in the sampling can be replaced by an isotropy assumption for the particle process. The marked point process $\Xi = \{[y_i; Z_i]\}$ is isotropic if

$$B\Xi = \{[By_i; BZ_i]\}$$

has the same distribution as Ξ for all rotations B in $\mathbf{R}^3$. (The rotation B can be represented by a 3×3 orthogonal matrix, satisfying $BB^* = B^*B = I_3$.) For some purposes, it will be enough to assume invariance under a subgroup of rotations but below we assume full isotropy in order to present the principles of model-based estimation. Isotropy of Ξ implies that the mark distribution is invariant under rotations

$$\mathbf{P}_m(BK) = \mathbf{P}_m(K), \quad B \text{ rotation}, K \subseteq M_d.$$

In the model-based setting, we will also assume that φ is invariant under rotations and translations, i.e.

$$\varphi(B(y + Z)) = \varphi(Z),$$

for all $B \in SO(3)$, $y \in \mathbf{R}^3$ and $Z \in M_d$.

In the following subsections we will explain how the design-based methods of estimating particle number, mean particle size and particle distribution, described in the previous section, can be given a model-based justification, using the theory of marked point processes [388, 543, 545].

11.3.1 Particle number

Since the marked point process is stationary, the particles extend all over space. We will therefore seek a model-based justification of the methods of estimating particle number, described in the *Extended* case. Recall that in the *Extended* case, particles were sampled by a FUR box Q with reference position Q_0. The designs described in Section 11.2.1 had all the property that for an arbitrary particle $Y \in M_d$

$$V(\{u \in \mathbf{R}^3 : Y \text{ sampled in } u + Q_0\}) = V(Q_0). \tag{11.8}$$

In the model-based setting, the particles are random while the sampling box Q is fixed, say

$$Q = y_0 + Q_0.$$

Let

$$\mathcal{S} = \{i : Y_i \text{ sampled in } Q\}.$$

Then, using (11.7) and (11.8), we find

$$\begin{aligned}
&\mathbf{E}N(\mathcal{S}) \\
&= \mathbf{E}\sum_i \mathbf{1}\{Y_i \text{ sampled in } Q\} \\
&= N_V \int_{\mathbf{R}^3} \int_{M_d} \mathbf{1}\{y+Z \text{ sampled in } Q\}\, \mathbf{P}_m(\mathrm{d}Z)\,\mathrm{d}y \\
&= N_V \int_{M_d} \int_{\mathbf{R}^3} \mathbf{1}\{Z \text{ sampled in } Q-y\}\, \mathrm{d}y\, \mathbf{P}_m(\mathrm{d}Z) \\
&= N_V \int_{M_d} \int_{\mathbf{R}^3} \mathbf{1}\{Z \text{ sampled in } y+Q_0\}\, \mathrm{d}y\, \mathbf{P}_m(\mathrm{d}Z) \\
&= N_V V(Q_0). \qquad (11.9)
\end{aligned}$$

An unbiased estimator of N_V is therefore

$$\widehat{N_V} = N(\mathcal{S})/V(Q_0).$$

11.3.2 Mean particle size

Unweighted mean particle size

The aim is here to estimate the parameter $\mathbf{E}_N[\varphi]$ which in the model-based setting is defined as

$$\mathbf{E}_N[\varphi] = \int_{M_d} \varphi(Z)\mathbf{P}_m(\mathrm{d}Z).$$

In the model-based case, particles are sampled, using a fixed box Q. The sample is

$$\mathcal{S} = \{i : y_i \in Q\}.$$

For each sampled particle Y_i, we estimate its size $\varphi(Y_i)$, using a local geometric design with a local probe passing through the associated point y_i of the sampled particle Y_i. Such an estimator is often of the form

$$\widehat{\varphi}(Y_i) = g(Y_i \cap (y_i + T_{\omega_0})),$$

where T_{ω_0} is a probe (line, plane or slab) through the origin with fixed orientation. The function g is invariant under translations and rotations and for any $Z \in M_d$

$$\int_{S_+^2} g(Z \cap T_\omega) \frac{\mathrm{d}\omega}{2\pi} = \varphi(Z).$$

The resulting estimator of $\mathbf{E}_N[\varphi]$

$$\widehat{\mathbf{E}_N[\varphi]} = \sum_{i \in \mathcal{S}} \widehat{\varphi}(Y_i)/N(\mathcal{S})$$

is ratio-unbiased. Thus for any rotation B, we have using that also $\mathbf{P}_m$ is invariant under rotations,

$$\begin{aligned}
&\mathbf{E} \sum_{i \in \mathcal{S}} \widehat{\varphi}(Y_i) \\
&= N_V \int_Q \int_{M_d} g((y+Z) \cap (y+T_{\omega_0}))\mathbf{P}_m(\mathrm{d}Z)\,\mathrm{d}y \\
&= N_V \int_Q \int_{M_d} g(Z \cap T_{\omega_0})\mathbf{P}_m(\mathrm{d}Z)\,\mathrm{d}y \\
&= N_V V(Q_0) \int_{M_d} g(BZ \cap T_{\omega_0})\mathbf{P}_m(\mathrm{d}Z) \\
&= N_V V(Q_0) \int_{M_d} g(Z \cap B^* T_{\omega_0})\mathbf{P}_m(\mathrm{d}Z) \\
&= N_V V(Q_0) \int_{M_d} \int_{S^2_+} g(Z \cap T_{\omega}) \frac{\mathrm{d}\omega}{2\pi} \mathbf{P}_m(\mathrm{d}Z) \\
&= N_V V(Q_0) \int_{M_d} \varphi(Z)\mathbf{P}_m(\mathrm{d}Z) \\
&= N_V V(Q_0) \mathbf{E}_N[\varphi].
\end{aligned}$$

Combining with (11.9), we finally get

$$\frac{\mathbf{E}\sum_{i \in \mathcal{S}} \widehat{\varphi}(Y_i)}{\mathbf{E}N(\mathcal{S})} = \mathbf{E}_N[\varphi].$$

Volume-weighted mean particle size

The interest is here in the volume-weighted version of the mark distribution

$$\mathbf{P}_v(\mathrm{d}Z) \propto V(Z)\mathbf{P}_m(\mathrm{d}Z).$$

The parameter to be estimated is

$$\mathbf{E}_V[\varphi] = \int_{M_d} \varphi(Z)\mathbf{P}_v(\mathrm{d}Z).$$

As argued in [280], the estimators of $\mathbf{E}_V[\varphi]$ used in the design-based case can be given a model-based justification if the particle aggregate can be modelled by a stationary and isotropic marked point process, see also [283]. The intuitive reason is, of course, that the uniformity and isotropy in the sampling can be replaced by the stationary and isotropy in the particle model.

Let us explain in more detail how the point-sampled intercept estimator (11.3) can be justified in the model-based case. As discussed in [280], there are various ways of doing this, depending on the specific assumptions made on the particle model. Let us here concentrate on the case where particles are assumed to be non-overlapping. We consider a fixed point $u \in \mathbf{R}^3$ and a fixed line T through the origin O. If the particle aggregate hits u, we let $Y(u)$ be the particle hitting u, otherwise $Y(u) = \emptyset$, say. In [280], it is shown that, given the particle aggregate hits u, the point-sampled intercept

$$2\pi \int_{Y(u)\cap(u+T)} d(x,u)^2 \, \mathrm{d}x$$

is distributed as

$$2\pi \int_{Z\cap(U+T)} d(x,U)^2 \, \mathrm{d}x,$$

where Z is a random particle from P_v and, given Z, U and T are independent, $U \sim \mathsf{Unif}(Z)$ and T is isotropic through O. It follows that

$$\begin{aligned}
&\mathbf{E}[2\pi \int_{Y(u)\cap(u+T)} d(x,u)^2 \, \mathrm{d}x \mid u \in \cup_i Y_i] \\
&= \int_{M_d} \int_Z \int_{S_+^2} [2\pi \int_{Z\cap(u+T_\omega)} d(x,u)^2 \, \mathrm{d}x] \frac{\mathrm{d}\omega}{2\pi} \frac{du}{V(Z)} \mathbf{P}_v(\mathrm{d}Z) \\
&= \int_{M_d} \int_Z \int_{S_+^2} [2\pi \int_{(Z-u)\cap T_\omega} d(v,O)^2 \, \mathrm{d}v] \frac{\mathrm{d}\omega}{2\pi} \frac{du}{V(Z)} \mathbf{P}_v(\mathrm{d}Z) \\
&= \int_{M_d} \int_Z V(Z-u) \frac{du}{V(Z)} \mathbf{P}_v(\mathrm{d}Z) \\
&= \int_{M_d} V(Z) \mathbf{P}_v(\mathrm{d}Z) \\
&= \mathbf{E}_V[V].
\end{aligned}$$

11.3.3 The particle size distribution

Under a stationary marked point process model $\Xi = \{[y_i; Z_i]\}$ with $Y_i = y_i + Z_i$, we have for a bounded Borel subset Q of $\mathbf{R}^3$ that

$$\{\varphi(Y_i) : y_i \in Q\}$$

can be regarded as a sample from the particle size distribution with distribution function

$$F(x) = \int_{M_d} \mathbf{1}\{\varphi(Z) \leq x\} \mathbf{P}_m(\mathrm{d}Z).$$

Using (11.7), we thus have

$$\mathbf{E} \sum_i \mathbf{1}\{\varphi(Y_i) \leq x, y_i \in Q\}$$

$$= N_V \int_Q \int_{M_d} \mathbf{1}\{\varphi(y+Z) \leq x\} \mathbf{P}_m(\mathrm{d}Z)\,\mathrm{d}y$$
$$= N_V \int_Q \int_{M_d} \mathbf{1}\{\varphi(Z) \leq x\} \mathbf{P}_m(\mathrm{d}Z)\,\mathrm{d}y$$
$$= N_V V(Q_0) F(x).$$

As in the design-based case, the crucial point is that the sizes of the sampled particles are usually not determined exactly, but instead estimated. The information available is therefore

$$\{\widehat{\varphi}(Y_i) : i \in Q\}. \tag{11.10}$$

Only in cases where the estimates are precise compared to the variability in the particle size distribution, (11.10) can be regarded as a sample from the particle size distribution.

11.4 Exercises

Exercise 11.1 In this exercise, we prove the key relation (11.1), $N_A = \mathbf{E}_N[H] \cdot N_V$, in the *Restricted* case. We suppose that the reference space Z is sectioned by an unbounded FUR plane T hitting Z.

The different terms in (11.1) are defined as follows. The parameter N_V is simply $N/V(Z)$, $\mathbf{E}_N[H]$ is defined as $\sum_{i=1}^N H(Y_i)/N$ where $H(Y_i)$ is the length of the projection of Y_i onto the normal direction of the section, and

$$N_A = \frac{\mathbf{E}\sum_{i=1}^N \mathbf{1}\{Y_i \cap T\}}{\mathbf{E}A(Z \cap T)}.$$

Show that

$$\mathbf{E}\sum_{i=1}^{N} \mathbf{1}\{Y_i \text{ sampled}\} = \sum_{i=1}^{N} H(Y_i)/H(Z)$$
$$\mathbf{E}A(Z \cap T) = V(Z)/H(Z).$$

Combining these two results, (11.1) follows directly.

Exercise 11.2 This exercise contains a proof of the relation (11.1) in Miles' extended case. The section available for observations is assumed to be a (bounded) FUR quadrat Q hitting Z. The parameters N_V, $\mathbf{E}_N[H]$ and N_A are defined as in Exercise 11.1 with T replaced by Q. Show that

$$\mathbf{E}\sum_{i=1}^{N} \mathbf{1}\{Y_i \cap Q \neq \emptyset\} = \sum_{i=1}^{N} H(Y_i) A(Q_0)/V(Z \oplus \check{Q}_0)$$

and

$$\mathbf{E}A(Z \cap Q) = V(Z) A(Q_0)/V(Z \oplus \check{Q}_0),$$

Once more the relation (11.1) has been given a rigorous stochastic interpretation.

11.5 Bibliographic notes

For references on particle counting, see the Bibliographic Notes for Chapter 10. For references on the theory of particle size estimation, see the Bibliographic Notes for Chapter 8.

A valuable discussion of particle number and size estimation in the presence of shrinkage is [164].

Applications and implementations of the techniques described here include (for cell number) [11, 15, 41, 53, 54, 55, 60, 186, 277, 321, 353, 416, 430, 456, 453, 473, 474, 551, 595, 598, 599] and (for cell size) [17, 19, 52, 277, 340, 416, 533]. Cell size variability has been studied using the selector in [385].

CHAPTER 12

Design of stereological experiments

Here we describe strategies for designing and implementing stereological experiments. Practical stereological experiments range in complexity from the analysis of a single plane section to intricate designs with multiple levels of subsampling and processing.

The sampling protocol must be designed with a view to ensuring the validity of the estimators and to minimising the variance contribution from sampling variation. In most applications, we have the freedom to choose from various alternative experimental designs and estimation techniques, as long as the fundamental principles of sampling and inference are followed. A judicious choice of technique helps to resolve practical difficulties and to optimize the efficiency of estimation.

Section 12.1 describes the constraints imposed on the experimental design by the nature of the material and the experiment. Section 12.2 discusses the definition of parameters and their estimators. Section 12.3 lists the main strategies for designing a stereological experiment. Section 12.4 discusses implementation of stereological technique. Section 12.5 discusses possible sources of error in stereological sampling experiments. Section 12.6 discusses the optimal design of stereological experiments.

12.1 Design constraints

Constraints are imposed on the experimental design by the nature of the material, its accessibility, the mode of microscopy, and the scale of the features of interest.

The **nature of the material** determines the cutting operations that are feasible. Rocks and metals may usually be sawn and polished to yield a single plane section at a time, observable in reflected light. Some minerals may be sliced and mounted on a glass slide to yield a translucent slice observable in transmitted polarised light. Other materials may require processing before they can be cut. Frangible materials such as soil samples may be soaked in resin, after which they can be sawn, giving us access to one section plane at a time. Biological tissues are usually fixed by chemical treatment to increase their rigidity, then embedded in a supporting medium such as wax or plastic, after which they can be cut to yield translucent slices.

Translucent materials allow us to observe the contents of a three-dimensional region in the material (under some modes of microscopy). In opaque materials, a substitute for three-dimensional microscopy is to observe two or more section planes cut successively from the material.

The operations of "cutting a thick section of known thickness" and "cutting two sections a known distance apart" are more difficult in the less rigid materials, such as biological tissues, because section thickness is difficult to control, to maintain, and to measure accurately.

The **accessibility of the specimen**, and of the reference space, also influence the sampling design. If the specimen is small enough to fit inside our cutting device in its entirety, then it may be sectioned completely in one operation. If it is too large for this, then it must first be cut into pieces before it may be sectioned. A very small specimen must be embedded in a supporting medium in order that it can be handled. For example, if the goal is to cut an IUR stack of serial sections from a biological organ, the isector technique (see Section 12.4.1) may be used if the organ is small enough to fit in the enclosing spherical mould, while other techniques (the orientator, slabs, bars, etc) must be used if it is larger.

Our ability to observe the material depends on the **mode of microscopy** (optical reflected light, optical transmission, optical confocal, backscattered electron, transmitted electron, acoustic, etc), the visualisation principle (reflection, absorption, fluorescence, interference) and the preparation of the material for microscopic visualisation (staining, etching, immunolabelling). For example in biological tissues, a cell stain is a chemical which renders a desired cellular protein visible. The stain is applied to sections after cutting. The depth of penetration of the stain into the tissue determines the maximum thickness of a thick section if it is desired to observe the tissue in three dimensions.

Finally the **scale of the features of interest** determines the magnification and subsampling required. This was explained in Section 9.2.2.

12.2 Parameters and estimators

12.2.1 Parameters

It is worth investing considerable time — before the experiment is conducted — to focus on the ultimate goal: the estimation of parameters of the material.

First, the desired parameters should be carefully elicited and defined in a three-dimensional context. It should especially be clarified whether the target parameters are absolute quantities (like V), relative quantities (like S_V) or particle population means (like $\mathbf{E}_V[V]$).

Sampling design is also strongly influenced by the inferential context, as explained in Chapter 6.

12.2.2 Estimation technique and probe type

For a given stereological parameter, there are usually several alternative estimators available, which involve different test probes of different geometries and dimensions.

The volume fraction V_V or absolute volume V of a three-dimensional object can be estimated using random 2-dimensional plane sections (e.g. Cavalieri estimator), random 1-dimensional linear probes, or random test points. As we have shown elsewhere, it is not necessarily true that lower-dimensional probes yield estimators with a higher variance.

Surface area S or surface density S_V may be estimated in several ways. We may take IUR plane sections and measure the lengths of section profile curves ($S_V = 4/\pi B_A$). Instead of measuring profile length we may take IUR test lines in the section planes, and count intercepts ($S_V = 2I_L$). Alternatively we may take vertical random planes (Chapter 8), and within these planes, take test lines with a weighted distribution which ensures that the test lines are isotropic in three dimensions, again using $S_V = 2I_L$. All three techniques satisfy the requirement that the test lines be isotropic in space.

Length L and length density L_V may be estimated by taking IUR plane sections and counting intersection profiles ($L_V = 2Q_A$) or — in translucent materials — using vertical slices or total vertical projections (Chapter 8).

Estimators of moments of particle volume were described in Chapter 11.

12.2.3 Requirements of each estimator

Each stereological estimator imposes conditions on the experiment and/or the material, corresponding to the statistical assumptions under which the estimator is valid. As we have explained in Chapter 6, these assumptions may be satisfied either by randomising the sample, or by assuming homogeneity of the material.

Different estimators impose different conditions on the experiment. For full details of the requirements of each estimator, refer to the description of the relevant estimator in Chapters 5–11. The following is an overview of differences between them.

Random rotation of the test probe or the specimen, or a model assumption of isotropy, is required for the estimation of surface area S and length L, and relative densities S_V, L_V, as explained in Section 4.1.5. Estimation of volume V and volume fraction V_V requires only uniformity in location.

Knowledge of section spacing or section thickness is required for the estimation of the absolute quantities V, S, L using methods like the Cavalieri estimator, while estimation of the relative quantities V_V, S_V, L_V does not.

Knowledge of the magnification is required for all techniques, except for the estimation of dimensionless quantities such as V_V and N.

Spatial uniformity of sampling, for example, uniformly random translation of the position of a stack of serial sections, is required for all estimators, with the exception of discrete sampling methods (Chapter 10). Estimation of particle number N by discrete sampling does not require spatial uniformity of sampling, nor even preservation of metric properties of the material.

Three-dimensional observation, that is, observation of the contents of a three-dimensional region in the material, is required for the estimation of particle number N using the disector method, and for the estimation of connectivity parameters such as the Euler-Poincaré characteristic. Under some conditions it may be possible to approximate three-dimensional observation by comparing pairs of successive plane sections (as in the "physical disector" and serial 3D reconstruction).

Thus, a sample which is adequate for some stereological purposes is inadequate for others. This point cannot be stressed too often. In particular, microscope images that were obtained by arbitrary selection (without following any sampling protocol) cannot afterwards be analysed stereologically, without making heroic assumptions. Good stereological software packages carry numerous warnings about applying stereological estimators to a microscope image of unspecified provenance.

An urban legend of stereology states that surface area S and S_V can be estimated from plane sections of arbitrary spatial orientation using the Merz grid, a test system of circular arcs (Figure 7.7). This is false. Although the test system is isotropically oriented in two dimensions, it is not isotropic in three dimensions when the test system is placed at random in a section plane of arbitrary spatial orientation. It is unfortunately impossible to estimate surface area S or S_V from plane sections if the sampling distribution of section plane orientation is unknown.

Given that different estimators have widely varying requirements, it may be worthwhile to relinquish a difficult goal, such as the estimation of particle population fractions N_N from single sections, in favour of a surrogate, such as the estimation of the volume fraction V_V of the particle subpopulation.

12.3 Design strategies

This section describes general sampling strategies which can be used to develop good sampling protocols. Apart from ratio estimation, which we have dealt with in detail in Chapter 9, good sampling protocols may involve stratification and nesting.

If nothing else is stated we consider the *Restricted* case where X, Y are fixed bounded subsets of $\mathbf{R}^3$ and $Y \subset X$.

12.3.1 Stratification

Stratification of a population is a standard technique for reducing variance in survey sampling. In stereology this method can be applied when the parameter of interest is an absolute geometrical quantity $\beta(Y)$ which is additive in the sense that $\beta(Y_1 \cup Y_2) = \beta(Y_1) + \beta(Y_2)$ when Y_1, Y_2 are disjoint.

We may stratify the specimen X by physically dividing X into disjoint pieces $X_1, \ldots, X_k$ (with consequent unseen division of Y into pieces $Y_i = Y \cap X_i$, $i = 1, \ldots, k$). Then we treat each piece X_i as a separate specimen and estimate the desired *absolute* geometrical property $\beta(Y_i)$ by sampling X_i. Finally we form an estimate of $\beta(Y) = \sum_{i=1}^{k} \beta(Y_i)$ by summing the estimates of each $\beta(Y_i)$.

Advantages of stratification include the ability to sample with different intensity or sample sizes in each stratum or piece X_i, and the fact that it allows us to randomise over the section orientation (the common orientation of the section planes through X_i is random and different in different pieces X_i).

Figure 12.2 sketches an application of the first kind: a biological organ which is known anatomically to consist of a cortex (outer part) and medulla (inner part) with very different densities of the cells of interest (dots in the Figure). The material may be stratified by cutting it macroscopically into two pieces X_1 and X_2 which are *approximately* the cortex and medulla, then sampling the two strata with different sampling fractions. Note that a *precise* division of the material into cortex and medulla is only required if we need to estimate stereological parameters for the cortex or medulla separately. Otherwise the stratification need only approximate the anatomical division of the organ, as the stratification serves merely to reduce the variance of the estimation of stereological parameters for the entire organ.

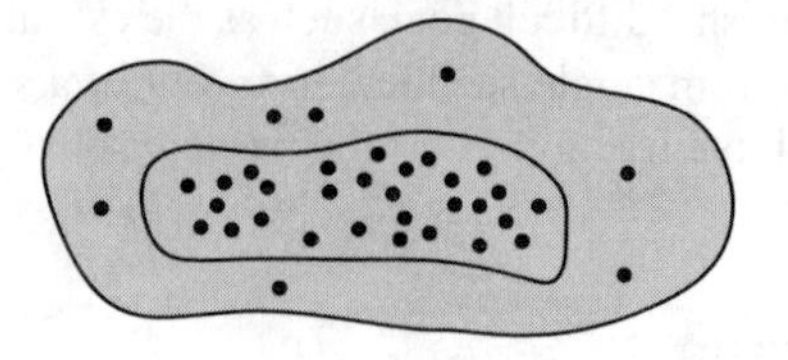

Figure 12.1 *A candidate for stratification. A biological organ containing a cortex (outer part) and medulla (inner part) with very different densities of the cells of interest (dots).*

An application of the second kind is sketched in Figure 12.2. The estimator of curve length using a grid of parallel lines has high variance when the curve has a strong preferential orientation (a very non-uniform orientation distribution). Instead we may stratify the containing space, and apply the same estimator to each stratum using independent rotations in each stratum. This can reduce the variance substantially.

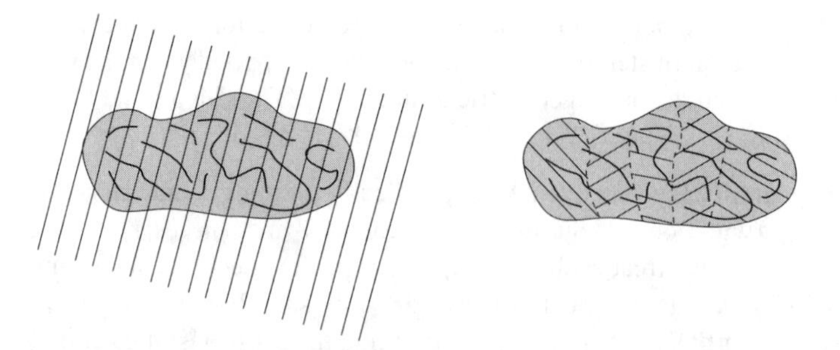

Figure 12.2 *Variance reduction by stratified sampling.* Left: *Estimation of curve length using an isotropic uniform grid of parallel test lines. This has a high variance because of strong preferential orientation of the curve.* Right: *Curve length estimated by dividing the material arbitrarily into strata and applying an independent IUR grid of test lines to each stratum.*

Figure 8.8 in Chapter 8 shows a real application of stratification to estimate surface area using vertical sections.

12.3.2 Subsampling and nesting

Subsampling is an efficient technique for dealing with the widely different scales of organization that are typical in microscopy. Indeed it is often a practical necessity, in order to reduce the amount of material to a manageable size at the scale of magnification used.

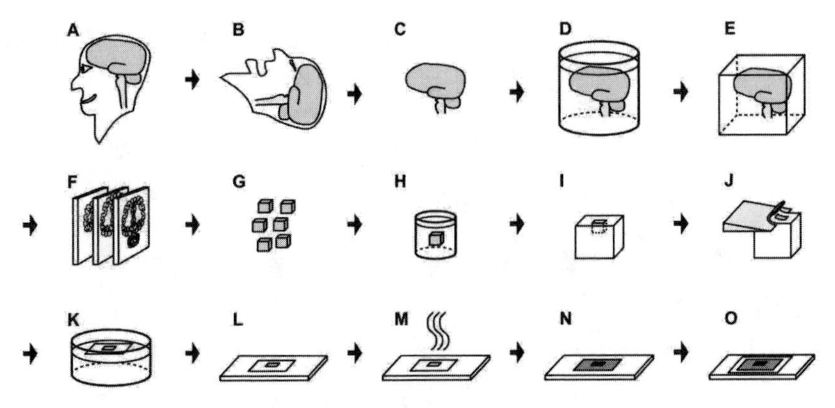

Figure 12.3 *Processing stages and subsampling levels in a stereological study. (A) Organ in living subject; (B) organ in dead subject; (C) organ removed from subject; (D) formaldehyde fixed organ; (E) fixed organ embedded in agar; (F) slabs cut from fixed organ; (G) blocks of fixed tissue; (H) dehydration of tissue block; (I) embedding of dehydrated tissue block; (J) sectioning of embedded tissue; (K) section of tissue floating on liquid; (L) section mounted on glass slide; (M) section fixed on glass slide – typically by drying; (N) histological staining of section; (O) section under cover slip. Note difference in scale between (A)–(F) and (G)–(O). Reproduced from [161] by kind permission of Dr K.-A. Dorph-Petersen (University of Pittsburgh) and Blackwell Publishing. Copyright © 2003 Blackwell Publishing.*

Figure 12.3 sketches the many stages of processing and successive subsampling in a stereological investigation of the human brain. Stages (A)–(E) are preparatory. At stage (F) the brain tissue is cut into large slabs and a sample of slabs is selected. A change in scale occurs at stage (G) in which the slabs are further cut into tissue blocks small enough to be embedded, and a sample of blocks is selected. After further processing (H)–(I), the tissue blocks are sliced (J) and further processed (K)–(O) for microscopy. The tissue available for viewing in these microscope slides would be further subsampled (not shown; further change of scale) by systematic sampling of fields of vision, and finally subsampled by placing test systems in these fields.

There are many examples of such multilevel or 'cascade' designs in the stereological literature. In particular, as touched upon in Chapter 9, the 'smallness' of the feature of interest Y may suggest a series of such sampling levels whereby the feature of interest at one level becomes the reference space at the next level.

Repeated subsampling results in a *nested* or hierarchical sampling design [140, 214]. For example, in many biological applications, organs are taken from each of several animals; several tissue blocks are sampled from each organ; several thin sections are cut from each block; and several sampling windows are photographed on each section. A typical nested structure is sketched in Figure 12.4.

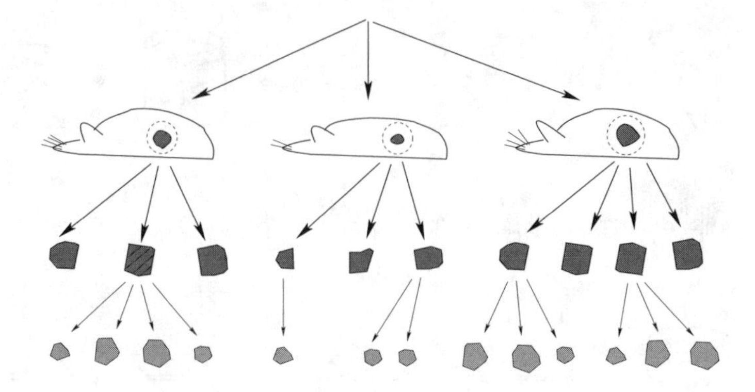

Figure 12.4 *Nested structure of a typical stereological experiment in biological science. At the first (top) level, individual animals are sampled from some population. At the second level, several tissue blocks are taken from each animal. At the third level, several plane sections are taken from each tissue block. After an original drawing by L.M. Cruz-Orive.*

Gundersen and Østerby [214] proposed that standard techniques of analysis of variance (ANOVA) for nested experimental designs [522, chap 13], [523, Chap 10], [617, Section 11.8] be applied in this context. See Section 13.5.

In brief, each level of subsampling is regarded as a source of variation. Nested ANOVA can then be applied to estimate variances and variance contributions empirically. Variances of estimators based on systematic designs require, however, special attention, see below.

12.4 Implementation

12.4.1 Sectioning

Sections may be obtained 'physically' by cutting the experimental material (as in optical microscopy of metals) or 'virtually' by nondestructive visualisation of a plane or volume within the material (as in confocal optical microscopy of living tissue).

There may be several stages of physical cutting prior to the visualisation of the 'actual' section plane (the plane which contains the material visualised in the microscope image and which provides the data for stereological estimators). The statistical requirements of stereological estimators refer to the location and orientation of the 'actual' section plane and microscope field of view. There are often many ways of designing the intermediate cutting and sampling steps so as to satisfy the statistical requirements at the final stage.

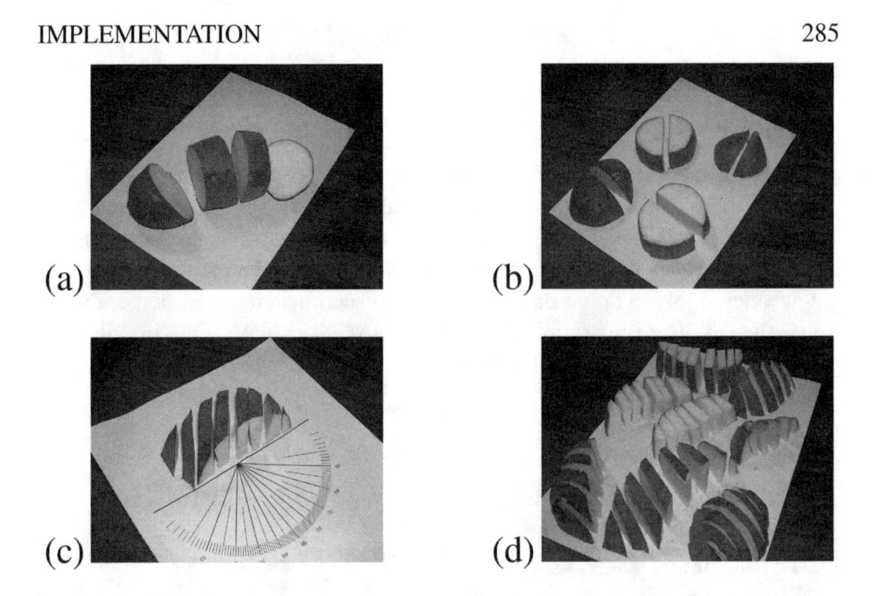

Figure 12.5 *Stratified orientator design. (a) the specimen is cut into slabs; (b) each slab is rotated uniformly about an axis perpendicular to the cut face C of the slab, and then divided into two pieces by a plane E perpendicular to C; (c) each piece is placed on the* sin*-weighted protractor with the division plane E placed face down, and the cut face C aligned with the base of the protractor. A random integer uniformly distributed from 1 to 100 is selected. Planes are cut parallel to the corresponding direction; (d) This process is applied to all pieces.*

Randomisation of the orientation of the section plane can be achieved in many ways, but care must be taken to avoid subtle probability paradoxes (Chapter 5). The following designs all provide correct IUR serial sections through a specimen:

1. isector design [441]: the specimen is embedded in a sphere (by placing it in a spherical mould and filling the mould with a hardening medium such as wax or resin). The sphere is rolled a few times across the laboratory bench to ensure random rotation, and then cut in parallel plane sections at constant spacing in an arbitrary common orientation.

2. stratified orientator design [373]: see Figure 12.5. The specimen is stratified by cutting it into slabs. In each slab, the orientator method (Section 5.5.3) is applied to generate an IUR direction, and serial sections of the slab are taken parallel to this direction.

3. virtual IUR sections [327, 342]: the specimen is divided arbitrarily into slabs. The material and the mode of microscopy must allow us to visualise the three-dimensional contents of each slab. Computer software randomly generates an IUR stack of parallel planes in three dimensional space, and

these mathematical planes are superimposed graphically on the video image of the contents of the slab.

For vertical section techniques (Section 8.1), there is even more freedom, since we have many ways of defining the vertical direction. Figure 12.6 shows one design for obtaining vertical sections from an object. First the object is cut into parallel slices (a). Each selected slice is then placed on a bench (panel (b) shows view looking down on the bench) and sections are taken perpendicular to the bench. The illustration shows vertical sections with three orientations. The vertical direction is the direction perpendicular to the bench.

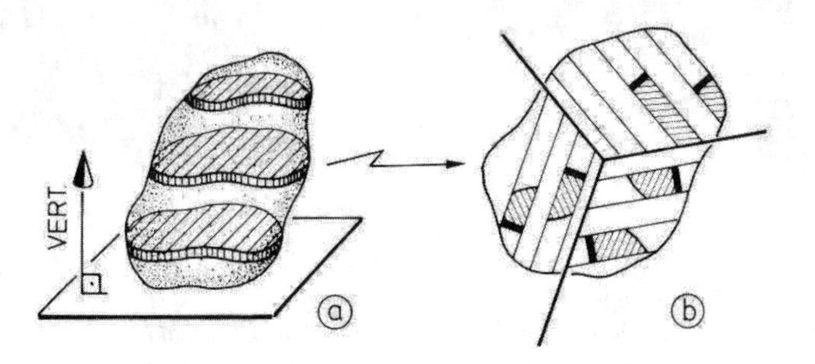

Figure 12.6 *Vertical sampling design for an arbitrary object. See text. Reproduced from [33] by kind permission of Blackwell Publishing. Copyright © 1986 Blackwell Publishing.*

Two different vertical sampling designs for the same tubular organ are sketched in Figure 12.7. In one design, the tube is first cut into flat rings, which are then slit open (a), opened flat on a table and chopped into pieces (c). In the other design, the tube is first cut into long segments, which are then slit open (b), opened flat on a table, and chopped into pieces (c). In both designs, the pieces are then placed on a table and sectioned perpendicular to the table (d). The result in either case is that the vertical direction (perpendicular to the table) corresponds to the radial direction in the original tube.

Figure 12.8 shows another convenient design which yields vertical sections. After cutting the object into thick slabs (as in Figure 12.6(a)) and selecting some slabs, we cut the slabs into elongated bars and select some bars. The bars are rotated uniformly randomly about their long axis, then places in an embedding medium. They are then serially sectioned by planes parallel to their long axis. The vertical direction is aligned with the long axis of each bar.

Of course, the sectioning technique must enable us to measure or observe the quantities involved in our stereological estimators. For an estimator which requires knowledge of section thickness or section spacing (such as the Cavalieri

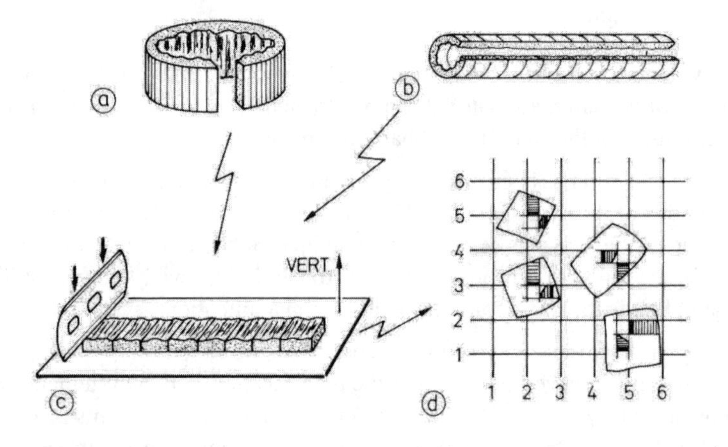

Figure 12.7 *Two vertical section designs for sampling a tubular organ. See text. Reproduced from [33] by kind permission of Blackwell Publishing. Copyright © 1986 Blackwell Publishing.*

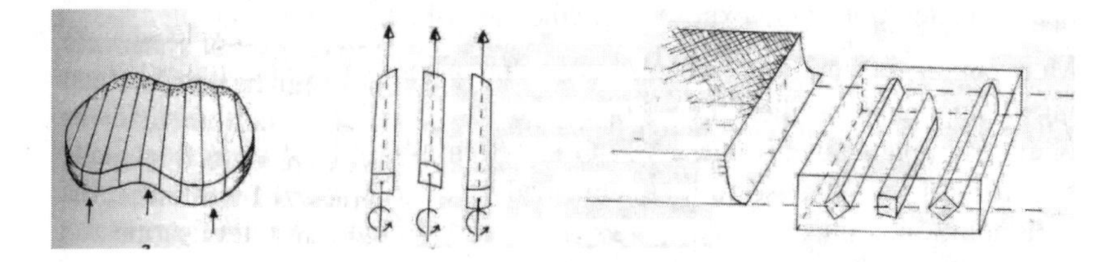

Figure 12.8 *Vertical bars design. See text. Reproduced from [225] by kind permission of the authors and Blackwell Publishing. Copyright © 1988 Blackwell Publishing.*

estimator of absolute volume), the sectioning technique must permit the accurate measurement of this quantity. For an estimator which requires the comparison of adjacent sections (such as the physical disector) we must be able to collocate or register successive sections.

Section thickness measurement is important. In metallography, a simple strategy for measuring the distance between two successive planes of polish is to strike a dent in the metal, using a pyramid with a known apex angle. The reduction in the observed size of the dent between the two microscope images then indicates the separation between the planes of polish. In biological microscopy, it is much more difficult to measure section thickness accurately, and the appropriate methods depend on the material. One practical technique can be seen in [322].

12.4.2 Subsampling

We saw that subsampling of the material is a practical necessity. Random subsampling of the sampled material can be effected in numerous ways. It is usually required that the sampling probabilities be uniform.

When the material has been physically cut into a manageable number of pieces, the methods of discrete sampling (Chapter 10) may be applied: we simply select some of these pieces at random. Techniques for random selection from a finite population with uniform sampling probability are well known. Systematic sampling is the most common technique in applications.

Spatially uniform sampling of microscope fields of vision on a plane section is traditionally achieved by covering the plane section with a systematic two-dimensional array of sampling tiles, as sketched in Figure 10.9. This is feasible using a transparent sheet or a video graphics overlay, only if the entire section is visible in a single microscopic field of vision. Otherwise, alternative techniques involve shifting the position of the microscope objective relative to the section. A traditional technique in electron microscopy is to physically 'mask' those areas of the section that are not to be sampled, by covering them with a grid of opaque material, leaving exposed only the tiles which are to be sampled.

More complex sampling protocols in two dimensions can be implemented using a computer-controlled microscope stage. Figure 12.9 shows 'meander sampling' as implemented in the CAST-Grid system [446]. First the approximate boundary of the plane section is delineated by the user using point-and-click tools at low magnification. Then the system switches to high magnification and performs systematic sampling across each horizontal row of the section, with independent starting position in each row. A computer-controlled stepper motor shifts the microscope stage to each new sampling position, and the image is presented to the user.

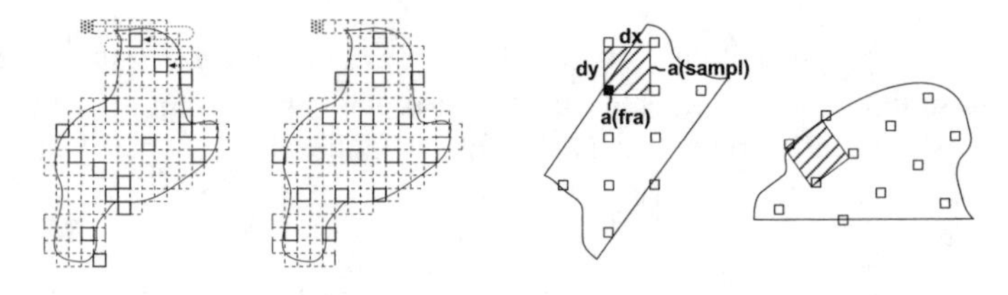

Figure 12.9 *Meander sampling.*

12.4.3 Test systems

To implement stereological methods involving test systems (arrays of points, lines, curves, sampling frames etc) we have to (a) superimpose the test system on the plane section image and (b) count or measure intersections between the test system and the features of interest.

Test systems can be superimposed on any image using optical techniques, such as drawing the test system in black ink on a transparent sheet which is then superimposed on a photographic print (yielding black lines) or superimposed on the photographic paper during exposure (yielding white lines); projecting the image optically onto a piece of white paper bearing the test system; and etching the test system onto glass placed in front of the microscope eyepiece.

Test systems can be superimposed on a digital image or video signal using a computer. Software generates the cartesian coordinates of the test system elements and converts this numerical representation to a vector graphics or pixel image format for display.

Intersections between the test system and the features of interest must usually be recognised by human observers. In most applications of stereology, the features of interest are not easy to isolate by simple image filtering, because of poor image contrast, high false positive rates, and complex definition criteria. Advanced image analysis can often recognise the desired features, but its application requires expertise and fine tuning.

Intersections can be counted and measured manually or semi-automatically. In semi-automatic operation the test system elements (individual test points, test lines, etc) are displayed one at a time and the microscope user is prompted for a response. In the case of point counting, the user clicks "hit" or "no hit" for each test point. In the case of measuring the length of intersection between an object and a test line, the software displays an unbounded test line and prompts the user to indicate (with the mouse) the locations of the two endpoints of the intersection. Software/hardware systems for semiautomatic stereology include CAST-Grid [446], Stereo Investigator [394] and Stereologer [537].

While the ability to capture digital images makes it theoretically possible to analyse all information in the microscope field-of-view in software without human intervention, in practice this is usually too hard. It is also usually unnecessary, because of the paradoxical fact that coarser test systems may yield more accurate parameter estimates.

12.5 Sources of non-sampling error

Discrepancies between the stereological findings of different laboratories are common. This suggests that potential sources of bias and variability external to the sampling design also need careful study.

Methodological bias occurs when an estimation method intrinsically yields a biased estimator. Some classical stereological methods (notably, classical methods for estimating particle number) have methodological bias. This has also been demonstrated empirically in a few studies (notably [391]) by applying competing methods to the same experimental material. We cannot measure bias empirically, of course, but a systematic difference between two methods in this context indicates that at least one of them is biased.

Observation artefacts include *lost caps* (the physical loss of small end pieces of particles which fall out of a thick section, and are therefore not observed); the *Holmes effect* of overprojection and underprojection (the failure to observe correct three-dimensional geometry in a projected two-dimensional image because some features are partly obscured by others); and *contrast artefacts* (the failure to observe some features at some angles because they do not show sufficient optical contrast).

Shrinkage of biological tissues during processing is a considerable problem. Sections cut on a cryostat may shrink by 75%. Absolute and relative parameters may be severely affected by shrinkage, with the exception of V_V. For particle number and size estimation in the presence of shrinkage, see Chapter 11 and [164].

Definition of the feature of interest is not a simple matter. In practice there must be an operational definition (a set of criteria for recognizing the features of interest on plane sections), which may depend on the preparation (staining, etching, labelling) and on the microscopic technique (mode of microscopy, magnification, immersion). Estimates of surface area and length (S, S_V, L, L_V) in biological tissues depend on the magnification used, and results from different laboratories have been reconciled by interpreting the feature as having fractional dimension [460, 183], [580, p. 156].

Operator effects include systematic differences between the judgement of different operators, systematic differences in the decisions made in marginal cases (e.g. always including marginal cases or always excluding them), time-dependent changes such as training effects, fatigue and impatience, and subconscious anticipation.

12.6 Optimal design

12.6.1 Variance reduction in systematic sampling

In systematic sampling of a finite population, the units are initially placed in a sequence. The ordering of units is arbitrary, and does not affect the validity of the sampling technique. To be precise, systematic sampling has constant sampling probability for each unit, and the associated estimator of the population total is unbiased, without regard to the ordering of units.

However, the variance of the estimator of population total does depend on this ordering. This is studied at length in Chapter 13. Some orderings of the sampling units yield far more efficient estimators than other orderings.

A comparatively efficient estimator is obtained by arranging the units in a 'unimodal' sequence. Suppose that the sampling units are numbered $1,\ldots,n$ and that unit i would (if it were sampled) yield a response y_i contributing to the estimator of population total. Rearrange the units in ascending order of response y_i: thus they are arranged in the order $i_1, i_2, \ldots, i_n$ where i_k is the unit with the k-th smallest response, that is, $y_{i_k} = y_{[k]}$ where $y_{[1]} \leq y_{[2]} \leq \ldots \leq y_{[n]}$ are the ranked responses. Now rearrange them in an ascending-then-descending sequence by setting $i'_k = i_{2k-1}$ for $k \leq n/2$ and $i'_k = i_{2n-2k}$ for $k > n/2$. Then the ordering $i'_1, \ldots, i'_n$ yields a highly efficient estimator of population total.

In stereological applications of systematic sampling it is often possible to rearrange the units (tissue blocks, slabs, bars) which are about to be systematically sampled. Hence, it is worth trying to reorder these units so as to minimise the variance due to systematic sampling. This is possible in practice if we have auxiliary or covariate information. Suppose that covariate values x_i are observable such that y_i is positively correlated with x_i. For example, y_i might be the cell count from a tissue bar and x_i might be the bar's overall length. Then the 'unimodal' ordering of the units based on the covariate values x_i may be hoped to lead to an efficient estimator of the population total of the responses y_i.

Variance reduction using non-uniform systematic sampling is developed in [163].

12.6.2 Use of prior information

Artful choices of the sampling design based on prior information about the material may substantially reduce the variance of estimators. Examples of prior information include the knowledge that there is a gradient in a particular direction; knowledge of the spatial organisation (such as anatomical knowledge of a biological organ); and covariate information.

A **gradient** in the material is a spatial trend in the quantity being measured. The presence of a gradient tends to inflate the sampling variability of estimators. Two strategies for reducing sampling variability [222] are sketched in Figure 12.10. In biological science the most common situation (A) is an elongated biological organ with a lengthwise gradient. It is usually necessary to cut the organ into slabs by slicing perpendicular to the gradient (upper drawing). A systematic sample of slabs is taken. The sampled slabs are cut into bars (lower drawing) by slicing slabs parallel to the gradient. A systematic subsample of the bars is taken (every fourth bar has been sampled, as indicated by displacing

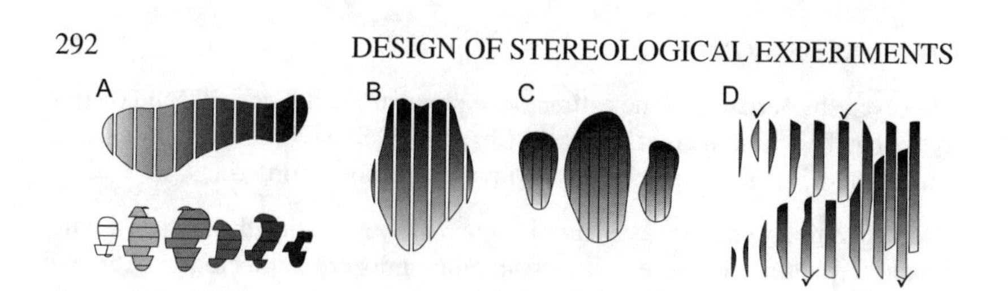

Figure 12.10 *Strategies for sampling an organ which contains a gradient, indicated by the shading. (A): Common situation of an elongated organ with lengthwise gradient which must be cut perpendicular to the gradient (upper drawing). The slabs should then be sliced into bars (lower drawing) parallel to the gradient. (B)–(D): Desirable situation of an organ which can be cut parallel to the gradient. A small sample of slabs, shown by the tick marks, is efficient here. Reproduced by kind permission of Prof. H.J.G. Gundersen.*

it from the others). A more desirable situation (B) is where it is feasible to cut the organ into slabs parallel to the gradient. Systematic sampling of these slabs yields efficient estimators. Note that in both situations, estimators are still unbiased (under randomisation of the sample) and these considerations only affect the estimator's variance.

Knowledge of the **spatial organisation** of the material, such as anatomical knowledge about a biological organ, can be used to stratify the material efficiently, as explained in Section 12.3.1.

12.6.3 Pilot experiments

A pilot experiment is a trial in which the intended experimental protocol is carried out with a smaller sample size. The results obtained from pilot experiments enable us to assess the performance of the design and to identify improvements to the design.

The variability inherent in each sampling level can be estimated (albeit roughly) from the results of the pilot experiment. The cost per sampling unit (in elapsed time or other measures) can also be determined. This can be used to calculate the optimal allocation of effort to each sampling level, by standard methods of design of experiments.

The sampling design may also be adjusted in other ways. For example, the test systems may be rescaled using elementary scaling principles. Suppose that in the pilot experiment a test point grid yielded an average count of 320 hits per field of vision, while the results suggest that an average count of about 20 hits per field of vision would be more efficient. If the test grid is rescaled by expanding each linear dimension by a factor of 4 (the spacing between test points is quadrupled) then the expected number of hits per field is reduced by $1/4^2 = 16$ so that it becomes about $320/16 = 20$ as desired.

The trial also helps us to identify potential sources of non-sampling error. This may help us to eliminate such errors, for example, by refining the operational definition for recognizing the features of interest, or by identifying operator differences and eliminating them by retraining. Operators also learn the protocol during this pilot phase so that the final experiment is less susceptible to training effects.

12.6.4 Do more less well

In biological experiments, the results of a nested ANOVA frequently indicate that the between-animal variance is substantially larger than other sources of variation. The optimal allocation of sampling effort is then to sample a large number of animals and spend relatively little effort on measuring the data in each sampling window ('do more less well'). See Section 13.5.

CHAPTER 13

Variance of stereological estimators

This chapter presents theoretical results about the variance of stereological estimators, and describes practical techniques for estimating the variance from sample data.

The precision of stereological estimators is a question of great practical importance, and has been investigated actively by stereologists, statisticians, geologists, materials scientists, cartographers, computer scientists and others since the 1940's. See the Bibliographic Notes.

Variance is a complex problem for several reasons. Firstly the basic theory of stereology, like survey sampling theory, only guarantees that parameter estimates are unbiased, and has little to say about variances in general.

Secondly, there are no simple, exact formulae for the variance of a stereological estimator; the variance is not simply a function of sample size. Just as the variance of a sum of random variables

$$\mathsf{var}[\sum_i A_i] = \sum_i \mathsf{var}[A_i] + \sum_i \sum_{j \neq i} \mathsf{cov}(A_i, A_j)$$

involves the *covariance* $\mathsf{cov}(A_i, A_j)$ between each pair of summands, so also the variance of a point count or area fraction involves a covariance term associated with each pair of points in the sample. These second-order properties depend on the spatial organisation of the material.

Thirdly, there may be insufficient information in the sample to estimate the variance. Typically the contents of a systematic sample do not contain sufficient information to enable us to estimate the variance of the sample total.

Section 13.1 discusses the variance of stereological estimators in a model-based setting. Section 13.2 covers design-based estimators, principally estimators based on systematic sampling. Section 13.3 presents a version of the Rao-Blackwell theorem which allows some comparisons of variance between different estimators. Section 13.4 is a brief mention of the concept of prediction variance. Section 13.5 discusses the technique of nested analysis of variance which is widely accepted in stereological practice. Section 13.6 discusses the variance of estimators of particle number N and numerical density N_V.

13.1 Variances in model-based stereology

13.1.1 Estimators of volume fraction

We start by considering the variances of classical stereological estimators of V_V in the model-based setting. Suppose the feature of interest Y is a stationary random set in $\mathbf{R}^3$ with volume fraction $V_V = p$. Any plane section of Y is again a stationary random set, with area fraction $A_A = V_V = p$.

Two unbiased estimators of V_V from the plane section are the observed area fraction $A(Y \cap F)/A(F)$, where F is a fixed quadrat of positive area, and the point count fraction $N(Y \cap G)/N(G)$ where G is a *fixed* finite set of test points. See Figure 13.1.

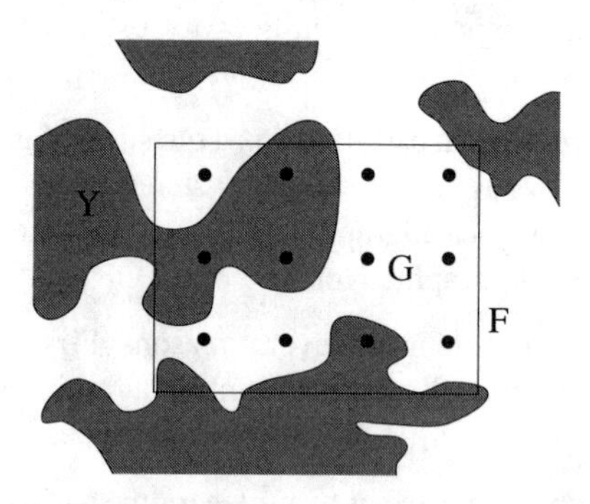

Figure 13.1 *Estimation of the volume fraction of a random set Y by point-counting on a grid G of test points, or by measuring the area in a quadrat F. Copyright © Applied Probability Trust. Reproduced from [32] by permission.*

For any fixed point v in $\mathbf{R}^3$, let

$$I_v = \mathbf{1}\{v \in Y\}$$

be the indicator of the event that v hits the set Y. Since Y is stationary, each indicator has the same success probability $\mathbf{E}[I_v] = V_V = p$. Writing

$$N(Y \cap G) = \sum_{v \in G} I_v$$

and applying the familiar expression for the variance of a sum, we get

$$\begin{aligned} \mathsf{var}[\frac{N(Y \cap G)}{N(G)}] &= \frac{1}{N(G)^2} \left\{ \sum_{v \in G} \mathsf{var}[I_v] + \sum_{v \in G} \sum_{u \in G \setminus v} \mathsf{cov}(I_v, I_u) \right\} \\ &= \frac{1}{N(G)^2} \left\{ N(G) p(1-p) + \sum_{v \in G} \sum_{u \in G \setminus v} C(v-u) \right\}. \quad (13.1) \end{aligned}$$

since $\mathsf{var}[I_v] = p(1-p)$ for all v. Here C is the 'spatial covariance function'

$C : \mathbf{R}^3 \to [-1,1]$ defined by $C(x) = \mathrm{cov}(I_0, I_x) = \mathrm{cov}(I_u, I_{u+x})$ for all $u, x \in \mathbf{R}^3$. The spatial covariance of a random set is analogous to the autocovariance function of a time series. It satisfies $C(0) = p(1-p)$ and $C(-u) = C(u)$ for $u \in \mathbf{R}^3$. Note that many writers use the term 'spatial covariance' for the uncentred covariance $D(x) = \mathbf{E}[I_0 I_x] = C(x) + p^2$.

It would often be assumed that the random set Y is also *isotropic*, i.e. that its probability distribution is invariant under rotations in $\mathbf{R}^3$. For a stationary and isotropic random set, the covariance function is also isotropic, say $C(x) = C_1(||x||)$. This scalar covariance function C_1 for the three-dimensional random set Y is also the scalar covariance for two-dimensional plane sections of Y, and for one-dimensional linear transects of Y. Thus, C_1 can be determined from plane sections or linear transects of Y.

Similarly for the quadrat sample we may write

$$A(Y \cap F) = \int_F I_v \, \mathrm{d}v$$

so that

$$A(Y \cap F)^2 = \int_F \int_F I_v I_u \, \mathrm{d}v \, \mathrm{d}u$$

which yields

$$\mathrm{var}[\frac{A(Y \cap F)}{A(F)}] = \frac{1}{A(F)^2} \int_F \int_F C(v-u) \, \mathrm{d}v \, \mathrm{d}u. \tag{13.2}$$

The same technique yields a similar formula for the variance of the length fraction $L(Y \cap T)/L(T)$ where T is a test line or system of test lines.

These formulae have a long history in statistics. See the Bibliographic Notes. They are also related to the 'change of support' problem in geostatistics [16, 360, 365], [102, Section 5.2, pp. 284–289], and to the theory of stationary random fields [435, 518, 570].

The right side of (13.2) can be interpreted as $\mathbf{E}[C(U-V)]$ where U and V are independent random points, uniformly distributed in F. Intuitively this says that the variance of the observed area fraction depends on the size of the image field F relative to the scale of variability in the material. The latter is usually measured by the *integral range*

$$R = \frac{1}{p(1-p)} \int_0^\infty C_1(r) \, \mathrm{d}r.$$

Roughly speaking, when the diameter of F is much larger than R, we may expect an accurate estimate of V_V. Precise results were obtained by Lantuéjoul [339].

The derivative of the covariance function at the origin is related to the boundary of the random set Y. For a one-dimensional random set Z with covariance function C, a simple probabilistic argument (cf. [543, p. 205]) shows that the expected number of boundary points of Z per unit length is $-2C'(0+)$ where $C'(0+)$ is the derivative of C from the right at 0, assuming that this derivative is finite. For a three-dimensional, stationary and isotropic, random set Y, if the boundary ∂Y has finite surface area density S_V, then from $S_V = 2I_L$ we obtain

$$C_1'(0+) = -(1/4)S_V. \tag{13.3}$$

13.1.2 Variance estimation

Assuming stationarity, we can estimate values of the spatial covariance function $C(x)$ from the microscope image [444, chap. 5], [514, chap. IX] and plug this estimate into (13.1) and (13.2) to obtain estimates of the variances of the two estimators.

Since $\mathbf{E}[I_u I_{u+x}] = C(x) + p^2$ for any point u, we may estimate $C(x) + p^2$ by taking the sample mean of $J_u = I_u I_{u+x}$ over all points u at which J_u is observable. For any set B define the erosion

$$B \ominus \{0, -x\} = \{u : \; u \in B, \; u + x \in B\} = B \cap (-x + B).$$

In the point-counting case, if we recorded I_u for each $u \in G$ rather than simply the total point count, then an unbiased estimator of $D(x) = C(x) + p^2$ is

$$\widetilde{D}(x) = \frac{N((Y \cap G) \ominus \{0, -x\})}{N(G \ominus \{0, -x\})} \tag{13.4}$$

while for the area fraction, if we recorded I_u for all $u \in F$ then an unbiased estimator is

$$\widehat{D}(x) = \frac{A((Y \cap F) \ominus \{0, -x\})}{A(F \ominus \{0, -x\})}. \tag{13.5}$$

This is analogous to the sample covariance of a time series. Computing these estimators is straightforward using the techniques of mathematical morphology [514]. Under the additional assumption of isotropy, $D_1(r) = C_1(r) + p^2$ would be estimated by pooling estimates of $D(x)$ for all vectors x with $||x|| = r$.

The distributions of these estimators of the covariance function depend on higher order properties of Y. However it is clear that when $||x||$ is large (close to the diameter of G or F) the estimates of $C(x)$ are based on small samples and will be inaccurate. This parallels the poor performance of the sample covariance of a random field at large lags (cf. [458, p. 116], [102]). Nonetheless, the resulting estimates of variance of the volume fraction estimators, obtained by substituting the estimated covariance function into (13.1) and (13.2), may still be accurate. A particular value $C(x)$ appears in (13.1) with multiplicity $N(G \ominus \{0, -x\})$ equal to the sample size of its estimator, so that the inaccurate estimates for large $||x||$ are downweighted. Similarly for (13.2).

Alternatively, one can fit parametric models to the covariance function. The methods of geostatistics [16, 82, 360, 362] provide general methodology for modelling and estimating these variances. A frequent choice in geological applications is the exponential covariance

$$C(x) = p(1-p)\exp(-\alpha\|x\|) \tag{13.6}$$

where $\alpha > 0$ is a scale parameter of dimensions length^{-1}.

Assuming an exponential covariance model, the variance formulae (13.1) and (13.2) can be evaluated as a function of α. A rough approximation to the variance (13.2) of the observed area fraction for large domains F is [243], [514, chap. IX], [543, p. 213]

$$\mathrm{var}[\frac{A(Y\cap F)}{A(F)}] \approx \frac{2\pi p(1-p)}{\alpha^2 A(F)} = \frac{32\pi(V_V)^3(1-V_V)^3}{A(F)(S_V)^2} \tag{13.7}$$

where again S_V is the surface area density of the boundary ∂Y of the feature of interest Y. The expression on the right can be estimated stereologically from large samples.

13.1.3 Variance paradoxes

Intuitively, point counting should be much less accurate than quadrat area measurement. That is, we expect the point count estimator of volume fraction $N(Y\cap G)/N(G)$ to have much larger variance than the empirical area fraction $A(Y\cap F)/A(F)$.

Surprisingly, this turns out to be false, in general. When the test grid G has a **fixed** position relative to the quadrat F, it is possible for the point counting estimator to be more precise than quadrat area measurement.

Jensen and Gundersen [282, sec. 6] constructed such a paradox, and it was later shown [32] that this is not an isolated phenomenon. Figure 13.2 shows a modified version of the Jensen-Gundersen paradox [32]. The random set Y in the plane is a Boolean model [415] formed by generating a uniform Poisson point process of intensity λ, constructing a disc of radius r centred on each point of the process, and taking the union of all discs. The two competing estimators are $A(Y\cap F)/A(F)$, where F is a square of side s, and $N(Y\cap G)/N(G)$, where G is the set of four vertices of the unit square. Since $N(G) = 4$ the latter estimator is very coarse indeed. Yet, for certain values of the parameters r, s and λ, the point count estimator is more accurate.

This paradox may be explained by the covariance contributions to the variance. There is large positive covariance $C(u)$ at small lags u. Hence the double integral (13.2) may well be larger than the double sum over pairs of vertices (13.1).

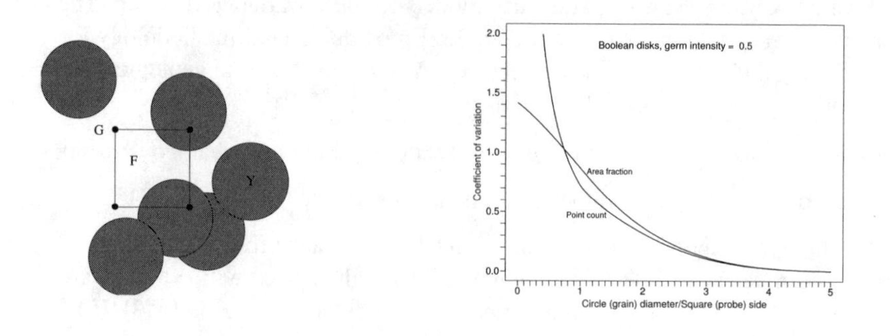

Figure 13.2 *Jensen-Gundersen paradox, stationary version [32].* Left: *Y is a Boolean model of discs. The subsample G consists of the four corner points of the window F.* Right: *Coefficients of variation (=standard deviation/mean) of the point-counting estimator and of the empirical area fraction are plotted against scale parameter* $w = 2r/s$. *For* $w > 0.7$, *the point-counting estimator has lower variance. First published in [32]. Copyright © Applied Probability Trust. Reproduced by permission of the Applied Probability Trust.*

Similar paradoxes occur with other stereological estimators. Hennig [255] and Hilliard and Cahn [261] showed that point counting with a coarse grid of test points may be more efficient than lineal intercept measurement for estimating V_V.

Paradoxes of this type are apparently well-known in the theory of wide-sense stationary random fields [435]. Consider a second-order stationary random function $Z : \mathbf{R} \to \mathbf{R}$. *Smit's paradox* [518, 570] is the fact that, under certain conditions, there exist equi-spaced sampling designs $\{x_1, \ldots, x_n\} \subset [0, T]$ for which the sample mean $\bar{Z} = \frac{1}{n}\sum_{i=1}^{n} Z(x_i)$ has lower variance than $\tilde{Z} = \frac{1}{T}\int_0^T Z(x)\,dx$. Grenander [206] showed that, among all 'linear' estimators of $\mu = \mathbf{E}\,Z(0)$ of the form $\widehat{\mu}_G = \int_0^T Z(x)\,dG(x)$ where G is a probability measure, the best linear unbiased estimator is not necessarily $\tilde{Z}$. *Grenander's theorem* [206, p. 236], [435, pp. 55-57] characterises the optimal G as the solution of

$$\int_0^T c(x-z)\,dG(x) = \text{constant} \tag{13.8}$$

where $c(u) = \mathsf{cov}(Z(0), Z(u))$ is the autocovariance function of Z. For the particular case of exponential covariance $c(t) = a\exp\{-\beta|t|\}$, the BLUE is [206, 435]

$$\widehat{\mu} = \frac{[Z(0) + Z(T) + \beta\int_0^T Z(t)\,\mathrm{d}t]}{2 + \beta T}. \tag{13.9}$$

The above results have been generalised to random fields on $\mathbf{R}^k$ in [435].

Optimal sampling design for a given random field has been discussed by many authors, e.g. [43, 470, 614, 618], see [102, §5.6.1, pp. 313–324].

Schladitz [506] applied these results to optimal estimation of area fraction in stereology. Schwandtke et al. [511] study the estimation of stereological parameters using linear combinations of standard estimators, analogous to the combination in (13.9).

In conclusion, the optimal sampling design for stereological estimation of volume fraction depends on the second-order properties of the material; there is no uniformly optimal sampling design; and the empirical area fraction is not always optimal.

13.1.4 Binomial approximation

For a coarse grid, the variance of the point-counting estimator of volume fraction may be approximated by the binomial variance.

Assuming that the covariance $C(x)$ converges to zero as $||x|| \to \infty$, two test points u, v separated by a large distance will have approximately uncorrelated indicators I_u, I_v. Hence the variance of the point count $N(Y \cap G)$ from a coarse grid of n points should be approximately $np(1-p)$ where $p = V_V$. This coincides with the variance of the binomial (n, p) distribution, although we are not intending to approximate the distribution of $N(Y \cap G)$ itself but only its variance. The variance of the estimator of $V_V = p$ is then $p(1-p)/n$.

Chayes [78] used this approximation for the variance of point counting estimators in geology, and showed [78, p. 42] that the binomial approximation is in good agreement with the sample variance of point counts obtained in replicated experiments in various rocks. Similar studies have been performed by metallographers [199, 261] and cartographers [350, pp. 420–424], [38]. Chayes [78, chap. 8] also investigated the influence of the grain size and measurement area on the (empirical) variances of point counting estimators in rocks.

When the denominator of the point counting estimator is also random (i.e. in Cruz-Orive's Model II, see Section 6.4.2) the natural bivariate extension of this approximation is the bivariate binomial. Chia [80, 81] developed this and fitted the bivariate binomial distribution to point count data in neuroanatomy.

13.1.5 Other stereological estimators

The variances of stereological estimators of S_V, L_V, N_V and other quantities require the methods of stochastic geometry [543] rather than the simple use of indicator variables. Consider the total 'amount' $\nu(Y \cap T)$ of the feature of interest observable in the sample T, where ν is an appropriate geometrical measure. We have

$$\begin{aligned} \mathsf{var}[\nu(Y \cap T)] &= \mathbf{E}\left[\nu(Y \cap T)^2\right] - \mathbf{E}[\nu(Y \cap T)]^2 \\ &= \mu^{(2)}(T \times T) - \mu^{(1)}(T)^2 \end{aligned} \tag{13.10}$$

where

$$\mu^{(1)}(A) = \mathbf{E}\nu(Y \cap A)$$

is the *first moment measure*, which is estimated by the stereological technique in question, and

$$\mu^{(2)}(A \times B) = \mathbf{E}\nu(Y \cap A)\nu(Y \cap B)$$

is the *second moment measure*. Techniques for estimating the second moment measure are discussed in [543, Section 11.6].

In section 13.1.1 we saw that the spatial covariance function of a stationary isotropic random set Y is identical to the covariance function of its plane section $Y \cap T$. Things are more complicated in the case of surfaces and curves, where the second order properties of Y and $Y \cap T$ are not identical, and their relationship involves the random angle of incidence between the feature and the section plane.

13.2 Variances in design-based stereology

Attention focuses mainly on the variance due to systematic sampling because so many stereological estimators are based on systematic samples.

13.2.1 Elementary example

A simple one-dimensional example will help to show the main features of variance under systematic sampling [127].

Example 13.1 *Suppose we want to estimate the length of a line segment Y on the real line by point-counting. We use a systematic grid of test points on the real line with spacing $t > 0$,*

$$\mathcal{G} = \{U + jt : \ j \in \mathbf{Z}\},$$

where $U \sim \mathsf{Unif}([0,t))$. An unbiased estimator of the length $L(Y)$ is

$$\widehat{L}(Y) = t\, N(Y \cap \mathcal{G})$$

See Figure 13.3.

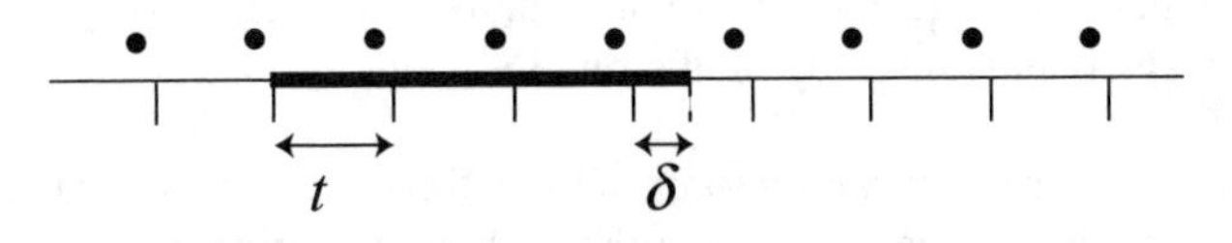

Figure 13.3 *Estimating the length of a line-segment from counting the number of points in a uniform random point grid, hitting the line-segment.*

Now the possible values of $N(Y \cap \mathcal{G})$ *are* k *and* $k+1$ *where* k *is the largest integer such that* $kt \leq L(Y)$. *Writing* $L(Y) = kt + \delta$ *where* $0 \leq \delta < t$, *we have*

$$N(Y \cap \mathcal{G}) = \begin{cases} k+1 & \text{with probability } \delta/t \\ k & \text{with probability } 1 - \delta/t \end{cases}$$

It follows that the variance of $\widehat{L}(Y)$ *is*

$$\mathsf{var}[\widehat{L}(Y)] = (t - \delta)\delta. \tag{13.11}$$

Figure 13.4 shows the variance of $\widehat{L}(Y)$ *for a line segment of unit length, or equivalently the relative variance* $\mathsf{var}[\widehat{L}(Y)]/L(Y)^2$ *for any line segment, plotted against the mean point count* $L(Y)/t$.

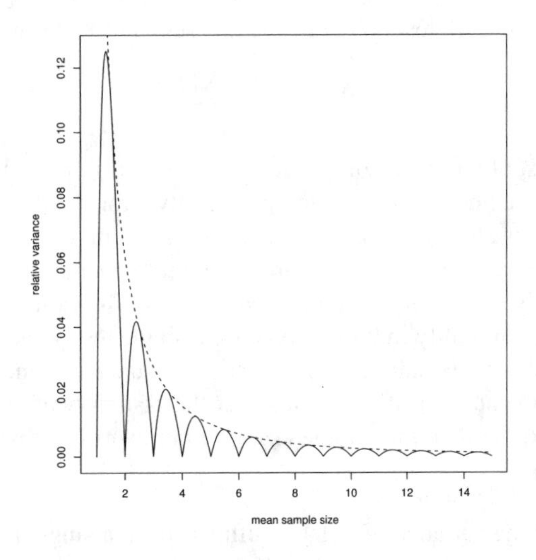

Figure 13.4 *Relative variance (= variance/mean*^{2}*) of estimator of line segment length from systematic sampling, plotted against mean sample size. Dashed curve shows upper bound* $1/(4 \cdot mean^2)$.

A striking feature of the graph in Figure 13.4 is the oscillation in variance, rather than a monotone decrease in the variance, as the sample size increases. Indeed the variance drops to zero when the grid spacing t is an integer divisor of the true length $L(Y)$. The oscillation is traditionally referred to by the German term *Zitterbewegung*, used by Matheron [360] and others.

Another feature is the *superefficiency* of the estimator compared to independent random sampling. Since the mean number of points hitting Y is

$$\begin{aligned} n &= \mathbf{E}N(Y \cap \mathcal{G}) \\ &= L(Y)/t, \end{aligned}$$

we can regard n as the expected sample size. Rewriting (13.11) as

$$\text{var}[\widehat{L}(Y)] = (1-\Delta)\Delta \cdot L(Y)^2 \cdot n^{-2}, \tag{13.12}$$

where $\Delta = \delta/t$, we find that the variance of $\widehat{L}(Y)$ is of asymptotic order n^{-2} as $n \to \infty$. Under independent random sampling the variance is of smaller order n^{-1}.

One geometrical insight is that *variance contributions are associated with the boundary.* Let I_j be the indicator

$$I_j = \mathbf{1}\{U + jt \in Y\}$$

that equals 1 if the jth test point hits the line segment Y. Then

$$N(Y \cap \mathcal{G}) = \sum_j I_j.$$

However, most of the indicators I_j are constant. Taking $Y = [0, \ell]$ without loss of generality, we find that $I_j = 1$ with probability 1 for $j = 0, 1, \ldots, k-1$, since the corresponding test points are always inside Y; while $I_j = 0$ with probability 1 for all $j < 0$ and all $j > k+1$, since the corresponding points are always outside Y. Only I_k is random, with $\mathbf{P}[I_k = 1] = \delta/t$. The variance of $\widehat{L}(Y)$ is due entirely to the variability in the outcome for a single test point at the boundary of Y. This helps to explain the phenomenon of superefficiency: the effective sample size (the number of test points near the boundary of the line segment) is of smaller order than the sample size (the number of test points hitting the line segment).

Estimation of the variance of $\widehat{L}(Y)$ is difficult from a single realisation of the experiment. If the only observable information is the point count $N(Y \cap \mathcal{G})$, then δ is unknown and cannot be estimated. However, since $x(1-x) \leq 1/4$ for all $0 \leq x \leq 1$, we have the upper bound

$$\text{var}[\widehat{L}(Y)] \leq \frac{t^2}{4}. \tag{13.13}$$

Of course this applies only to estimation of the length of a single line segment.

13.2.2 Systematic sampling in one dimension

The variance due to systematic sampling in one dimension was treated by Moran [419]. Suppose we wish to estimate the integral

$$\theta = \int_{-\infty}^{\infty} f(x)\,\mathrm{d}x,$$

where $f : \mathbf{R} \to \mathbf{R}$ is an integrable function, called the *measurement function*. Using systematic sampling with period $t > 0$, an unbiased estimator of θ is

$$\widehat{\theta} = t \sum_{j=-\infty}^{\infty} f(U + jt),$$

where $U \sim \mathsf{Unif}([0,t))$. For example, this has direct application to the Cavalieri estimator of volume (Sections 3.3.3 and 7.1.1) where the measurement function is $f(z) = A(Y \cap T_z)$ for a solid region Y where T_z is the horizontal plane at height z.

If f is square-integrable, then the variance of $\widehat{\theta}$ is finite and can be expressed as follows. Writing

$$\begin{aligned} (\widehat{\theta})^2 &= t^2 \left[\sum_{j=-\infty}^{\infty} f(U + jt) \right]^2 \\ &= t^2 \sum_j \sum_i f(U + jt) f(U + it) \end{aligned}$$

we have

$$\begin{aligned} \mathbf{E}\left[(\widehat{\theta})^2 \right] &= t \sum_j \sum_i \int_0^t f(u + jt) f(u + it)\,\mathrm{d}u \\ &= t \sum_m \sum_j \int_0^t f(u + jt) f(u + jt + mt)\,\mathrm{d}u \\ &= t \sum_m \int_{-\infty}^{\infty} f(x) f(x + mt)\,\mathrm{d}x \\ &= t \sum_m g(mt) \end{aligned}$$

where

$$g(y) = \int_{-\infty}^{\infty} f(x) f(x + y)\,\mathrm{d}x \tag{13.14}$$

is called the *covariogram*. Hence

$$\mathsf{var}[\widehat{\theta}] = t \sum_{m=-\infty}^{\infty} g(mt) - \theta^2. \tag{13.15}$$

The covariogram is symmetric around 0, i.e. $g(y) = g(-y)$, and satisfies

$$g(0) = \int_{-\infty}^{\infty} f(x)^2 \, dx \tag{13.16}$$

$$\int_{-\infty}^{\infty} g(y)\, dy = \int\int f(x) f(x+y)\, dx\, dy = \left[\int f(x)\, dx\right]^2 \tag{13.17}$$

so that (13.15) can be rewritten

$$\text{var}[\widehat{\theta}] = t \sum_{j=-\infty}^{\infty} g(jt) - \int_{-\infty}^{\infty} g(y)\, dy. \tag{13.18}$$

Thus the variance of $\widehat{\theta}$ can be interpreted as the difference between the integral of g and a discrete approximation to this integral. It follows that the variance should be small when t is small.

Example 13.2 *Reformulate Example 13.1 as the problem of integrating the rectangular function*

$$f(x) = \mathbf{1}\{a \leq x \leq a + \ell\}$$

say, where ℓ is the true length of the line segment. The covariogram is

$$g(y) = \begin{cases} \ell - |y| & \text{if } |y| \leq \ell \\ 0 & \text{otherwise} \end{cases}$$

It can be checked directly that (13.18) reduces to (13.11).

Equivalent expressions in the Fourier domain are sometimes useful [419]. If f^* is the Fourier transform of f, then the Fourier transform of g is

$$g^*(s) = \sqrt{2\pi}\, |f^*(s)|^2 \tag{13.19}$$

so that

$$\text{var}[\widehat{\theta}] = 2\pi \sum_{m=-\infty}^{\infty}{}' \left| f^*\left(\frac{2\pi m}{t}\right)\right|^2 \tag{13.20}$$

where the prime indicates that the term for $m = 0$ is omitted.

Recent work by Kiêu and collaborators [315] has identified an important relation between the smoothness of the measurement function and the covariogram. For integers m and p, say that $f : \mathbf{R} \to \mathbf{R}$ is $(m, p)-$*piecewise smooth* if

- f has bounded support;
- all derivatives of f of order less than m are continuous on $\mathbf{R}$;
- the derivatives of f of order $j \in \{m, m+1, \ldots, m+p\}$ have a finite number of jumps of finite size.

Thus m is the lowest order such that the mth derivative of f may have jumps.

For an (m,p)-piecewise smooth measurement function with $p \geq 1$, it can be shown [312] that the covariogram g is $(2m+1,p)-$piecewise smooth. Furthermore, $\mathsf{var}[\widehat{\theta}]$ can be decomposed as

$$\mathsf{var}[\widehat{\theta}] = \mathsf{var}_E[\widehat{\theta}] + Z(t) + o(t^{2m+2}), \quad t \to 0 \tag{13.21}$$

where $\mathsf{var}_E[\widehat{\theta}]$ is the *extension term*

$$\mathsf{var}_E[\widehat{\theta}] = -2t^{2m+2} \frac{B_{2m+2}}{(2m+2)!} g^{(2m+1)}(0^+), \tag{13.22}$$

and $Z(t)$ is called the *Zitterbewegung* or *fluctuation term*. The constant B_α is the Bernoulli number of order α, see e.g. [4]. In particular,

$$B_2 = \frac{1}{6} \text{ and } B_4 = -\frac{1}{30}.$$

The extension term $\mathsf{var}_E[\widehat{\theta}]$ represents the overall trend in the variance, while the Zitterbewegung term oscillates around zero. For small t, the Zitterbewegung term can be neglected, and the extension term can be used as an approximation to the variance.

Example 13.3 *Continuing the example of estimating the length of a line segment (Examples 13.1 and 13.2), we note that the measurement function f is a $(0,\infty)-$piecewise smooth function. With $m = 0$ the extension term is*

$$\begin{aligned} \mathsf{var}_E[\widehat{\theta}] &= -2t^2 \frac{B_2}{2!} g'(0+) \\ &= \frac{1}{6} t^2. \end{aligned}$$

Comparing this with the exact variance $\mathsf{var}[\widehat{\theta}] = (t-\delta)\delta$ we find that the approximation is obtained by averaging the exact variance uniformly over $\delta \in [0,t]$. The analogue of (13.12) is

$$\mathsf{var}_E(\widehat{L}(Y)) = \frac{1}{6} \cdot L(Y)^2 \cdot n^{-2}.$$

Example 13.4 (Cavalieri estimator for a sphere) *Suppose we estimate the volume of the unit sphere in $\mathbf{R}^3$ using the Cavalieri estimator. The measurement function is the area of the plane section at height x,*

$$f(x) = \pi(1-x)^2$$

for $-1 \leq x \leq 1$, and $f(x) = 0$ otherwise. The covariogram $g(y)$ is a polynomial for $y \in [-2,2]$ which can easily be evaluated and differentiated. Figure 13.5 shows the exact value of $\mathsf{var}[\widehat{\theta}]$ computed using (13.18) plotted on a log-log *scale against t. Since f is continuous and has piecewise continuous derivatives of all orders $k \geq 1$, it is $(1,\infty)$-smooth, so the variance is of order $t^{2m+2} = t^4$ and the extension term is*

$$\mathsf{var}_E[\widehat{\theta}] = -2t^4 \frac{B_4}{4!} g^{(3)}(0+) = \frac{\pi^2}{90} t^4. \qquad (13.23)$$

The extension term is also plotted in Figure 13.5.

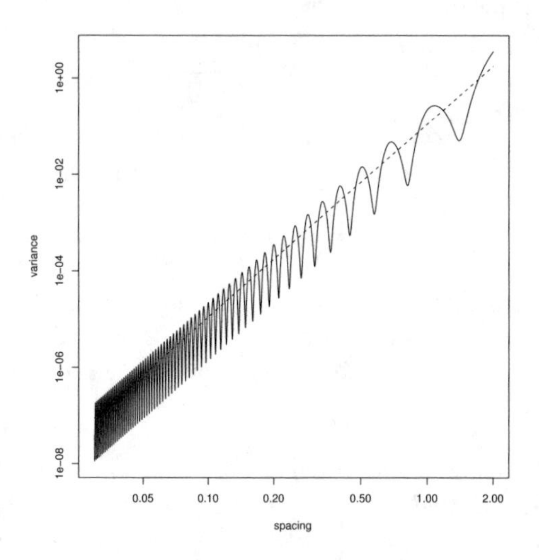

Figure 13.5 *Variance of Cavalieri estimator of volume of the unit sphere (solid lines) and the extension term approximation (dotted lines) plotted against section spacing t on a log-log scale.*

Notice that in Figure 13.5 the variance drops below 0.01 when the section spacing t is below 0.5. In other words, the Cavalieri estimator of sphere volume based on 4 plane sections has a coefficient of variation (SD/mean) of only 2.4%.

13.2.3 Variance of Cavalieri estimator

For a three-dimensional solid of general shape, the variance of the Cavalieri estimator of volume can also be derived along the lines of Example 13.4. However the computations can be quite lengthy, cf. [189, 190].

Alternatively, there is a simple asymptotic approximation to the variance, which holds under randomisation of the section orientation. Suppose that the stack of serial section planes

$$\mathcal{T} = \{T_{\Omega,U+mt} : m \in \mathbf{Z}\}$$

is given an isotropic random orientation $\Omega \sim \mathsf{Unif}(S^2_+)$ as well as a uniform initial position $U \sim \mathsf{Unif}([0,t))$ where U, Ω are independent. Then assuming the boundary ∂Y of Y is compact and differentiable,

$$\mathsf{var}[\widehat{V}(Y)] \approx \frac{\pi}{360} S(\partial Y) t^4, \quad t \to 0. \tag{13.24}$$

For example, if Y is the unit sphere, this coincides with (13.23). This was obtained by Matheron [360, p. 105], see [120, Eq. (7.3), (7.7), (7.8)].

The number $N(Y \cap \mathcal{T})$ of planes hitting Y has expectation

$$n = \mathbf{E} N(Y \cap \mathcal{T}) = \frac{1}{t} \overline{H}(Y)$$

where $\overline{H}(Y)$ is the mean projected height (caliper diameter) of Y discussed in Section 5.6.2. Hence the variance approximation (13.24) is of the order n^{-4} which is 'super-efficient' compared to the order n^{-1} obtained from a sample of i.i.d. planes with any common distribution.

Moran [422] analysed the equivalent two-dimensional problem of estimating the area of a region by measuring the lengths of intersection with a grid of parallel lines. For a convex set of maximum caliper diameter D, the variance of the Cavalieri estimator of area is bounded by $\frac{1}{4} t^2 D^2$. This bound also applies in higher dimensions, so that the variance of the Cavalieri estimator of volume (for a convex set) is also bounded by $\frac{1}{4} t^2 D^2$.

13.2.4 Variance of point-counting in the plane

The variance of the point-counting estimator of area of a fixed set, based on a randomly-translated grid of test points, was first investigated by Kendall and Rankin [306, 308]. There is now an immense literature on this topic: see the Bibliographic Notes.

For a compact set Y in $\mathbf{R}^2$, define the *geometric covariogram*

$$g(u) = A(Y \cap (Y + u))$$

for all $u \in \mathbf{R}^2$, where $Y + u$ denotes the set obtained by translating Y by the vector u. That is, $g(u)$ is the area of the intersection between Y and a copy of Y translated by u. This function is the analogue of the covariogram defined in the one-dimensional case. We have

$$g(0) = A(Y) \tag{13.25}$$

$$\int_{R^2} g(u)\, \mathrm{d}u = A(Y)^2. \tag{13.26}$$

Consider the point-counting estimator of area

$$\widehat{A}(Y) = a N(Y \cap \mathcal{G})$$

based on a randomly-translated grid of test points

$$\mathcal{G} = U + \mathcal{G}_0$$

where $\mathcal{G}_0$ is a periodic set of points with fundamental domain D_0, where $a = A(D_0)$ is the area per test point, and where $U \sim \mathsf{Unif}(D_0)$. Then we get

$$\mathsf{var}[\widehat{A}(Y)] = a \sum_{u \in \mathcal{G}^*} g(u) - \int_{R^2} g(u)\,\mathrm{d}u. \tag{13.27}$$

An elegant result is obtained when the grid is also randomly rotated [306, 308]. Let

$$\mathcal{G} = \{R_\Theta(U + s(k,l)) : k,l \in \mathbf{Z}\}$$

be the square grid of mesh $s > 0$ subjected to a uniformly random translation $U \sim \mathsf{Unif}([0,s)^2)$ followed by an isotropic random rotation R_Θ with $\Theta \sim \mathsf{Unif}([0,\pi])$. Then the point-counting estimator of area

$$\widehat{A}(Y) = s^2 N(Y \cap \mathcal{G})$$

has asymptotic variance

$$\mathsf{var}[\widehat{A}(Y)] \approx \frac{1}{6\pi}(\frac{3}{\pi^2}\zeta(3) + 1)L(\partial Y)s^3, \quad s \to 0 \tag{13.28}$$

where ζ is the Riemann zeta function

$$\zeta(\alpha) = \sum_{k=1}^{\infty} \frac{1}{k^\alpha}.$$

This can be evaluated numerically to yield

$$\mathsf{var}[\widehat{A}(Y)] \approx 0.0724\, L(\partial Y)\, s^3, \quad s \to 0. \tag{13.29}$$

This result holds for all sets Y in the plane with finite area and rectifiable boundary. Since the 'expected sample size' $n = \mathbf{E}N(Y \cap \mathcal{G})$ is proportional to s^{-2}, this says that the variance is of order $n^{-3/2}$.

13.2.5 Data-based variance estimation

It is also possible to estimate the variance due to systematic sampling from data, by estimating and modelling the covariogram.

Consider the theory for systematic sampling in one dimension (Section 13.2.2). From the observations in a systematic sample of period t,

$$X_i = f(U + it), \quad i \in \mathbf{Z}$$

we can estimate the values of the covariogram $g(y)$ at integer multiples of t by the sample cross moments,

$$\widehat{g}(kt) = t\sum_i X_i X_{i+k}. \tag{13.30}$$

The asymptotic variance depends only on the behaviour of the covariogram g near the origin. If g has an expansion of the form

$$g(h) = \sum_{j=0}^{\infty} b_j |h|^j, \tag{13.31}$$

in a neighbourhood of the origin, then g is (m,∞)-smooth where $m = 2k$ is the lowest even integer such that $b_{2k+1} \neq 0$, and

$$\mathsf{var}_E[\widehat{\theta}] = -t^{2m+2}\frac{B_{2m+2}}{m+1}b_{2m+1}. \tag{13.32}$$

The coefficient b_{2m+1} can be estimated by fitting a truncated version of the model (13.31) to the empirical estimates of g. For $m = 0$ we may assume that

$$g(h) \approx b_0 + b_1|h| + b_2 h^2. \tag{13.33}$$

The relevant coefficient b_1 can be estimated from estimates of g at $g(0)$, $g(t)$ and $g(2t)$ by solving the linear equations

$$g(jt) = b_0 + b_1(jt) + b_2(jt)^2, \quad j = 0,1,2.$$

Similarly, for $m = 1$, we can consider the cubic model

$$g(h) \approx b_0 + b_1|h| + b_2 h^2 + b_3|h|^3. \tag{13.34}$$

For $m = 1$, smoothness of g implies that $b_1 = 0$. The interesting coefficient b_3 can again be estimated from estimates of $g(0)$, $g(t)$ and $g(2t)$. Under the relevant models (13.33) and (13.34), we get for $m = 0,1$

$$\mathsf{var}_E[\widehat{\theta}] = \alpha_m t[3g(0) - 4g(t) + g(2t)], \tag{13.35}$$

where $\alpha_0 = 1/12$ and $\alpha_1 = 1/240$. Plugging in the moment estimates (13.30) of $g(0)$, $g(t)$ and $g(2t)$ we obtain an empirical estimate of the right hand side of (13.35). The estimation approach described above was developed in [125, 229]. This technique is now widely accepted in applied stereology as an estimator of the variance of the Cavalieri estimator of volume (Section 7.1.1).

Refinements and further developments have been presented in a number of subsequent papers, mainly from the group around Cruz-Orive. In [189], it is pointed out that for some measurement functions the derivatives may not have *finite* jumps as required for a $(m, p)-$piecewise smooth function. A generalised estimation method is developed in order to deal with these cases, see [189, Section 5.1] and [190, 191].

Recently derived formulae for the variance of estimators of absolute volume, including the contributions from point counting of areas (Section 7.1.2) can be found in [187, sect. 3.2], see also [125]. Variance formulae relating to volume estimation from parallel thick sections or slabs (cf. Exercise 7.6) can be found in [210, Sections 5.4 and 5.5].

13.2.6 Systematic sampling of orientations; curve length estimation

Systematic sampling on the circle is used in stereological methods that require isotropic rotation.

For example, curve length in the plane can be estimated by counting the number of intersections with test lines, provided the orientation of the test lines is isotropically random. In Steinhaus' estimator (Section 7.2.1) a grid of parallel lines at constant spacing t is repeatedly superimposed on the curve at n different orientations $\theta_j = \Psi + j\pi/n$ for $j = 1, \ldots, n$ where $\Psi \sim \mathsf{Unif}([0, \pi/n])$. The angles θ_j constitute a systematic sample of $[0, \pi)$ with fixed sample size n.

Steinhaus [535] derived optimal *absolute* bounds on the relative error of the estimator $\widehat{L}(Y)$:

$$\frac{\varepsilon}{\tan\varepsilon} \leq \frac{\widehat{L}(Y)}{L(Y)} \leq \frac{\varepsilon}{\sin\varepsilon} \tag{13.36}$$

where $\varepsilon = \pi/(2n)$. For $n = 2$, corresponding to a test system containing two perpendicular sets of parallel lines, the bound is $0.78 \leq \widehat{L}(Y)/L(Y) \leq 1.11$.

Moran [420] studied the variance of $\widehat{L}(Y)$. The variance is a sum of two contributions, the first from systematic sampling of orientations on the circle, and the second from systematic sampling of lines parallel to a given orientation. Each variance contribution can be evaluated using the theory for one-dimensional systematic sampling. An upper bound for the first contribution is

$$\frac{L(Y)^2}{2\pi n \sin^2\varepsilon}[2\varepsilon + \sin(2\varepsilon) - \frac{4}{\varepsilon}\sin^2\varepsilon] \tag{13.37}$$

and an upper bound for the second contribution is

$$\frac{1}{3}k^2 t^2 \tag{13.38}$$

where

$$k = \max_{\theta \in [0,\pi)} k(\theta)$$

and $k(\theta)$ is the number of jumps in the piecewise constant function $h(p) = N(Y \cap T_{\theta,p})$ for fixed θ. For example, any closed convex curve or convex arc has $k = 2$.

Periodic systematic sampling has been treated in detail in [132, 211, 212]. As an example, let us here briefly consider the case of systematic sampling on the circle. The parameter to be estimated is then of the form

$$\theta = \int_0^{2\pi} f(x)\,\mathrm{d}x.$$

In contrast to systematic sampling on the real line, the sample size is here fixed. For sample size n, the estimator of θ becomes

$$\widehat{\theta} = \frac{2\pi}{n} \sum_{i=0}^{n-1} f(U + \frac{i2\pi}{n}),$$

where $U \sim \mathsf{Unif}([0, \frac{2\pi}{n}))$. Again, if f is square integrable, the variance can be expressed in terms of the covariogram

$$g(h) = \int_0^{2\pi} f(x)f(x+h)\,\mathrm{d}x, \quad h \in [0, 2\pi), \tag{13.39}$$

where f is extended periodically. Instead of following the classical approach described above for one-dimensional systematic sampling, Cruz-Orive and Gual Arnau [132] postulate a global model for the covariogram, which exploits specific symmetry properties due to the periodicity in the measurement function such as

$$g(h) = g(2\pi - h), \quad h \in [0, 2\pi).$$

If the covariogram model fits the data, its parameters can be estimated and inserted into the expression for the variance of $\widehat{\theta}$. The case where the measurement function is determined with error is also treated.

This approach can be used to estimate the variance of the Steinhaus estimator of planar curve length (Section 7.2.1). The estimator requires repeated use of line grids at n systematic angles. For $n = 2$, two perpendicular line grids are used, one of them with uniform angle in $[0, \frac{\pi}{2})$.

The results in [132] also include variance estimators relating to surface area estimators, based on systematic vertical sections (Section 8.1.4), curve length estimators in 3D from total vertical projections (Section 8.2), local area estimators (Exercise 8.6) and vertical rotator volume estimators (Section 8.4).

13.2.7 Estimating variance by curve estimation

The data-based estimation of variance presented in Section 13.2.5 is based on asymptotics as the grid spacing t tends to zero. It might be inaccurate for coarsely-spaced grids.

An alternative technique for variance estimation was developed by Chia [80].

In the one-dimensional case, given the observed values

$$X_i = f(U + it), \quad i \in \mathbf{Z}$$

of the measurement function f, we use nonlinear smoothing or curve estimation techniques to interpolate the values of $f(s)$ for all s. Plugging this estimate $f^\dagger$ into the definition of the covariogram (13.14) we compute an estimated covariogram $g^\dagger$ by numerical integration. We can then compute an estimate of the variance of the systematic sampling estimator at any sample spacing $t > 0$ by evaluating (13.18).

13.3 Variance comparisons and Rao-Blackwell

Intuition suggests that taking a subsample or lower-dimensional sample of the available information ought to result in an increase in the variance of stereological estimators. Thus, estimating volume fraction by point-counting ought to be less accurate than estimating it using one-dimensional transect lengths, which in turn ought to be less accurate than using two-dimensional plane section areas.

Paradoxical counterexamples, where this intuition is not correct, were described in Section 13.1.3.

However, there are also very many examples where this principle does work. Davy and Miles [147, sec. 6] proved that, for IUR and WUR probes in a design-based setting, estimators based on lower-dimensional probes have higher variances. Lantuéjoul [338] presented a general derivation with many classes of examples.

The principle here is the Rao-Blackwell theorem. In basic statistical theory, this theorem [56, 472], [101, p. 258] can be invoked to prove that an unbiased estimator A based on a subsample of the data must have larger variance than the corresponding unbiased estimator B based on the entire data set. Given the first estimator A, define the second estimator B by

$$B = \mathbf{E}[A \mid S]$$

where S is a sufficient statistic. Then the variance decomposition

$$\mathsf{var}[A] = \mathbf{E}\left[\mathsf{var}[A \mid S]\right] + \mathsf{var}[\mathbf{E}[A \mid S]] \tag{13.40}$$

becomes

$$\mathsf{var}[A] = \mathbf{E}\left[\mathsf{var}[A \mid S]\right] + \mathsf{var}[B]$$

implying that

$$\mathsf{var}[A] \geq \mathsf{var}[B].$$

If additionally the statistical model is ‘complete’, then the Rao-Blackwell theorem implies that B is the best (minimum variance) unbiased estimator.

The Rao-Blackwell theorem was adapted to stereology by Baddeley and Cruz-Orive [32] as follows. See also [338].

Theorem 13.1 (Stereological Rao-Blackwell) *Let X, T_1, T_2 be random closed sets in $\mathbf{R}^d$ with joint probability distribution P belonging to some class $\mathcal{P}$ of distributions. Assume that*

- *$T_2 \subseteq T_1$ with probability 1;*
- *the conditional distribution of T_2 given T_1 is fixed, and does not depend on P, for $P \in \mathcal{P}$.*

If $A = a(X \cap T_2, T_2)$ is any unbiased estimator of a parameter $\theta = \theta(P)$, then

$$B = \mathbf{E}[A \mid X \cap T_1,\, T_1] \qquad (13.41)$$

is a well-defined statistic $B = b(X \cap T_1, T_1)$, is unbiased for θ, and $\mathsf{var}(S) \leq \mathsf{var}(R)$ under any $P \in \mathcal{P}$.

Theorem 13.1 encompasses both design-based estimation (where X is a fixed but unknown compact set and T_1, T_2 are random with known distribution) and model-based estimation (where X is random with postulated invariance properties, and T_1 may as well be fixed).

Note the theorem requires that T_2 be a **randomised** subsample of T_1. Taking the conditional expectation of A given T_1 and $X \cap T_1$ amounts to averaging over the conditional distribution of T_2 given T_1.

Example 13.5 *Consider the estimators of the absolute area $A(X)$ of a set X in the plane based on point counting and on line intercept measurement,*

$$\begin{aligned} \widehat{A}(X) &= tL(X \cap \mathcal{L}) \\ \widetilde{A}(X) &= t^2 N(X \cap \mathcal{G}) \end{aligned}$$

where $\mathcal{L}$ is an array of test lines with spacing t,

$$\mathcal{L} = \left\{T_{0,U_1+kt} : k \in \mathbf{Z}\right\}$$

and $\mathcal{G}$ is a square grid of test points with the **same** *spacing t,*

$$\mathcal{G} = \{(U_1 + kt, U_2 + mt) : m, k \in \mathbf{Z}\}$$

where $U_1, U_2 \sim \mathsf{Unif}([0,t])$ are independent. Notice that $\mathcal{G} \subset \mathcal{L}$ since both test systems use the same random shift variable U_1. The conditions of Theorem 13.1 apply. We have

$$\mathbf{E}[\widetilde{A}(X) \mid \mathcal{L}] = \mathbf{E}[\widetilde{A}(X) \mid U_1] = \widehat{A}(X)$$

so that

$$\mathsf{var}[\widehat{A}(X)] \leq \mathsf{var}[\widetilde{A}(X)],$$

that is, the point-counting estimator cannot be more precise. Similar arguments hold in the model-based case.

Notice that Example 13.5 requires the two grids to have equal spacing, or at least that the grid geometry allow $\mathcal{G}$ to fit inside $\mathcal{L}$. Variance comparisons for grids with incompatible geometries are unresolved.

Example 13.6 *Consider Delesse's principle in the design-based, weighted sampling version,*

$$\mathbf{E}\left[\frac{A(Y \cap T)}{A(X \cap T)}\right] = \frac{V(Y)}{V(X)}$$

where X, Y are fixed sets in $\mathbf{R}^3$ with $Y \subset X$, and T is a random plane with the area-weighted distribution that has probability density

$$f_T(T) = \frac{A(X \cap T)}{2\pi V(X)}$$

with respect to the invariant measure for planes in $\mathbf{R}^3$. Given T, let T' be a random line in T with the length-weighted density

$$f_{T'|T}(T' \mid T) = \frac{L(X \cap T')}{2\pi V(X)}$$

with respect to the invariant measure for lines in T. We have

$$\mathbf{E}\left[\frac{L(Y \cap T')}{L(X \cap T')} \mid T\right] = \frac{A(Y \cap T)}{A(X \cap T)}.$$

It follows that

$$\text{var}[\frac{A(Y \cap T)}{A(X \cap T)}] \le \text{var}[\frac{L(Y \cap T')}{L(X \cap T')}].$$

Davy and Miles [147, sec. 6] proved that the standard unbiased stereological estimators based on IUR and WUR probes are connected to one another by the Rao-Blackwell relation (13.41), and hence that lower-dimensional estimators have higher variance. Lantuéjoul [338] proved that stereological estimators based on isotropic random projections onto k-planes are also related in this way. Jensen and Gundersen [295, eq. (5.1)] showed that the estimate of the volume of a body from coaxial sections ('rotator') is the conditional expectation of the 'nucleator' estimate given the section plane information. See Section 8.6. This is also a consequence of Theorem 13.1.

On the other hand, Theorem 13.1 does not imply that a lower-dimensional sample will always yield an estimator with higher variance. If A and B are two estimators based on samples of dimension a and b, respectively, where $a < b$, the key relationship (13.41) does not necessarily hold. It will hold only in situations where the subset or lower-dimensional probe T_2 is randomised, and where we can use results from geometrical probability to establish (13.41). Other variance paradoxes are discussed in [255, 274, 368], [442, Appendix 2], [443, 506].

We cannot expect a uniform minimum variance unbiased estimator to exist in most stereological applications of Theorem 13.1. For example in model-based stereology, the attractive assumption of spatial homogeneity implies that the underlying σ-field is not complete, so that a UMVUE does not exist [32].

13.4 Prediction variance

In the model-based framework, where the feature of interest Y is a realisation of a random set, we may also be interested in predicting the realised value $h(Y)$ of some functional h, rather than the expectation $\mathbf{E}[h(Y)]$. For example, in ore reserve estimation, the actual volume $V(Y)$ of an ore body Y is of primary interest, rather than its expectation $\mathbf{E}[V(Y)]$ under the stochastic model.

Using a random subsample T independent of Y, a predictor $\widehat{h}(Y \cap T)$ of $h(Y)$ is constructed. Unbiasedness with respect to the sampling design can be formulated as

$$\mathbf{E}[\widehat{h}(Y \cap T) \mid Y] = h(Y).$$

The theory of variances for design-based stereology applies to this situation when we condition on Y. In Section 13.2 we presented estimators $\widehat{\sigma}^2_Y(Y \cap T)$ of the conditional variance

$$\sigma^2_Y = \mathsf{var}[\widehat{h}(Y \cap T) \mid Y]$$

which are design-unbiased.

In a model-based setting, the prediction error

$$\mathbf{E}[\widehat{h}(Y \cap T) - h(Y)]^2$$

is typically the important quantity to study. Since $\widehat{h}(Y \cap T)$ and $\widehat{\sigma}^2_Y(Y \cap T)$ are design-unbiased, it follows from (13.40) that

$$\mathbf{E}[\widehat{h}(Y \cap T) - h(Y)]^2 = \mathbf{E}[\widehat{\sigma}^2_Y(Y \cap T)].$$

Therefore, $\widehat{\sigma}^2_Y(Y \cap T)$ or an average of such estimators over a sample of objects can be regarded as an unbiased estimate of the prediction error.

Estimation of the prediction error is important in geostatistics, but is not often studied in stereology. In [266], model-based statistical inference for data based on systematic sampling on the circle is discussed, see also [80]. Here, $Y(x)$, $x \in [0, 2\pi)$ is a stationary, cyclic stochastic process and

$$h(Y) = \int_0^{2\pi} Y(x)\,\mathrm{d}x.$$

A parametric model for Y with a covariance structure similar to that suggested in [211] is studied. The parameters of the model are estimated using maximum likelihood and the estimates are inserted into a closed form parametric expression for the prediction error.

13.5 Nested ANOVA

Many stereological experiments are nested sampling designs. For example, in many biological applications, several animals participate in the experiment; several tissue blocks are sampled from each animal; several thin sections are cut from each block; and several sampling windows are photographed on each section. See Figure 12.4 in Section 12.3.2.

Each level of the design can be regarded as a source of sampling variability. It is desirable to quantify the variability at each level. This can be used to calculate the standard error of the stereological estimates obtained in the experiment. An analysis of variance also would allow us to optimise the design of future experiments [214]. In particular, the variability between experimental animals, or 'biological' variability, is of scientific interest in its own right, and is important when considering the sample size.

13.5.1 General theory

The basic principles of analysis of variance can be used. The variance splitting formula (13.40) can be applied at each level of the design. Suppose there are n levels and let $\mathcal{S}_k$ be the sample obtained at level k (where $k = 1$ is the top level). For a stereological estimator A obtained from the lowest subsample $\mathcal{S}_n$, we have

$$\begin{aligned} \mathsf{var}[A] &= \mathsf{var}[\mathbf{E}[A \mid \mathcal{S}_{n-1}]] + \mathbf{E}[\mathsf{var}[A \mid \mathcal{S}_{n-1}]] \\ &= \mathsf{var}[A_{n-1}] + \mathbf{E}[\mathsf{var}[A \mid \mathcal{S}_{n-1}]] \end{aligned}$$

where $A_{n-1} = \mathbf{E}[A \mid \mathcal{S}_{n-1}]$; and so recursively

$$\mathsf{var}[A] = \mathsf{var}[A_1] + \mathbf{E}[\mathsf{var}[A_2 \mid \mathcal{S}_1]] + \mathbf{E}[\mathsf{var}[A_3 \mid \mathcal{S}_2]] + \ldots + \mathbf{E}[\mathsf{var}[A \mid \mathcal{S}_{n-1}]] \tag{13.42}$$

where $A_k = \mathbf{E}[A \mid \mathcal{S}_k]$. The n summands on the right hand side of (13.42) are the variance contributions, from the corresponding levels $1, \ldots, n$ of the nested design, to the overall variance of the estimator A.

The variance decomposition (13.42) holds in all nested sampling designs and for all estimators A. The difficulty is to estimate the variance contributions from the sample data. The constructed variables A_k may not be observable or estimable, and we will have only one realisation of the experiment, so that the incremental variances $\mathbf{E}[\mathsf{var}[A_k \mid \mathcal{S}_{k-1}]]$ may not be estimable.

For nested cluster sampling, *assuming the clusters at each level are selected by simple random sampling*, the variance contributions can be estimated using the corresponding observed sums of squares from an analysis of variance of the sample data [83, Chapter 9], [555, pp. 121–123]. However, this is generally inappropriate, because systematic sampling is generally used to subsample the experimental material in stereology.

Otherwise, estimation of the variance contributions in (13.42) requires statistical model assumptions.

13.5.2 Random effects ANOVA

Gundersen and Østerby [214] proposed that the data be analysed using the standard variance components analysis for a nested random effects model [522, chap 13], [523, Chap 10], [617, Section 11.8].

For example, suppose there are a animals, b tissue blocks per animal and c section fields per block. Let X_{ijk} denote the observation from the kth section field in the jth tissue block in the ith animal. The random effects model is

$$X_{ijk} = \mu + A_i + B_{ij} + C_{ijk} \tag{13.43}$$

where μ is constant and A_i, B_{ij}, C_{ijk} for all i, j, k are independent random variables with mean zero and variances $\mathsf{var}[A_i] = \sigma_A^2$, $\mathsf{var}[B_{ij}] = \sigma_B^2$ and $\mathsf{var}[C_{ijk}] = \sigma_C^2$, respectively. Suppose the stereological estimator is the grand mean

$$\widehat{\mu} = \overline{X}_{\cdots} = \frac{1}{abc} \sum_i \sum_j \sum_k X_{ijk}.$$

Then the variance decomposition (13.42) becomes

$$\mathsf{var}[\widehat{\mu}] = \frac{\sigma_A^2}{a} + \frac{\sigma_B^2}{ab} + \frac{\sigma_C^2}{abc}. \tag{13.44}$$

Unbiased estimators of the three variance components are the corresponding sample variances

$$\widehat{\sigma_A^2} = \frac{1}{a-1} \sum_i (\overline{X}_{i\cdot\cdot} - \widehat{\mu})^2 \tag{13.45}$$

$$\widehat{\sigma_B^2} = \frac{1}{a(b-1)} \sum_i \sum_j (\overline{X}_{ij\cdot} - \overline{X}_{i\cdot\cdot})^2 \tag{13.46}$$

$$\widehat{\sigma_C^2} = \frac{1}{ab(c-1)} \sum_i \sum_j \sum_k (X_{ijk} - \overline{X}_{ij\cdot})^2 \tag{13.47}$$

of the sample means at each level,

$$\overline{X}_{i\cdot\cdot} = \frac{1}{bc} \sum_j \sum_k X_{ijk} \tag{13.48}$$

$$\overline{X}_{ij\cdot} = \frac{1}{c} \sum_k X_{ijk}. \tag{13.49}$$

The estimators are more complicated when the design is not balanced (i.e. when the number of blocks per animal, or the number of sections per block, is not constant), see [523, Chap. 10].

Examples of these calculations for nested stereological sampling designs are presented in [214]. This approach has been widely adopted in biological stereology, and is now taught in basic courses, cf. [427, Chap. 10]. However, very little is known about the validity and statistical properties of the method.

The assumptions of the random effects model (13.43) are probably wrong in most applications. Since the subsampling at each level (apart from the top level in some cases) is almost always performed by systematic sampling, the independence assumption is not appropriate. Indeed if systematic sampling is used because it is more efficient than simple random sampling, then (cf. Appendix A) the sample variance of the observations within a systematic sample is greater than the sample variance in a simple random sample, suggesting that the random effect variances σ_B^2, σ_C^2 etc will be overestimated. We might argue that sparse sampling and stratification together ensure that contributions from different parts of the sample are approximately independent. However the random effects model may still be inappropriate because of heteroscedasticity and non-constant mean, both stemming from spatial inhomogeneity. However, we might hope that these departures from the model would all lead to overestimation of the variance components, making the method conservative.

Since the observations at the lowest sampling level are often integer counts, it is inappropriate to model them by a normal distribution, and the statistical properties of the procedure are unknown. Classical hypothesis tests based on the normal distribution should be treated with caution in this context.

13.5.3 Optimal sampling design: "Do more less well!"

Under the random effects model, we may also optimise the sampling design. Performing a pilot experiment with trial values of a, b, c, we estimate the three variances. In the example given, suppose there is a fixed cost k_A associated with each animal, a cost k_B for each block, and a cost k_C for each section. Then the optimal design a, b, c can be obtained by integer quadratic programming, minimising variance (13.44) for a given total cost.

One of the main findings obtained with the Gundersen-Østerby approach is that, in most biological experiments, the inter-animal or 'biological' variability is quite large. Estimates of relative variance (= variance$/$mean^2 = CV^2) of around 10% are common. Since we have seen that the other ('stereological') sources of variability have comparable magnitude, the contribution of stereological sampling variability to the final variance of the estimator is much lower than the contribution of biological variability. For example, in (13.44) if $\sigma_A^2 \approx \sigma_B^2 \approx \sigma_C^2$ then the stereological contribution to the estimator variance, $\sigma_B^2/ab + \sigma_C^2/abc$, is a factor $1/b + 1/bc$ smaller than the biological contribution σ_A^2/a. In practice it is often found that the optimal value of a is higher, and the optimal values of b and c are lower, than the values that were used in the experiment. This has the important implication that we ought to take more animals, to attenuate the biological variance contribution, and perform less sampling per animal. E.R. Weibel paraphrased this as **'Do more, less well'**.

13.5.4 Go forth and replicate

A possible remedy for the statistical weaknesses of the nested ANOVA technique is to include independent replication at each level of the sampling design.

Suppose the units available for sampling at a given level are numbered $1,\ldots,n$. Typically we take a systematic sample of period k

$$\mathcal{S} = \{u,\ u+k,\ u+2k,\ldots\}$$

where u is one of the integers $1,\ldots,k$ selected at random with equal probability. Instead of this, let us take a simple random sample $\{u_1,\ldots,u_m\}$ of size m without replacement from the integers $1,\ldots,k$, and form the m systematic subsamples

$$\mathcal{S}_i = \{u_i,\ u_i+k,\ u_i+2k,\ldots\}$$

If $H_i = h(\mathcal{S}_i)$ is an unbiased estimator of a parameter θ obtained from the ith such sample, then the variance due to systematic sampling $\mathsf{var}[H_i]$ may be estimated from $H_1,\ldots,H_m$ by the standard formulae for a simple random sample without replacement. This is the context in which ANOVA may be applied to nested cluster sampling [83, Chapter 9], [555, pp. 121–123].

13.6 Particle number

Stereological estimators of the total number N or numerical density N_V of a particle population were presented in Chapters 10 and 11.

There are very few theoretical results for the variance of such estimators. For those estimators which use the principles of 'discrete sampling', such as the fractionator, the variance may not even be a well-defined quantity, because the design requires the experimenter to divide the material arbitrarily into discrete pieces for random selection at each sampling stage. The only viable approach seems to be statistical modelling of the data.

At the lowest sampling level, where individual particles are counted by the disector and other unbiased counting rules, familiar statistical models for count data may be applied. Suppose the unbiased estimator of total particle number is

$$\widehat{N} = a \sum_i Q_i \tag{13.50}$$

where Q_i are the counts from individual disectors, and a is the appropriate inverse sampling fraction. A reasonable starting-point is to model the Q_i as Poisson random variables, conditionally independent given the outcome of the sampling design $\mathcal{S}$. Then we immediately have

$$\begin{aligned} \mathsf{var}[\widehat{N}] &= \mathsf{var}[\mathbf{E}[\widehat{N} \mid \mathcal{S}]] + \mathbf{E}\left[\mathsf{var}[\widehat{N} \mid \mathcal{S}]\right] \\ &= a^2 \mathsf{var}[\sum_i \mathbf{E}[Q_i \mid \mathcal{S}]] + a^2 \mathbf{E}\left[\sum_i \mathbf{E}[Q_i \mid \mathcal{S}]\right] \\ &= a^2 \mathsf{var}[\sum_i \mathbf{E}[Q_i \mid \mathcal{S}]] + aN \end{aligned} \tag{13.51}$$

since $\mathsf{var}[Q_i \mid \mathcal{S}] = \mathbf{E}[Q_i \mid \mathcal{S}]$.

Given the sample, the expected count $\mu_i = \mathbf{E}[Q_i \mid \mathcal{S}]$ can be regarded as the value of the 'measurement function' for the ith sampled item, and Q_i as an inaccurate estimate of μ_i. Thus the first term on the right-hand side of (13.51) is analogous to the variance due to sampling of a measurement function, while the second term is due to observation error. The variance formulae given in section 13.2 can be applied to the first term.

Variances of estimators of particle number based on systematic sampling have been discussed in [80, 125]. Suppose we take a systematic sample of parallel slabs $S_1, S_2, \ldots$ of constant thickness t and constant spacing t/a through the specimen X, and let Q_i be the total disector count for slab i. Suppose it is reasonable to assume a constant average density of particles. Then the expected disector count is proportional to the slab's volume,

$$\mathbf{E}[Q_i \mid S_i] = N_V \, V(S_i \cap X).$$

When t is small relative to the diameter of X we may approximate $V(S_i \cap X) \approx t A(T_i \cap X)$ where T_i is one of the two boundary planes of the slab S_i. Then (13.51) becomes

$$\begin{aligned} \mathsf{var}[\widehat{N}] &= (N_V)^2 \mathsf{var}[at \sum_i A(T_i \cap X)] + aN \\ &= \frac{N^2}{V(X)^2} \mathsf{var}[at \sum_i A(T_i \cap X)] + aN \end{aligned}$$

or

$$\frac{\mathsf{var}[\widehat{N}]}{N^2} = \frac{\mathsf{var}[at \sum_i A(T_i \cap X)]}{V(X)^2} + \frac{a}{N}. \tag{13.52}$$

The first term is the relative variance of the Cavalieri estimator of $V(X)$, which can be predicted by the methods of Section 13.2. The second term is the variance due to Poisson variability, which may be estimated by $a/\widehat{N} = a^2/\sum_i Q_i$.

The assumption of homogeneity is not necessary in the analysis above. In the general case, suppose there is a spatially-variable density of particles $\mu(x), x \in X$. Then

$$\mathbf{E}[Q_i \mid S_i] = \int_{S_i} \mu(x)\,\mathrm{d}x.$$

This becomes the measurement function and a representation similar to (13.52) holds.

Chia [80] developed a completely data-based approach to variance estimation, by combining the Poisson model with nonlinear curve estimation (Section 13.2.7). Standard software for generalised linear models (say) is used to fit the model

$$Q_i \sim \mathsf{Poisson}(f(ati)) \tag{13.53}$$

where the measurement function $f(x)$ is a smooth, positive real-valued function of position x, perhaps given in parametric form, or a spline with a smoothing penalty. An estimate of the covariogram g is obtained by plugging the fitted measurement function $f^\dagger$ into (13.14), and then we estimate the systematic sampling variance for arbitrary section spacing t by the usual relation (13.15).

Figure 13.6 from [80] shows the disector counts Q_i from a systematic subsample of serial sections, plotted against the original section number, in the hippocampus for each of several rats. To each dataset the model (13.53) has been fitted, where $\log f$ is quadratic. The fitted mean curve is shown for each dataset.

Figure 13.7 shows the result of variance prediction for a typical dataset. The *Zitterbewegung* in the predicted variance may be erroneous, because it depends unstably on the exact diameter of the specimen. Instead, the upper envelope of the predicted variance is computed by taking the convex hull of the graph. Note that the variance contribution from Poisson variability far exceeds that from systematic sampling.

The validity of the Poisson model should be checked. Chia [80] developed goodness-of-fit tests for Poisson models with constant and non-constant mean, and applied them to disector counts of neurons in the hippocampus. The Poisson model with constant mean was appropriate for some datasets. Other datasets showed evidence of overdispersion (which could be explained by a non-constant mean) or underdispersion relative to the Poisson distribution.

In the very recent paper [129], variance estimation for estimators from fractionator designs (Section 10.6) is studied. A new formula, see [129, (2.27)], incorporating the known sampling fraction and local error effects is proposed. This formula is meant as a replacement of a formula available from [122].

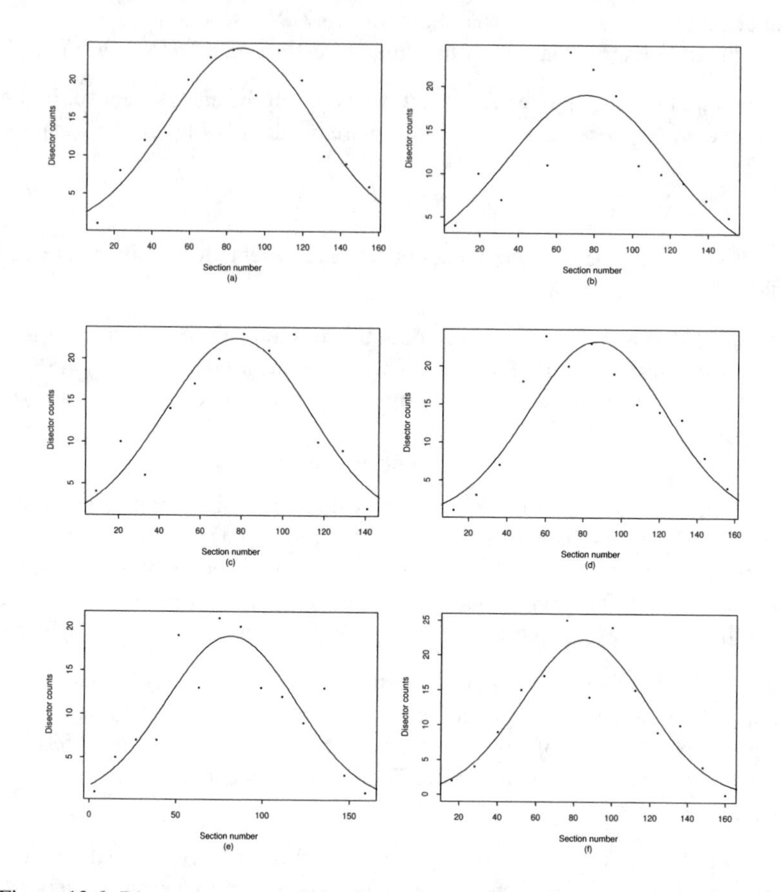

Figure 13.6 *Disector counts and fitted mean count according to model (13.53), from [80]. Reproduced by kind permission of Dr. J.L.C. Chia. Copyright © 2002 J.L.C. Chia.*

13.7 Advice to consultants

Variance is the most widely misunderstood issue in applications of stereology. There is often an expectation that a simple variance formula must exist (analogous to σ^2/n for independent replication) and the consultant will need to explain why it does not exist in general. It is also important to explain the distinction between the true variance and an estimator of the variance, and to stress that a variance estimate may be inaccurate.

In experimental design it is wise to encourage replication at all levels of sampling. Replication is the only reliable way to obtain variance estimates without

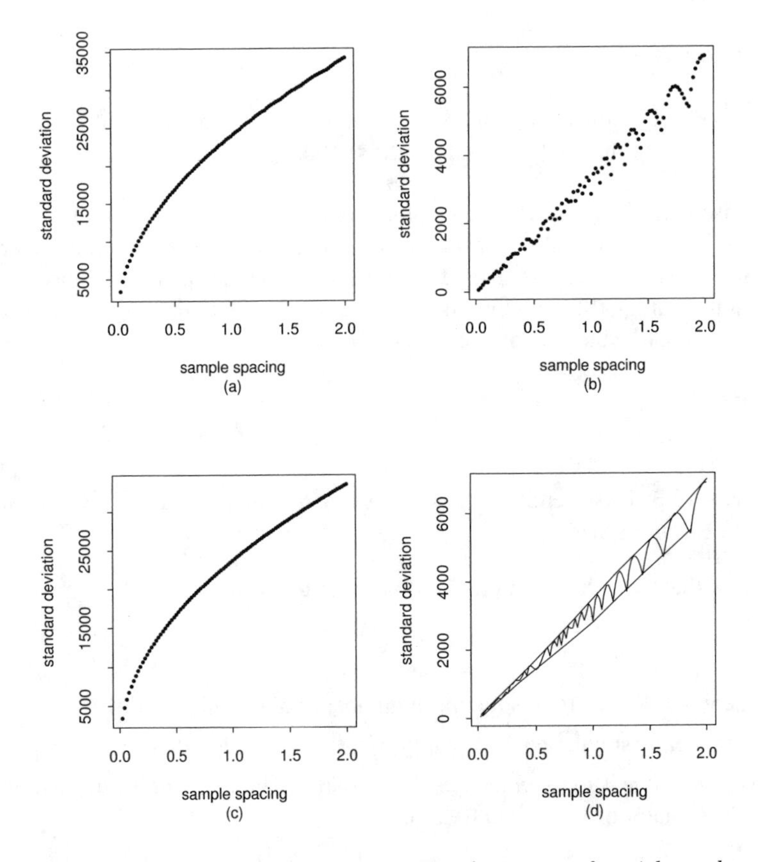

Figure 13.7 *Estimated contributions to variance of estimator of particle number, plotted against section spacing. (a): total variance; (b): systematic sampling contribution; (c): Poisson contribution; (d) upper and lower envelopes of (b). Reproduced by kind permission of Dr. J.L.C. Chia. Copyright © 2002 J.L.C. Chia.*

some model assumptions (of course, independence of replicates is also an assumption).

If stereological estimates of a parameter are computed for each of several animals, these estimates can often be regarded as independent replications, but of course the sample variance of the individual animal estimates is an overestimate of the biological variance. The observed variability between estimates for different animals is the true biological variability plus the sampling variance contribution.

13.8 Exercises

Practical exercises

Exercise 13.1 Estimate V_V and S_V in Figure 2.5 by hand, and calculate the variance approximation (13.7) using $32\pi \approx 100$.

Exercise 13.2 In Exercise 3.2 the volume of a banana was estimated from serial sections using Cavalieri's principle and point counting. Repeat this experiment and estimate the variance of the estimator (a) ignoring the variance due to point-counting, (b) including the variance due to point-counting, calculated by approximating the banana sections by circles.

Theoretical exercises

Exercise 13.3 This exercise concerns the variance approximation (13.24). We let Y be the unit sphere.

1) Show that the variance approximation takes the form

$$\mathrm{Var}\widehat{V}(Y) \approx \frac{8}{45}\pi^2 n^{-4},$$

where $n = \mathbf{E}N(Y \cap \mathcal{T})$ is the mean number of planes hitting Y.

2) Examine, by simulation, the quality of the variance approximation.

3) An alternative is to use n independent parallel planes with uniform position. The estimator of $V(Y)$ then becomes

$$\tilde{V}(Y) = \frac{1}{n}\sum_{i=1}^{n} \widehat{V}_i(Y),$$

where

$$\widehat{V}_i(Y) = 2 \cdot \pi(1 - U_i^2)$$

and $U_1, \ldots, U_n$ are independent and uniformly distributed in $[0,1]$. Find an expression for $\mathrm{Var}\tilde{V}(Y)$ and compare with $\mathrm{Var}\widehat{V}(Y)$.

Exercise 13.4 Find an approximation to the squared coefficient of variation for the area estimator $\widehat{A}(Y)$ based on point-counting. Express the result in terms of the mean number of points hitting Y, $n = \mathbf{E}N(Y \cap \mathcal{G})$.

Exercise 13.5 Suppose Y is the union of 3 convex curves in the plane with total length 1 metre. Evaluate the upper bound, (13.37) plus (13.38), for the variance of the Steinhaus estimator of the length $L(Y)$ length based on the intersection count with two perpendicular sets of test lines at spacing 5 cm.

13.9 Bibliographic notes

Delesse [156] and Rosiwal [486] offered theoretical and empirical statements about the precision of their methods. They derived absolute bounds for the error by deterministic arguments. Rosiwal [486] concluded that a high density of lines was required. Early studies of the number of lattice points in a plane region were also purely deterministic, see [609].

The calculation of moments of A_A and V_V using the indicator function I originates with Kolmogorov [318, p. 41] but is often attributed to Robbins [480, 481, 482]. See also [64, 404, 524, 525], [309, pp. 109-114], [514, chap. IX]. Similar formulae appear in [479, chap. 3], [102, 358], [435, pp. 52–53]. An important paper by Lantuéjoul [339] relates the sampling variance to the integral range.

Cartographers use point counting to estimate the (absolute) area of regions or 'parcels' on a map [350, Chap. 18]. The earliest description appears to be in forestry applications [2]. The accuracy of point-counting methods was investigated theoretically and empirically by cartographers from the 1950's onward [38, 58, 184, 263, 346, 355, 615]. Frolov and Maling [184] related the variance to the boundary length.

Geologists and metallographers have used the binomial approximation to the variance of point counts since the 1950's and also assessed its validity [78, 199, 261]. Chayes [78, p. 42] showed that these estimates are in good agreement with the sample variance of point counts X_i obtained in replicated experiments in various rocks. Chayes [78] gives an extensive critical discussion of the accuracy of volume fraction estimates in rocks, including the influence of the grain size and measurement area on the (empirical) variance [78, chap. 8].

Empirical and theoretical studies as far back as the 1940's [78, 255, 261, 274] suggested that point counting may be more accurate than lineal intercept measurement, in some situations. Experimental comparisons of efficiency between point counting and area measurement include [274, 368]. Gundersen et al [226] compared point-counting with digitizer tablet estimates of area and found a systematic difference.

Statisticians have studied the variance of point-counting estimates of the area of a fixed set in two dimensions since the 1940's (notably B. Matérn [356, 357, 358], D.G. Kendall [306, 308], P.A.P. Moran [419, 420, 422, 424] and others [470, 179, 180, 181, 144, 145, 414, 448, 459]; and more recently G. Bach [22], D. Stoyan [539] and [42, 112, 160, 240, 282, 410], [358, chap. 3] [479, chap. 3]).

R.E. Miles and P.J. Davy's rigorous theory of stereological estimation [146, 147, 396, 401, 404, 408, 409, 410, 411] included a consideration of efficiency issues. See also [422, 424, 90, 514]. Efficiency questions have been pursued in statistics PhD theses by Davy [146], Dhani [158], Downie [165], Ohser [443] and Chia [80]. The variance in ratio estimation was studied by numerous writers, surveyed in Chapter 9.

The founder of geostatistics, G. Matheron, provided theoretical tools which allow the computation of variances for different types of geometric probes (lines, planes, slabs, quadrats, etc) in [360, 362, 363]. Matheron's transitive methods were 'rediscovered' in stereology in the 1980's prompting a flurry of papers on systematic sampling, including [120, 227, 305, 359, 370, 554]. During the last fifteen years, the two groups around Kiêu and Cruz-Orive have been main contributors to the design-based theory and practice of variance estimation for stereological estimators based on systematic sampling. Significant progress in the foundation of the methods were later made in the nineties by Kiêu and collaborators [312, 314, 315, 528]. See also [229]. Cruz-Orive and co-workers have contributed substantially to the literature on systematic sampling [119, 121, 122, 123, 125, 129, 131, 132, 188, 189, 190, 191, 187, 208, 209, 210, 211, 212, 386].

The role of the Rao-Blackwell theorem and Cartier's formula emerged from findings of Miles and Davy, especially [147, sect. 6], and was elucidated by Lantuéjoul [337, 338] and Baddeley & Cruz-Orive [32].

P. Gy [230] developed variance models in the sampling of particulate materials.

Estimators of variance for particle counting are still controversial, especially in neuroscience [253, 507, 596].

In finite population survey sampling, the variance of systematic sampling is discussed thoroughly in [83, pp. 207 ff.]. A systematic sample gives more precise estimators than a simple random sample iff the average sample variance of the unit response *within* systematic samples is *greater* than the population variance as a whole. That is, systematic sampling is precise when the units within the sample are heterogeneous.

The variance of systematic sampling in natural populations (including spatial systematic sampling) has been studied [83, pp. 221–223] and suggests that systematic sampling gives a consistent gain in efficiency over simple random sampling. The problem of (unexpected) periodicity has also been discussed in real examples [83, pp. 217–219].

CHAPTER 14

Frontiers and open problems

This chapter presents some current research issues in stereology, with emphasis on topics where further statistical research is expected to make a difference. The selection of material is clearly affected by our own personal views and engagement in stereological issues.

14.1 Variances

14.1.1 Needs

The research in design-based variance estimation in systematic sampling, cf. Chapter 13, has been focused on developing automatic procedures that are generally applicable without model checking. There is a need for further investigation of these variance estimation procedures and comparison with what can be obtained by probabilistic modelling (e.g. Gy's work [230, 465]), or statistical modelling and data analysis (Sections 13.5 and 13.6). Notice that there are different levels of modelling. One approach is to model the covariances directly, another to model the spatial structure and derive the resulting model for the covariances. The last approach has been discussed in a stereological context in [266].

14.1.2 Non-normal ANOVA

As emphasised in [80], the modelling of structured experiments has not yet been fully explored in stereology. It has earlier been suggested to model data obtained from stereological experiments by classical ANOVA for nested designs [214]. The assumption that the error term is normally distributed and has constant variance is, however, not satisfied for count data. An alternative approach [7], based on non-parametric maximum likelihood in experimental designs, has been tried out in [80].

14.1.3 Spatial covariance

Covariance modelling for spatial stochastic processes is an important issue that may imply new variance estimation procedures in stereology, see [102, Chap. 3] and [514, Chap. IX]. In recent years, models for nonstationary covariances have been suggested based on transformation of stationary covariances, cf. e.g. [496].

14.2 Inhomogeneous materials and tissues

14.2.1 Gradient structures

Gradient structures are inhomogeneous along a particular gradient direction but homogeneous perpendicular to that direction. For such structures, densities such as the volume or surface area density are spatially variable characteristics. In [236], it is shown that vertical sections parallel to the gradient direction can be used to estimate the surface area density for gradient surface processes. Gradient structures are studied more generally in [235] where a number of applications from materials science are discussed. Some of the spatial structures under investigation have a local scaling property, see also [234].

14.2.2 Modelling spatial inhomogeneity in organs

In [80], the modelling of inhomogeneity in organs is discussed. It is suggested to model data from structured stereological experiments, using generalised linear models. For instance, the experiments may involve sections perpendicular to an axis of inhomogeneity and the position of a section along the axis is used as explanatory variable in the generalised linear model.

14.3 Anisotropy and digital stereology

Stereological estimation of 3D anisotropy (e.g. surface orientation distribution, curve orientation distribution) has been studied using parametric methods from directional statistics, see [133, 369], and non-parametric methods, see [310] and references therein. A Bayesian method has recently been developed in [468]. Methods of estimating the orientation distribution of planar random sets from digitized versions of sets are discussed in [296, 311]. These methods belong to the emerging discipline of digital stereology, where stereological analysis of digital images in two and three dimensions is studied, see also [444, 514].

14.4 Spatial arrangement

A subfield of local stereology [293] is second-order stereology which in fact was developed before the general local methods, see e.g. [288]. Using second-order stereology, it is possible to get information about the spatial arrangement of the structure under study. One example is quantification of the phenomenon in neuroscience called satellitosis where small glial cells are distributed around neurons in the brain. Such a phenomenon has been described by a discretised version of the pair correlation function [174]. In statistical terms the correlation between two point processes is studied. In [371] and [379], second-order stereology has been applied in the study of the human mammary gland and the kidney. It is here of interest to quantify disease caused changes of the spatial arrangement of the tissue.

If it is possible to collect the required measurements, using optical sectioning, these methods of second-order stereology are superior to methods depending on specific assumptions about particle shape [244, 245].

14.5 Stereology of extremes

In recent years, there has been some interest in estimating stereologically the tail of a particle size distribution. From a practical point of view, this is clearly of interest since the property of a material may be related to the extreme particle sizes rather than the mean particle size. Some results are available for spherical and spheroidal particles [47, 264, 549]. It seems difficult to construct methods that do not rely on specific shape assumptions.

14.6 Fractal behaviour of surfaces

The general methodology relating to microscopy for estimating the dimension of a fractal sets [351] has been laid down by Rigaut [333, 478] and others [67, 258, 460]. Fractals have been used to describe trabecular bone texture [176, 349, 457, 593], see also [106].

14.7 Non-recognition artefacts

In microscopy, the observer is faced with the problem that some observations are lost. The situation is analogous to non-response in surveys. One example is the non-recognition of a sectioned surface at some angles because of lack of contrast. Examples from particle analysis are glancing sections and lost caps (small particle sections which are not observed). A review for spherical particles is given in [113]. The stereology of membranes is often affected by an inability to recognise the membrane at certain orientations to the section plane, for example because of poor contrast. Curves may also be difficult to recognise at some angles, e.g. [328, pp. 196–197]

14.8 Reconciliation with other areas

An important task for statisticians is to reconcile stereological techniques with those used in other areas, especially spatial sampling [66, 65] modern survey sampling [555] and spatial statistics [159], and with the work of Gy [230, 465].

14.9 Shape models

A feature of recent research in stereology has been the development of flexible semi-parametric models that provide a framework for constructing optimal estimators and assessing variability of estimators. Very recently, a flexible parametric model for rotation invariant spatial particles has been suggested in [265]. The parameters of the model can be estimated from observations on central sections through the particles. This approach opens up the possibility of making stereological inference about the shape variability in a population of spatial particles, see also [246] and [267].

APPENDIX A

Sampling theory

This appendix lists some key concepts of survey sampling theory. For further details see the textbooks of Thompson [555] and Cochran [83]. The accessible, non-mathematical essay by Stuart [548] communicates the key ideas very well.

A.1 Population and sample

A.1.1 Sampling from a finite population

In classical sampling theory we are interested in studying a finite 'population' $\mathcal{P}$ of separate units, such as the human population of India, the books in a library, or the registered businesses in Florida. These will be called *sampling units* and arbitrarily numbered $1, 2, \ldots, N$ where the population size N may not be known.

For practical reasons we investigate the population $\mathcal{P}$ by taking a sample from it. A *random sample* is a randomly-chosen subset $\mathcal{S}$ of the population $\mathcal{P}$. The goal is to draw conclusions about $\mathcal{P}$ from information that we can observe in the subset $\mathcal{S}$.

A.1.2 Sampling designs

A random mechanism for selecting the sample is called a *sampling design*. Two examples of sampling designs are the following.

In **simple random sampling** ('without replacement') we select a fixed number n of distinct units from the population, in such a way that each possible subset of size n is equally likely to be the sample selected. One may think of dealing a hand of n cards from a well-shuffled pack of N cards. A simple random sample is usually obtained by picking the units one-at-a-time. The first unit is chosen at random from $\mathcal{P}$ with equal probability $1/N$ for each unit. The subsequent units are chosen at random from amongst the *remaining* population (the units not already selected) with equal probability for each unit.

Figure A.1 *Simple random sample.*

In **systematic random sampling** we choose every kth item in the population where k is the 'sampling period'. We start by choosing one of the integers 1 to k at random with equal probability (the 'random start'). Then the sample consists of the units numbered $m, m+k, m+2k, \ldots$. See Figure A.2.

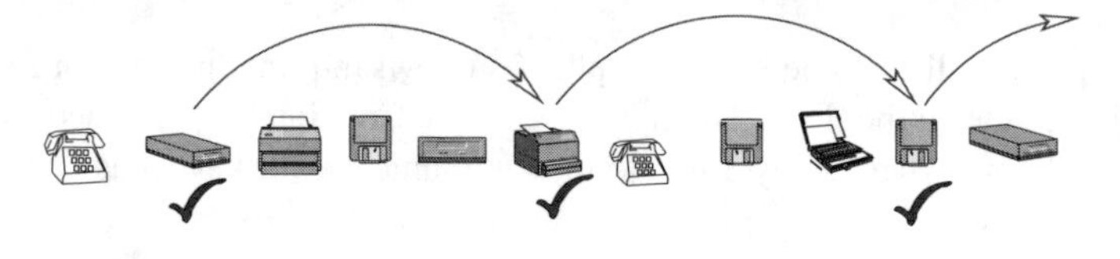

Figure A.2 *Systematic random sample of period* 4.

Other examples of sampling designs are given in Section A.3.

A.1.3 Sampling probability and sampling bias

The most important attribute of a sampling design is the *sampling probability* of each unit. The probability that the ith unit is selected in the sample is

$$p_i = \mathbf{P}[i \in \mathcal{S}].$$

For example, in simple random sampling of size n, the sampling probability for a particular unit i is

$$p_i = \binom{N-1}{n-1} / \binom{N}{n} = \frac{n}{N}$$

since there are $\binom{N}{n}$ possible samples of size n, all equally likely, and amongst these there are $\binom{N-1}{n-1}$ samples that contain unit i.

Note that the sampling probabilities p_i do not form a probability distribution on $\mathcal{P}$, since these events are not exclusive, nor do they specify the probability distribution of the random sample.

The sampling design is said to have *uniform sampling probability* if the sampling probabilities p_i are all equal. Otherwise the sampling design is *non-uniform*. For example, simple random sampling is uniform since $p_i = n/N$ for all i. Systematic random sampling also has uniform sampling probability, $p_i = 1/k$, since unit i will be sampled if and only if $i = m \pmod{k}$ where m

is uniform on $1, 2, \ldots, k$. We have found Figure A.3 useful for explaining this fact.

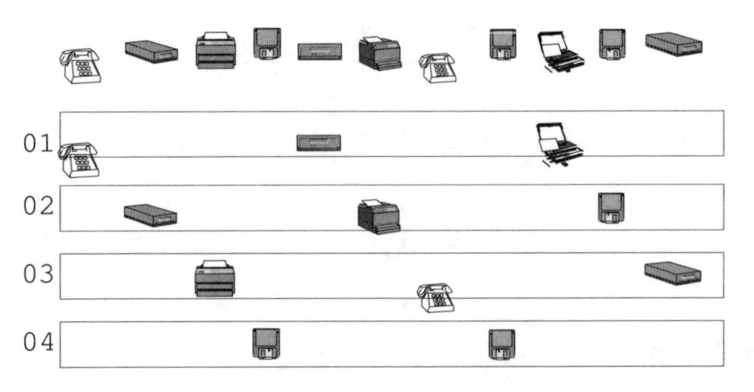

Figure A.3 *Graphical explanation for the fact that systematic sampling has uniform sampling probability. For a systematic random sample of period 4, all possible outcomes are shown. Each unit in the population is selected in precisely one of the 4 possible samples. So each unit has equal chance* $1/4$ *of being selected.*

A sampling design exhibits *sampling bias* if the actual sampling probabilities are different from the desired or intended probabilities. Most often the implicit intention is to sample with equal probability, so that sampling bias occurs if the sampling design is non-uniform. Causes of sampling bias are discussed by Cochran [83, chap 13].

A.2 Estimation from a sample

Suppose there is a quantity x_i associated with sampling unit i and we are interested in estimating the population total

$$X = \sum_{i=1}^{N} x_i \tag{A.1}$$

or the population mean $\bar{X} = X/N$.

If the random sample $\mathcal{S}$ has uniform sampling probability p, then an unbiased estimator of the population total X is the rescaled sample total,

$$\widehat{X} = \frac{1}{p} \sum_{i \in \mathcal{S}} x_i. \tag{A.2}$$

Note this requires that the sampling probability p be known and positive. The estimator is unbiased because

$$\mathbf{E}\left[\widehat{X}\right] = \frac{1}{p}\mathbf{E}\left[\sum_{i \in \mathcal{S}} x_i\right]$$

$$
\begin{aligned}
&= \frac{1}{p}\mathbf{E}\left[\sum_{i=1}^{N} \mathbf{1}\{i \in \mathcal{S}\} x_i\right] \\
&= \frac{1}{p}\sum_{i=1}^{N} x_i \mathbf{E}\left[\mathbf{1}\{i \in \mathcal{S}\}\right] \\
&= \frac{1}{p}\sum_{i=1}^{N} x_i \mathbf{P}\left[i \in \mathcal{S}\right] \\
&= \sum_{i=1}^{N} x_i \\
&= X
\end{aligned}
$$

since $\mathbf{P}[i \in \mathcal{S}] = p$ for all i.

The corresponding estimator of the population mean $\bar{X}$,

$$\bar{x} = \frac{1}{|\mathcal{S}|}\sum_{i \in \mathcal{S}} x_i,$$

(where $|\mathcal{S}|$ is the sample size) is generally a biased estimator of the population mean $\bar{X}$.

For example, a systematic random sample with period k, the sampling probability $p = 1/k$ is known and positive, so the rescaled sample total

$$\widehat{X} = k\sum_{i \in \mathcal{S}} x_i$$

is an unbiased estimator of the population total X. The sample mean $\bar{x}$ is biased for the population mean $\bar{X}$.

In a simple random sample of size n, since the sample size n is fixed, the sample mean $\bar{x}$ is an unbiased estimator of the population mean $\bar{X}$. To estimate the population total X, we need to know the population size N to determine the sampling probability $p = n/N$. In that case

$$\widehat{X} = \frac{N}{n}\sum_{i \in \mathcal{S}} x_i = N\bar{x}$$

is an unbiased estimator of X.

A.3 Sampling designs

A.3.1 Stratified sampling

Stratified sampling is a technique for reducing the variability of estimators of a population parameter. The population $\mathcal{P}$ is partitioned into disjoint subsets ('strata') $\mathcal{P}_1, \ldots, \mathcal{P}_m$. Each stratum $\mathcal{P}_i$ is then treated as a separate population, and sampled. Parameter estimates for each stratum, obtained from these samples, are combined to yield a parameter estimate for the original population $\mathcal{P}$.

The benefit of stratification is that we may use a different sampling fraction in each stratum. For example, to estimate the total wage bill of all factories in Spain, it may be prudent to stratify factories into large and small employers. The stratum of large employers has a comparatively large contribution to the total wage bill, so it should be sampled with a higher sampling fraction to ensure it is estimated more accurately.

Let $\mathcal{S}_i$ be the random sample from stratum $\mathcal{P}_i$ for $i = 1, \ldots, m$. It is generally assumed that $\mathcal{S}_1, \ldots, \mathcal{S}_m$ are independent. If the goal is to estimate the population total X then we first estimate each stratum total

$$X_{(i)} = \sum_{j \in \mathcal{P}_i} x_j$$

by an unbiased estimator $\widehat{X}_{(i)}$ based on the sample $\mathcal{S}_i$. Then

$$\widehat{X}^{\text{str}} = \sum_{i=1}^{m} \widehat{X}_{(i)}$$

is an unbiased estimator of X.

A.3.2 Cluster sampling

Cluster sampling is often a convenient way to generate a random sample with uniform sampling probability from a large or complex population. We divide the population $\mathcal{P}$ into a finite number of disjoint subsets or 'clusters' $\mathcal{P}_1, \ldots, \mathcal{P}_K$ in any convenient fashion. We then treat these clusters as the sampling units, and select a random sample of the clusters. Equivalently we choose a random sample $\mathcal{K}$ of the indices $1, 2, \ldots, K$. The clusters $\mathcal{P}_k$ for $k \in \mathcal{K}$ are sampled.

The cluster sample $\mathcal{S}$ of the original population $\mathcal{P}$ consists of all units belonging to the sampled clusters:

$$\mathcal{S} = \bigcup_{k \in \mathcal{K}} \mathcal{P}_k.$$

The basic principle of cluster sampling is that, if all clusters had equal probability of being selected, then each unit in the population has equal probability of being selected in the cluster sample $\mathcal{S}$. That is, if

$$\mathbf{P}[k \in \mathcal{K}] = p \quad \text{for all } k = 1, 2, \ldots, K$$

then

$$\mathbf{P}[x \in \mathcal{S}] = p \quad \text{for all } x \in \mathcal{P}.$$

To see this, observe that any unit x in the population $\mathcal{P}$ belongs to exactly one of the clusters $\mathcal{P}_k$, say $x \in \mathcal{P}_{k^*}$, since the clusters form a partition of $\mathcal{P}$. So x will be sampled if and only if $\mathcal{P}_{k^*}$ is selected, and this has probability p.

In the literature, cluster sampling is usually presented as a technique for sampling from a finite population $\mathcal{P}$. However, the arguments above did not assume that $\mathcal{P}$ was finite, and the cluster sampling principle applies to an arbitrary set $\mathcal{P}$ divided into a finite number of subsets.

A.4 Nonuniform sampling designs

A.4.1 PPS sampling

Probability proportional to size (PPS) sampling is often used as a remedy for estimation bias in the sample mean, and may also increase efficiency.

Consider a random sample $\mathcal{S}$ from the population $\mathcal{P}$ obtained by a sampling design with uniform sampling probability. As we saw in Section A.2, the sample mean $\bar{x}$ is generally a biased estimator of the population mean $\bar{X}$, unless the sample size is constant.

We modify the sampling design as follows. Let $p(S) = \mathbf{P}[\mathcal{S} = S]$ be the probability that the original sampling mechanism yields the sample $S \subset \mathcal{P}$. Then define

$$q(S) = \frac{|S|\, p(S)}{\sum_{T \subseteq \mathcal{P}} |T|\, p(T)}.$$

where $|S|$ denotes the size of the sample S. Then q is a probability distribution over the set of all possible subsets of $\mathcal{P}$. The original probabilities $p(S)$ have been weighted by sample size $|S|$ and renormalised. A random sampling design which has probability $q(S)$ of generating the sample S is called a PPS design relative to the original sampling design.

Under PPS sampling, the sample mean is unbiased:

$$\begin{aligned}
\mathbf{E}[\bar{x}] &= \sum_{S \subset \mathcal{P}} q(S) \frac{1}{|S|} \sum_{i \in S} x_i \\
&= \sum_{S \subset \mathcal{P}} \frac{|S|\, p(S)}{\sum_{T \subseteq \mathcal{P}} |T|\, p(T)} \frac{1}{|S|} \sum_{i \in S} x_i \\
&= \frac{\sum_{S \subset \mathcal{P}} p(S) \sum_{i \in S} x_i}{\sum_{T \subseteq \mathcal{P}} |T|\, p(T)} \\
&= \frac{pX}{pN} \\
&= \bar{X}.
\end{aligned}$$

A.4.2 Horvitz-Thompson estimator

The Horvitz-Thompson estimator [270] allows unbiased estimation of population totals from samples which have any non-uniform sampling probabilities, as long as the sampling probabilities are known. See [555, pp. 49–53], [83, pp. 259–261], [324, pp. 291, 313, 428], [62].

Suppose we observe a (non-independent, non-uniform) random sample $\mathcal{S}$ from $\mathcal{P}$. The sample size may be random; the only condition is that $\pi_i = \mathbf{P}[i \in \mathcal{S}]$ be known and positive for all i. Then we can estimate any population total by weighting each element i in the sample by $1/\pi_i$. The **Horvitz-Thompson estimator**

$$\widehat{X}_{HT} = \sum_{i \in \mathcal{S}} \frac{x_i}{\pi_i} \tag{A.3}$$

is unbiased for the population total X since

$$\begin{aligned} \mathbf{E}[\,]\widehat{X}_{HT} &= \mathbf{E}\left[\sum_{i=1}^{N} \mathbf{1}\{i \in \mathcal{S}\} \frac{x_i}{\pi_i}\right] \\ &= \sum_{i=1}^{N} \pi_i \frac{x_i}{\pi_i} \qquad = X. \end{aligned} \tag{A.4}$$

Actually the sampling probabilities π_i need only be known for the *sampled* items. They must be known exactly.

The poor performance of H-T estimator in adverse examples is shown by Basu's 'circus elephant' example [39, pp. 212–213].

A.5 Variances of estimators

It is well known that the variance of the sample mean $\bar{x}$, from a simple random sample with replacement, is equal to σ^2/n where σ^2 is the population variance and n is the sample size. The population variance σ^2 can be estimated from the sample: an unbiased estimator of σ^2 is the sample variance

$$S^2 = \frac{1}{n-1} \sum_i (x_i - \bar{x})^2.$$

Consequently we have an estimate of the variance of $\bar{x}$ from the sample, namely S^2/n. Indeed we can predict, from data in a sample of size n, the variance of the sample mean from a sample of size m.

For other sampling designs, the variances of estimators depend on the population in a more complex way. It may not be possible to estimate these variances from the sample, nor to predict the variance if the sample size is changed.

Under systematic sampling with period k, the variance of the estimator of the sample total is

$$\mathsf{var}[\widehat{X}] = \mathsf{var}[k\sum_{i\in S} x_i] = \frac{1}{k}\sum_{m=1}^{k}\left(k\sum_i x_{m+ik} - X\right)^2 .$$

In a simple random sample of size $n = N/k$ (the same average sample size), the variance of the estimator of the sample total is

$$\mathsf{var}[N\bar{x}] = N^2\frac{\sigma^2}{n} = kN\sigma^2,$$

so systematic sampling is more efficient than simple random sampling (with the same average sample size) if $\tau^2 < \sigma^2/n$ where

$$\tau^2 = \frac{1}{k}\sum_{m=1}^{k}\left(\frac{1}{n}\sum_i x_{m+ik} - \mu\right)^2$$

writing $\mu = X/N$ for the population mean. The expression for τ^2 is analogous but not identical to the variance of the sample mean in systematic samples. The total sum of squares for the entire population can be decomposed as

$$\sum_j (x_j-\mu)^2 = \sum_{m=1}^{k}\sum_i (x_{m+ik}-\bar{x}_m)^2 + n\sum_{m=1}^{k}(\bar{x}_m-\mu)^2$$

where $\bar{x}_m = \frac{1}{n}\sum_i x_{m+ik}$. The two terms on the right are the within-sample and between-sample sums of squares, respectively. Noting that τ^2 is proportional to the between-sample sum of squares, we find that systematic sampling is most efficient when the within-sample sum of squares is as *large* as possible. Thus, systematic sampling is most efficient when the units in a sample are as *heterogeneous* as possible. Similar arguments apply to cluster sampling [555, pp. 120–123].

Systematic sampling performs most poorly when the population responses x_i have inherent periodicity, the period being close to the period k of the systematic sample.

List of notation

Sets

$\mathbf{R}^d$	$d-$dimensional Euclidean space
$\mathbf{R}$	$= \mathbf{R}^1$, real numbers
O	The origin of $\mathbf{R}^d$
$\mathbf{R}_+$	Positive real numbers
$[a,b]$	Closed line-segment between $a \in \mathbf{R}^d$ and $b \in \mathbf{R}^d$
(a,b)	Open line-segment between $a \in \mathbf{R}^d$ and $b \in \mathbf{R}^d$
$[a,b)$	Half-closed-open line-segment between $a \in \mathbf{R}^d$ and $b \in \mathbf{R}^d$
$B_d(x,R)$	Closed ball in $\mathbf{R}^d$ with centre $x \in \mathbf{R}^d$ and radius $R \geq 0$
B^d	Unit ball in $\mathbf{R}^d$
S^{d-1}	Unit sphere in $\mathbf{R}^d$
S_+^{d-1}	$\{(\omega_1,\dots,\omega_d) \in S^{d-1} : \omega_d \geq 0\}$
T	m-dimensional plane in $\mathbf{R}^d$, the important examples are lines in $\mathbf{R}^2$ and lines and planes in $\mathbf{R}^3$
$V_{\theta,p}$	Vertical plane with orientation $\theta \in [0,\pi)$ and position $p \in \mathbf{R}$, see Chapter 8
E_ω	Plane through O with unit normal vector $\omega \in S^2$
Q	'Quadrat': closed bounded m-dimensional subset of $\mathbf{R}^d$.
$[Y]$	$= \{T : Y \cap T \neq \emptyset\}$, hitting set (page 115)
Convex	A set $Y \subseteq \mathbf{R}^d$ with the property $[y_1,y_2] \subseteq Y$ for all $y_1, y_2 \in Y$
Star-shaped	A set $Y \subseteq \mathbf{R}^d$ is said to be star-shaped relative to $y_0 \in Y$ if $[y_0,y] \subseteq Y$ for all $y \in Y$

Inner product and distances

$x \cdot y$	$= \sum x_i y_i, x,y \in \mathbf{R}^d$
$d(x,y)$	$= [(x-y)\cdot(x-y)]^{1/2}, \quad x,y \in \mathbf{R}^d$

Concepts relating to lines

$\omega^{\perp}$ The orthogonal complement of the line through O with direction $\omega \in S^2$

p_{ω} The orthogonal projection onto the line through O with direction $\omega \in S^2$

Matrices

I_d $d \times d$ identity matrix

$SO(d)$ The group of rotations in $\mathbf{R}^d$, special orthogonal transformations

Set operations

$X \oplus Y$ Minkowski sum $X \oplus Y = \{x + y : x \in X, y \in Y\}$ for $X, Y \subseteq \mathbf{R}^d$

$\check{X}$ reflection $\check{X} = (-1)X = \{-x : x \in X\}$ for $X \subseteq \mathbf{R}^d$

$X \oplus \check{Y}$ Dilation of X by Y

$X \ominus Y$ Minkowski subtraction $X \ominus Y = (X^c \oplus Y)^c$

$X \ominus \check{Y}$ Erosion of X by Y

$\mathsf{conv}(X)$ Convex hull, the smallest convex set containing X

∂X Boundary of X

$\mathsf{star}(X, O)$ star set: the part of X which can be seen directly from O (page 195)

Abbreviations

UR Uniform random

FUR Fixed orientation uniform random

IUR Isotropic uniform random

IR Isotropic random

Geometric quantities

V_k k-dimensional volume in $\mathbf{R}^d$

V Volume in $\mathbf{R}^3$

S Surface area in $\mathbf{R}^3$

M Integral of mean curvature in $\mathbf{R}^3$

L Length of space curve in $\mathbf{R}^3$

N Number of units in $\mathbf{R}^3$

A Area in $\mathbf{R}^2$

B Length of boundary or curve in $\mathbf{R}^2$

K Total curvature in $\mathbf{R}^2$

Q, N Number of profiles or units in $\mathbf{R}^2$

L Length in $\mathbf{R}^1$

I Number of (intersection) points

P Number of test points

H Caliper diameter or projected height in fixed direction

$\bar{H}$ The mean of H over all orientations

Functions

$\mathbf{1}\{\cdot\}$ Indicator function, equals 1 if statement is true and 0 otherwise

$\Gamma(\cdot)$ The gamma function

Statistics

$\mathbf{E}(X)$	Mean of random variable X
$\mathsf{var}[X]$	Variance of random variable X
$\mathsf{cov}(X, Y)$	Covariance between random variables X and Y
$\mathsf{Unif}(Z)$	Uniform distribution on Z (page 112)

Particle mean parameters

$\mathbf{E}_N[V]$ Mean particle volume

$\mathbf{E}_N[S]$ Mean particle surface area

$\mathbf{E}_N[H]$ Mean particle height

$\mathbf{E}_V[\varphi]$ Volume-weighted mean particle size φ, $= \mathbf{E}_N[V\varphi]/\mathbf{E}_N[V]$

References

[1] E.A. Abbott. *Flatland: a romance of many dimensions.* Blackwell, revised 6th edition, 1950.

[2] C.A. Abell. A method of estimating area in irregularly-shaped and broken figures. *Journal of Forestry*, 37:344–345, 1939.

[3] M. Abercrombie. Estimation of nuclear population from microtome sections. *Anatomical Record*, 94:239–247, 1946.

[4] M. Abramowitz and I.A. Stegun. *Handbook of mathematical functions.* Dover Publications, New York, 1965.

[5] H. Adam, G. Bernroider, and H. Haug, editors. *Fifth International Congress on Stereology*, Vienna and Munich, 1980. G Fromme. reprinted from Mikroskopie (Wien) vol 37 supplement 1980.

[6] H. Agarwal, A.M. Gokhale, S. Graham, and M.F. Horstemeyer. Quantitative characterization of three-dimensional damage evolution in a wrought Al-alloy under tension and compression. *Metallurgical and Materials Transactions A–Physical Metallurgy and Materials Science*, 33(8):2599–2606, 2002.

[7] M. Aitkin. A general maximum likelihood analysis of overdispersion in generalized linear models. *Statistics and Computing*, 6:251—262, 1996.

[8] G.C. Amstutz and H. Giger. Stereological methods applied to mineralogy, petrology, mineral deposits and ceramics. *Journal of Microscopy*, 95:145–164, 1972.

[9] B. Andersen. *Methodological errors in medical research.* Blackwell Scientific, Oxford, 1990.

[10] B.B. Andersen, H.J.G. Gundersen, and B. Pakkenberg. Aging of the human cerebellum: a stereological study. *J Comparative Neurol*, 466(3):356–365, November 2003.

[11] B.B. Andersen, L. Korbo, and B. Pakkenberg. A quantitative study of the human cerebellum with unbiased stereological techniques. *Journal of Comparative Neurology*, 326:549–560, 1992.

[12] R.S. Anderssen and A.J. Jakeman. Abel type integral equations in stereology, I. General discussion. *Journal of Microscopy*, 105:121–133, 1975.

[13] R.S. Anderssen and A.J. Jakeman. Abel type integral equations in stereology, II. Computational methods of solution and the random spheres approximation. *Journal of Microscopy*, 105:135–154, 1975.

[14] A. Antoniadis, J. Fan, and I. Gijbels. A wavelet method for unfolding sphere size distributions. *Canadian Journal of Statistics*, 29:251–268, 2001.

[15] S.M.G. Antunes. The optical fractionator. *Zoological Studies*, 34 Suppl I:180–183, 1995.

[16] M. Armstrong. *Basic linear geostatistics*. Springer Verlag, 1997.

[17] E. Artacho-Pérula and R. Roldán-Villalobos. Unbiased stereological estimation of the number and volume of nuclei and nuclear size variability in invasive ductal breast carcinomas. *Journal of Microscopy*, 186:133–142, 1997.

[18] E. Artacho-Pérula, R. Roldán-Villalobos, and L.M. Cruz-Orive. Application of the fractionator and vertical slices to estimate total capillary length in skeletal muscle. *Journal of Anatomy*, 195:429–437, 1999.

[19] E. Artacho-Pérula and R. Roldán-Villalobos. Volume-weighted mean particle volume estimation using different measurement methods. *Journal of Microscopy*, 173:73–78, 1994.

[20] ASTM E 112 – 96: Standard test methods for determining average grain size. American Society for Testing and Materials, 1996. Available at `www.astm.org`.

[21] G. Bach. Über die Bestimmung von charakteristichen Größen einer Kugelverteilung aus der unvollständigen Verteilung der Schnittkreise. *Metrika*, 9:228–233, 1965.

[22] G. Bach. Über die Auswertung von Schnittflächenverteilungen. *Biometrical Journal*, 18:407–412, 1976.

[23] A. Baddeley. Is stereology 'unbiased'? *Trends in Neurosciences*, 24:375–376, 2001.

[24] A.J. Baddeley. A fourth note on recent research in geometrical probability. *Advances in Applied Probability*, 9:824–860, 1977.

[25] A.J. Baddeley. Vertical sections. In W. Weil and R. V. Ambartzumian, editors, *Stochastic geometry and stereology (Oberwolfach 1983)*, Berlin, 1983. Teubner.

[26] A.J. Baddeley. An anisotropic sampling design. In W. Nagel, editor, *Geobild '85*, pages 92–97, Jena, DDR, 1985. Wissenschaftliche Publication, Friedrich-Schiller-Universität Jena.

[27] A.J. Baddeley. Stereology and image analysis for anisotropic plane sections. In *Proceedings of the 13th International Biometric Conference, Seattle 1987*, pages 1–9, 1987. section I-8.1.

[28] A.J. Baddeley. Stereology. In *Spatial Statistics and Digital Image Analysis*, chapter 10, pages 181–216. National Research Council USA, Washington DC, 1991.

[29] A.J. Baddeley. Stereology and survey sampling theory. *Bulletin of the International Statistical Institute*, 50, book 2:435–449, 1993.

[30] A.J. Baddeley. A crash course in stochastic geometry. In O.E. Barndorff-Nielsen, W.S. Kendall, and M.N.M. van Lieshout, editors, *Stochastic Geometry: Likelihood and Computation*, chapter 1, pages 1–35. Chapman and Hall, London, 1998.

[31] A.J. Baddeley. Spatial sampling and censoring. In O.E. Barndorff-Nielsen, W.S. Kendall, and M.N.M. van Lieshout, editors, *Stochastic Geometry: Likelihood and Computation*, chapter 2, pages 37–78. Chapman and Hall, London, 1998.

[32] A.J. Baddeley and L.M. Cruz-Orive. The Rao-Blackwell theorem in stereology and some counterexamples. *Advances in Applied Probability*, 27:2–19, 1995.

[33] A.J. Baddeley, H.J.G. Gundersen, and L.M. Cruz-Orive. Estimation of surface area from vertical sections. *Journal of Microscopy*, 142:259–276, 1986.

[34] V. Bagger. Direct estimation of volumes in ovarian follicles using the nucleator. *Acta Pathologica Microbiologica et Immunologica Scandinavica*, 101:784–790, 1993.

[35] F.A. Bahmer, S. Hantirah, and H.P. Baum. Rapid and unbiased estimation of the volume of cutaneous malignant melanoma using Cavalieri's principle. *Am J Dermatopath*, 18:159–164, 1996.

[36] G. Ballardini, S. Degli Esposti, F.B. Bianchi, L.B. de Giorgi, A. Faccani, L. Biolchini, C.A. Busachi, and E. Pisi. Correlation between Ito cells and fibrogenesis in an experimental model of hepatic fibrosis. A sequential stereological study. *Liver*, 3:58–63, 1983.

[37] E. Barbier. Note sur le problème de l'aiguille et le jeu du joint couvert. *J. Math. pures et appl.*, 5:273–286, 1860.

[38] J.P. Barrett and J.S. Philbrook. Dot grid area estimates: precision by repeated trials. *Journal of Forestry*, 68:141–151, 1970.

[39] D. Basu. An essay on the logical foundations of survey sampling, Part One. In V.P. Godambe and D.A. Sprott, editors, *Foundations of Statistical Inference*, pages 203–242. Holt, Rinehart and Winston, Toronto, 1971.

[40] S. Batra, M.F. Konig, and L.M. Cruz-Orive. Unbiased estimation of capillary length from vertical slices. *Journal of Microscopy*, 178:152–159, 1995.

[41] K.S. Bedi. Effects of undernutrition during early life on granule cell numbers in the rat dentate gyrus. *Journal of Comparative Neurology*, 211:425–433, 1991.

[42] D.R. Bellhouse. Area estimation by point-counting techniques. *Biometrics*, 37:303–312, 1981.

[43] D.R. Bellhouse. Systematic sampling. In P.R. Krishnaiah and C.R. Rao, editors, *Sampling*, volume 6 of *Handbook of statistics*, chapter 6. North-Holland, Amsterdam, 1988.

[44] T.F. Bendtsen and J.R. Nyengaard. Unbiased estimation of particle number using sections — an historical perspective with special reference to the stereology of glomeruli. *Journal of Microscopy*, 153:93–102, 1989.

[45] F.M. Benes and N. Lange. Reconciling theory and practice in cell counting. *Trends in Neurosciences*, 24:378–380, 2001.

[46] F.M. Benes and N. Lange. Two-dimensional versus three-dimensional cell counting: a practical perspective. *Trends in Neurosciences*, 24:11–17, 2001.

[47] V. Beneš, K. Bodlák, and D. Hlubinka. Stereology of extremes; bivariate models and computation. *Methodology and Computing in Applied Probability*, 5:289–308, 2003.

[48] V. Beneš and J. Rataj. *Stochastic geometry: selected topics*. Kluwer, 2004.

[49] J. Bertram, V. Nurcombe, and N. Wreford. *Stereological Methods for Biology*. Thomson Learning, 1997.

[50] J.F. Bertram. Counting in the kidney. *Kidney International*, 59(2):792–796, 2001.

[51] P.R. Billingsley and S.W. Ranson. On the number of nerve cells in the ganglion cervicale superius and of nerve fibers in the cephalic end of the truncus sympathicus in the cat and on the numerical relayionship of preganglionic and postganglionic neurones. *Journal of Comparative Neurology*, 29:359–366, 1918.

[52] M. Binder, A. Steiner, U. Mossbacher, M. Hunegnaw, H. Pehamberger, and K. Wolff. Estimation of the volume-weighted mean nuclear volume discriminates keratoacanthoma from squamous cell carcinoma. *American Journal of Dermatopathology*, 20:453–458, 1998.

[53] R. Bjugn. Estimation of the total number of cells in the rat spinal cord using the optical disector. *Micron and Microscopica Acta*, 22:25–26, 1991.

[54] R. Bjugn. The use of the optical disector to estimate the number of neurons, glial and endothelia cells in the spinal cord of the mouse — with a comparative note on the rat spinal cord. *Brain Research*, 627:25–33, 1993.

[55] R. Bjugn and H.J.G. Gundersen. Estimate of the total number of neurons and glial and endothelia cells in the rat spinal cord by means of the optical disector. *Journal of Comparative Neurology*, 328:406–414, 1993.

[56] D. Blackwell. Conditional expectation and unbiased sequential estimation. *Annals of Mathematical Statistics*, 18:105–110, 1947.

[57] W. von Blaschke. *Vorlesungen über Integralgeometrie.* Chelsea, New York, 1949.

[58] G.M. Bonnor. The error of area estimates from dot grids. *Can. J. For. Res.*, 5:10–17, 1975.

[59] A.E. Boycott. A case of unilateral aplasia of the kidney in a rabbit. *Journal of Anatomy and Physiology*, 4:20–22, 1911.

[60] H. Brændgaard, S.M. Evans, C.V. Howard, and H.J.G. Gundersen. The total number of neurons in the human neocortex unbiasedly estimated using optical disectors. *Journal of Microscopy*, 157:285–304, 1990.

[61] K.R. Brewer. Ratio estimation in finite populations: Some results deducible from the assumption of an underlying stochastic process. *Australian Journal of Statistics*, 5:93–105, 1963.

[62] K.R. Brewer and M. Hanif. *Sampling with unequal probabilities.* Number 15 in Lecture Notes in Statistics. Springer Verlag, New York, 1983.

[63] H.D. Brody. Organization of the cerebral cortex. III. A study of aging in the human cerebral cortex. *J. Comparative Neurology*, 1023:511–556, 1955.

[64] J. Bronowski and J. Neyman. The variance of the measure of a two-dimensional random set. *Annals of Mathematical Statistics*, 16:330–341, 1945.

[65] S.T. Buckland, D.R. Anderson, K.P. Burnham, J.L. Laake, D.L. Borchers, and L. Thomas. *Introduction to distance sampling: estimating abundance of biological populations.* Oxford University Press, July 2001.

[66] S.T. Buckland, S.R. Anderson, K.P. Burnham, and J.L. Laake. *Distance sampling: estimating abundance of biological populations.* Chapman and Hall, London, 1993. Book can be downloaded from: `http://www.ruwpa.st-and.ac.uk/distancebook/download.html`.

[67] S. Buczkowski, S. Kyriacos, F. Nekka, and L. Cartilier. The modified box-counting method: analysis of some characteristic parameters. *Pattern Recognition*, 31:411–418, 1998.

[68] Georges Louis Leclerc, Comte de Buffon. Essai d'arithmétique morale. In *Supplément à l'Histoire Naturelle*, volume 4. Imprimerie Royale, Paris, 1777.

[69] A.J. Cabo and A.J. Baddeley. Line transects, covariance functions and set convergence. *Advances in Applied Probability*, 27:585–605, 1995.

[70] A.J. Cabo and A.J. Baddeley. Estimation of mean particle volume using the set covariance function. *Stochastic Geometry and Statistical Applications (Advances in Applied Probability)*, 35:27–46, 2003.

[71] J.W. Cahn and R.L. Fullman. On the use of lineal analysis for obtaining particle size distribution functions in opaque samples. *Trans. AIME (Metallurgical)*, 197:610–612, 1956.

[72] H. Campbell and S.T. Tomkeieff. Calculation of the internal surface of a lung. *Nature*, 170:117, 1952.

[73] J.R. Casley-Smith. Expressing stereological results 'per cm^3' is not enough. *Journal of Pathology*, 156:263–265, 1988.

[74] A. Cauchy. *Oeuvres complètes.* Premier série, tome II. Gauthier-Villars, Paris, 1908.

[75] B. Cavalieri. *Geometria Indivisibilibus Continuorum.* Typis Clemetis Feronij, Bononi, 1635. Reprinted as Geometria degli Indivisibili. Torino : Unione Tipografico-Editorice Torinese, 1966.

[76] H.W. Chalkley. Method for quantitative morphological analysis of tissues. *J. National Cancer Institute*, 4:47–53, 1943.

[77] H.W. Chalkley, J. Cornfield, and H. Park. A method of estimating volume surface ratio. *Science*, 110:295–297, 1949.

[78] F. Chayes. *Petrographic modal analysis: an elementary statistical appraisal.* Wiley, New York, 1956.

[79] J.L. Chermant, editor. *Quantitative analysis of microstructures in biology, materials sciences and medicine.* Riederer, Stuttgart, 1978.

[80] J.L.C. Chia. *Data Models and Estimation Accuracy in Stereology.* PhD thesis, Department of Mathematics and Statistics, University of Western Australia, Perth, Australia, 2002.

[81] L.C.J. Chia and A.J. Baddeley. Accuracy of estimates of volume fraction. *Image Analysis and Stereology*, 19:199–204, 2000.

[82] J.-P. Chilès and P. Delfiner. *Geostatistics: modeling spatial uncertainty.* Wiley-Interscience, 1999.

[83] W. G. Cochran. *Sampling Techniques.* John Wiley and Sons, 3rd edition, 1977.

[84] R. Coggeshall. Commentary on the paper by Benes and Lange. *Trends in Neurosciences*, 24:376–377, 2001.

[85] R.E. Coggeshall. A consideration of neural counting methods. *Trends in Neurosciences*, 15(1):9–13, 1992.

[86] R.E. Coggeshall, K. Chung, D. Greenwood, and C.E. Hulsebosch. An empirical method for converting nucleolar counts to neuronal numbers. *Journal of Neuroscience Methods*, 12:125–132, 1984.

[87] R.E. Coggeshall, R. LaForte, and C.M. Klein. Calibration of methods for determining numbers of dorsal root ganglion cells. *Journal of Neuroscience Methods*, 35:187–194, 1990.

[88] R.E. Coggeshall and H.A. Lekan. Methods for determining numbers of cells and synapses: a case for more uniform standards of review. *Journal of Comparative Neurology*, 364:6–15, 1996. Erratum in J Comp Neurol 369 (1996) 16.

[89] R. Coleman. The stereological analysis of two-phase particles. In R.E. Miles and J. Serra, editors, *Geometrical Probability and Biological Structures: Buffon's 200th Anniversary*, Lecture Notes in Biomathematics, No 23. Springer Verlag, Berlin-Heidelberg-New York, 1978.

[90] R. Coleman. *An introduction to mathematical stereology.* Memoirs no 3. Department of Theoretical Statistics, University of Aarhus, Denmark, 1979.

[91] R. Coleman. The distribution of sizes of spheres from observations through a thin slice. In H Adam, G Bernroider, and H Haug, editors, *Fifth International Congress on Stereology.* G Fromme, Vienna and Munich, 1980. reprinted from Mikroskopie (Wien) vol 37 supplement 1980.

[92] R. Coleman. Size determination of transparent spheres in an opaque specimen

from a slice. *Journal of Microscopy*, 123:343–345, 1981.

[93] R. Coleman. The sizes of spheres from profiles in a thin slice. I. Opaque spheres. *Biometrical Journal*, 24:273–286, 1982.

[94] R. Coleman. The sizes of spheres from profiles in a thin slice. II. Transparent spheres. *Biometrical Journal*, 25:745–755, 1983.

[95] R. Coleman. Inverse problems. *Journal of Microscopy*, 153:233–248, 1989.

[96] C.B. Cordy. An extension of the Horvitz-Thompson theorem to point sampling from a continuous universe. *Statistics and Probability Letters*, 18:353–362, 1993.

[97] J. Cornfield and H.W. Chalkley. A problem in geometric probability. *J. Washington Acad. Sci*, 41:226–229, 1951.

[98] S. J. Corrsin. A measure of the area of a homogeneous random surface in space. *Quarterly of Applied Mathematics*, 12:404–408, 1955.

[99] D.R. Cox. Appendix to 'The dye sampling method of measuring fibre length distribution' by D.R. Palmer. *Journal of the Textile Institute*, 39:T8–T22, 1949.

[100] D.R. Cox. Some sampling problems in technology. In N.L. Johnson and H. Smith, editors, *New developments in survey sampling*, pages 506–527. John Wiley and Sons, 1969.

[101] D.R. Cox and D.V. Hinkley. *Theoretical statistics*. Chapman and Hall, London, 1974.

[102] N.A.C. Cressie. *Statistics for spatial data*. John Wiley and Sons, New York, 1991.

[103] M.W. Crofton. On the theory of local probability (etc.). *Philosophical Transactions of the Royal Society, London*, 158:181–199, 1869.

[104] M.W. Crofton. Sur quelques théorèmes du calcul intégral. *Comptes Rendus de l'Académie des Sciences de Paris*, 68:1469–1470, 1869.

[105] M.W. Crofton. Probability. In *Encyclopaedia Britannica*. 9th edition, 1885.

[106] S.S. Cross, S. Rogers, P.B. Silcocks, and D.W.K. Cotton. Trabecular bone does not have a fractal structure on light microscopic examination. *J. Pathology*, 170:311–313, 1993.

[107] L.M. Cruz-Orive. Particle size-shape distributions: the general spheroid problem. I. Mathematical model. *Journal of Microscopy*, 107:235–253, 1976.

[108] L.M. Cruz-Orive. Particle size-shape distributions: the general spheroid problem. II. Stochastic model and practical guide. *Journal of Microscopy*, 112:153–167, 1978.

[109] L.M. Cruz-Orive. Best linear unbiased estimators for stereology. *Biometrics*, 36:595–605, 1980.

[110] L.M. Cruz-Orive. On the estimation of particle number. *Mikroskopie*, 37 (Supplement):79–85, 1980.

[111] L.M. Cruz-Orive. On the estimation of particle number. *Journal of Microscopy*, 120:15–27, 1980.

[112] L.M. Cruz-Orive. The use of quadrats and test systems in stereology, including magnification corrections. *Journal of Microscopy*, 125:89–102, 1982.

[113] L.M. Cruz-Orive. Distribution-free estimation of sphere size distributions from slabs showing overprojection and truncation, with a review of previous methods. *Journal of Microscopy*, 131:265–290, 1983.

[114] L.M. Cruz-Orive. Estimating particle number and size. In L.F. Agnati and

K. Fuxe, editors, *Quantitative Neuroanatomy in Transmitter Research*, pages 11–24. MacMillan, London, 1985.

[115] L.M. Cruz-Orive. Estimating volumes from systematic hyperplane sections. *Journal of Applied Probability*, 22:518–530, 1985.

[116] L.M. Cruz-Orive. Particle number can be estimated using a disector of unknown thickness: the selector. *Journal of Microscopy*, 145:121–142, 1987.

[117] L.M. Cruz-Orive. Precision of stereological estimators from systematic probes. *Acta Stereologica*, 6(3):153–158., 1987.

[118] L.M. Cruz-Orive. Stereology: historical notes and recent evolution. *Acta Stereologica*, 6/Suppl. II:43–56, 1987. ISS Commemorative-Memorial Volume.

[119] L.M. Cruz-Orive. Stereology: recent solutions to old problems and a glimpse into the future. *Acta Stereologica*, 6(3):3–18.5, 1987.

[120] L.M. Cruz-Orive. On the precision of systematic sampling: a review of Matheron's transitive methods. *Journal of Microscopy*, 153:315–333, 1989.

[121] L.M. Cruz-Orive. Second-order stereology: estimation of second moment volume measures. *Acta Stereologica*, 8(2):641–646, 1989.

[122] L.M. Cruz-Orive. On the empirical variance of a fractionator estimate. *Journal of Microscopy*, 160:89–95, 1990.

[123] L.M. Cruz-Orive. Systematic sampling in stereology. *Bulletin of the International Statistical Institute*, 55(2):451–468, 1993. Proceedings 49th Session, Florence 1993.

[124] L.M. Cruz-Orive. Stereology of single objects. *Journal of Microscopy*, 186:93–107, 1997.

[125] L.M. Cruz-Orive. Precision of Cavalieri sections and slices with local errors. *Journal of Microscopy*, 193:182–198, 1999.

[126] L.M. Cruz-Orive. Stereology: a science between mathematical play and practical need. In E. Barcelo i Vidal, editor, *Homenatge al Profesor Lluis Santaló i Sors*, pages 223–225. 2002. Catedra Lluis Santalo, Universidad de Gerona.

[127] L.M. Cruz-Orive. Stereology: meeting point of integral geometry, probability, and statistics. In memory of Professor Luis A.Santaló (1911–2001). *Mathematicae Notae*, 41:49–98, 2002. Special issue (Homenaje a Santaló).

[128] L.M. Cruz-Orive. Estereologia: Punto de encuentro de la geometria integral, la probabilidad y la estadistica (En memoria del Profesor Luis A.Santaló (1911–2001)). *La Gaceta de la RSME*, 6:469–513, 2003.

[129] L.M. Cruz-Orive. Precision of the fractionator from Cavalieri designs. *Journal of Microscopy*, 213:205–211, 2004.

[130] L.M. Cruz-Orive, P. Gehr, A. Müller, and E.R. Weibel. Sampling designs for stereology. *Mikroskopie*, 37 (Supplement):149–155, 1980.

[131] L.M. Cruz-Orive and M. Geiser. Estimation of particle number by stereology: an update. *Journal of Aerosol Medicine*, 2004. in press.

[132] L.M. Cruz-Orive and X. Gual-Arnau. Precision of circular systematic sampling. *Journal of Microscopy*, 207:225–242, 2002.

[133] L.M. Cruz-Orive, H. Hoppeler, O. Mathieu, and E.R. Weibel. Stereological analysis of anisotropic structures using directional statistics. *Journal of the Royal Statistical Society C, (Applied Statistics)*, 34:14–32, 1985.

[134] L.M. Cruz-Orive and C.V. Howard. Estimating the length of a bounded curve in three dimensions using total vertical projections. *Journal of Microscopy*,

163:101–113, 1991.

[135] L.M. Cruz-Orive and C.V. Howard. Estimation of individual feature surface area with the vertical spatial grid. *Journal of Microscopy*, 178:146–151, 1995.

[136] L.M. Cruz-Orive and E.B. Hunziker. Stereology for anisotropic cells: application to growth cartilage. *Journal of Microscopy*, 143:47–80, 1986.

[137] L.M. Cruz-Orive, S.E. Larsen, M. Konig, and E.R. Weibel. Stereology of membrane loss in vertical sections: I. Theory. *Acta Stereologica*, 11(1):117–122, 1992.

[138] L.M. Cruz-Orive and A.O. Myking. A rapid method for estimating volume ratios. *Journal of Microscopy*, 115:127–136, 1979.

[139] L.M. Cruz-Orive and A.O. Myking. Stereological estimation of volume ratios by systematic sections. *Journal of Microscopy*, 122:143–157, 1981.

[140] L.M. Cruz-Orive and E.R. Weibel. Sampling designs for stereology. *Journal of Microscopy*, 122:235–257, 1981.

[141] L.M. Cruz-Orive and E.R. Weibel. Recent stereological methods for cell biology: a brief survey. *American Journal of Physiology*, 258:L148–L156, 1990.

[142] E. Czuber. *Geometrische Wahrscheinlichkeiten und Mittelwerte.* Leipzig, 1884.

[143] H.E. Daniels. A new technique for the analysis of fibre length distribution in wool. *Journal of the Textile Institute*, 33:T137–T150, 1942.

[144] A.C. Das. Two-dimensional systematic sampling and the associated stratified and random sampling. *Sankhyā*, 10:95–108, 1950.

[145] A.C. Das. Systematic sampling. *Bulletin of the International Statistical Institute*, 33(2):119–132, 1951.

[146] P.J. Davy. *Stereology: a statistical viewpoint.* Ph.D. thesis, Australian National University, 1978.

[147] P.J. Davy and R.E. Miles. Sampling theory for opaque spatial specimens. *Journal of the Royal Statistical Society, Series B*, 39:56–65, 1977.

[148] R.T. DeHoff. The determination of the size distribution of ellipsoidal particles from measurements made on random plane sections. *Transactions of the AIME*, 224:474–, 1962.

[149] R.T. DeHoff. The determination of the geometric properties of aggregates of constant size particles from counting measurements made on random plane sections. *Transactions of the Metallurgical Society of the AIME*, 230:764–769, 1964.

[150] R.T. DeHoff. Curvature and the topological properties of interconnected phases. In R.T.DeHoff and F.N. Rhines, editors, *Quantitative Microscopy*, pages 291–324. McGraw-Hill, New York, 1968.

[151] R.T. DeHoff. *Quantitative microstructural analysis*, pages 63–95. American Society for Testing and Materials, 1968.

[152] R.T. DeHoff. The geometric meaning of the integral mean curvature. In *Microstructural Science*, volume 5, pages 331–348, Amsterdam, 1977. Elsevier.

[153] R.T. DeHoff and F.N. Rhines. Determination of number of particles per unit volume from measurements made on random plane sections: the general cylinder and the ellipsoid. *Transactions AIME*, 221:975–982, 1961.

[154] R.T. DeHoff and F.N. Rhines, editors. *Quantitative Microscopy.* McGraw-Hill, New York, 1968.

[155] A. Delesse. Procédé mécanique pour déterminer la composition des roches. *Comptes Rendues de l'Académie des Sciences (Paris)*, 25:544–545, 1847.

[156] A. Delesse. Procédé mécanique pour déterminer la composition des roches. *Annales des Mines*, 13:379–388, 1848. Quatrième série.

[157] R. Deltheil. *Probabilités Géométriques*. Traité du Calcul des Probabilités et de ses Applications. Gauthier-Villars, Paris, 1926.

[158] A. Dhani. *Problèmes d'éstimation en stéréologie*. Mémoire, Licencié en sciences mathématiques, Faculté de Science, Facultés Universitaires de Notre Dame de la Paix, Namur, Belgium, 1979.

[159] P.J. Diggle. *Statistical analysis of spatial point patterns*. Arnold, second edition, 2003.

[160] P.J. Diggle and C.J.F. ter Braak. Point sampling of binary mosaics in ecology. In B. Ranneby, editor, *Statistics in theory and practice. Essays in honour of Bertil Matérn*, pages 107–122. Swedish University of Agricultural Sciences, Umeå, 1982.

[161] K.-A. Dorph-Petersen, J.R. Nyengaard, and H.J.G. Gundersen. Tissue shrinkage and unbiased stereological estimation of particle number and size. *Journal of Microscopy*, 204:232–246, 2001.

[162] K.A. Dorph-Petersen. Stereological estimation using vertical sections in a complex tissue. *Journal of Microscopy*, 195:79–86, 1999.

[163] K.A. Dorph-Petersen, H.J.G. Gundersen, and E.B.V. Jensen. Non-uniform systematic sampling in stereology. *Journal of Microscopy*, 200:148–157, 2000.

[164] K.A. Dorph-Petersen, J.R. Nyengaard, and H.J.G. Gundersen. Tissue shrinkage and unbiased stereological estimation of particle number and size. *Journal of Microscopy*, 204:232–246, 2001.

[165] A.S. Downie. *Efficiency of statistics in stereology*. Ph.D. thesis, Imperial College London, 1991.

[166] R.J. Duffin, Meissner, and Rhines. Report 32, Carnegie Institute of Technology, 1953. Cited by Saltykov [495].

[167] S.J. Edwards, S. Isaac, H.A. Collin, and N.J. Clipson. Stereological analysis of celery leaves infected by septoria apiicola. *Mycological Research*, 103:750–756, 1999.

[168] B.R. Eisenberg, A.M. Kuda, and J.B. Peters. Stereological analysis of mammalian skeletal muscle. I. Soleus muscle of the adult human guinea pig. *Journal of Cell Biology*, 60:732–754, 1974.

[169] H. Elias. Address of the President. In H. Haug, editor, *Proceedings of the First International Congress for Stereology*, page 2, Vienna, 1963. Congressprint.

[170] H. Elias, editor. *Second International Congress for Stereology*, New York, 1967. Springer Verlag.

[171] H. Elias. Identification of structure by the common sense approach. *Journal of Microscopy*, 95:59–68, 1972.

[172] H. Elias, A. Hennig, and P.M. Elias. Some methods for the study of kidney structure. *Z. wiss. Mikr.*, 65:70–82, 1961.

[173] H. Elias and D.M. Hyde. *A guide to practical stereology*. Number 1 in Karger Continuing Education Series. Karger, Basel, 1983.

[174] S.M. Evans and H.J.G. Gundersen. Estimation of spatial distributions using the nucleator. *Acta Stereologica*, 8:395–400, 1989.

[175] H.E. Exner and H.P. Hougardy, editors. *Quantitative Image Analysis of Microstructures*. DGM Metallurgy, Oberursel, Germany, 1998.

[176] N.L. Fazzalari and I.H. Parkinson. Fractal dimension and architecture of trabecular bone. *J. Pathology*, 178:100–105, 1996.

[177] H. Federer. *Geometric Measure Theory.* Springer Verlag, Heidelberg, 1969.

[178] W. Feller. *An introduction to probability theory and its applications*, volume 2. John Wiley and Sons, second edition, 1971.

[179] D.J. Finney. Random and systematic sampling in timber surveys. *Forestry*, 22:1–36, 1948.

[180] D.J. Finney. The efficiency of enumerations. I. Volume estimation of standing timber by sampling. II. Random and systematic sampling in timber surveys. *Indian Forest Bulletin*, page 146, 1949.

[181] D.J. Finney. An example of periodic variation in forest sampling. *Forestry*, 23:96–111, 1950.

[182] S. Floderus. Untersuchungen über den Bau der menschlichen Hypophyse mit besonderer Berücksichtigung der quantitativen mikromorphologischen Verhältnisse. *Acta Pathol. Microbiol. Immunol. Scand.*, 53 (Suppl 1):1–26, 1944.

[183] A.G. Flook. Fractal dimensions: their evaluation and their significance in stereological measurements. *Acta Stereologica*, 1:79–87, 1982.

[184] Y.S. Frolov and D.H. Maling. The accuracy of area measurement by point counting techniques. *Cartogr. J.*, 6:21–35, 1969.

[185] R.L. Fullman. Measurement of particle sizes in opaque bodies. *Journal of Metals*, 5:447–452, 1953.

[186] M. Gallagher, P.W. Landfield, B. McEwen, M.J. Meaney, P.R. Rapp, R. Sapolsky, and M.J. West. Hippocampal neurodegeneration in ageing. *Science*, 274:4–5, 1996.

[187] M. Garcia-Fiñana, L. M. Cruz-Orive, C.E. Mackay, E.B. Pakkenberg, and N. Roberts. Comparison of MR imaging against physical sectioning to estimate the volume of human cerebral compartments. *NeuroImage*, 18:505–516, 2003.

[188] M. Garcia-Fiñana and L.M. Cruz-Orive. Explanation of apparent paradoxes in Cavalieri sampling. *Acta Sterelogica*, 17:293–302, 1998.

[189] M. Garcia-Fiñana and L.M. Cruz-Orive. Fractional trend of the variance in Cavalieri sampling. *Image Analysis and Stereology*, 19:71–79, 2000.

[190] M. Garcia-Fiñana and L.M. Cruz-Orive. New approximations for the efficiency of Cavalieri sampling. *Journal of Microscopy*, 199:224–238, 2000.

[191] M. Garcia-Fiñana and L.M. Cruz-Orive. Improved variance prediction for systematic sampling on **R**. *Statistics*, 2004. in press.

[192] R.J. Gardner. *Geometric tomography.* Cambridge University Press, New York, 1995.

[193] J. Gaule and T. Lewin. Über die Zahlen der Nervenfasern und Ganglienzellen in den Spinalganglien des Kaninchens. *Zentralblatt f. Physiol.*, 10:437–440 and 465–471, 1896.

[194] S. Geuna. Appreciating the difference between design-based and model-based sampling strategies in quantitative morphology of the nervous system. *Journal of Comparative Neurology*, 427:333–339, 2000.

[195] S. Geuna. Cost-effectiveness of 3-D cell counting. *Trends in Neurosciences*, 24:374–375, 2001.

[196] H. Giger. Grundgleichungen der Stereologie I. *Metrika*, 16:43–57, 1970.

[197] H. Giger. Grundgleichungen der Stereologie II. *Metrika*, 18:84–93, 1972.

[198] H. Giger and H. Hadwiger. Über Treffzahlwahrscheinlichkeiten im Eikörperfeld. *Zeitschrift für Wahrscheinlichkeitstheorie und verwandte Gebiete*, 10:329–334, 1968.

[199] T. Gladman and J.H. Woodhead. The accuracy of point counting in metallographic investigations. *Journal of the Iron and Steel Institute*, 194:189–193, 1960.

[200] A.A. Glagolev. On geometrical methods of quantitative mineralogic analysis of rocks. *Trans. Inst. Econ. Min.*, 59:1–47, 1933.

[201] A.M. Gokhale. Unbiased estimation of curve length in 3-D using vertical slices. *Journal of Microscopy*, 159:133–143, 1990.

[202] A.M. Gokhale. Estimation of length density L_V from vertical slices of unknown thickness. *Journal of Microscopy*, 167:1–8, 1992.

[203] A.M. Gokhale and V. Benes. Estimation of average particle size from vertical projections. *Journal of Microscopy*, 191:195–200, 1998.

[204] A.M. Gokhale, W.J. Drury, and B. Whited. Quantitative microstructural analysis of anisotropic materials. *Materials Characterization*, 31:11–, 1993.

[205] A.M. Gokhale and E.E. Underwood. A general method for estimation of fracture surface roughness: Part I. Theoretical aspects. *Metallurgical Transactions*, 21A:1193–1199, 1990.

[206] U. Grenander. Stochastic processes and statistical inference. *Arkiv for Mathematik*, 1:195–277, 1950.

[207] P. Groeneboom and G. Jongbloed. Isotonic estimation and rates of convergence in Wicksell's problem. *Ann. Statist*, 23:1518–1542, 1995.

[208] X. Gual Arnau and L.M. Cruz-Orive. Consistency in systematic sampling. *Advances in Applied Probability*, 28:982–992, 1996.

[209] X. Gual Arnau and L.M. Cruz-Orive. Consistency in systematic sampling for stereology. *Advances in Applied Probability*, 28:329–330, 1996.

[210] X. Gual Arnau and L.M. Cruz-Orive. Variance prediction under systematic sampling with geometric probes. *Advances in Applied Probability*, 30:889–903, 1998.

[211] X. Gual-Arnau and L.M. Cruz-Orive. Systematic sampling on the circle and on the sphere. *Advances in Applied Probability*, 32:628–647, 2000.

[212] X. Gual-Arnau and L.M. Cruz-Orive. Variance prediction for pseudosystematic sampling on the sphere. *Advances in Applied Probability (SGSA)*, 34:469–483, 2002.

[213] R.W. Guillery and K. Herrup. Quantification without pontification: Choosing a method for counting objects in sectioned tissues. *Journal of Comparative Neurology*, 386(1):2–7, 1998.

[214] H.J. Gundersen and R. Østerby. Optimizing sampling efficiency of stereological studies in biology: or 'Do more less well!'. *Journal of Microscopy*, 121:65–74, 1981.

[215] H.J.G. Gundersen. Notes on the estimation of the numerical density of arbitrary profiles: the edge effect. *Journal of Microscopy*, 111:219–223, 1977.

[216] H.J.G. Gundersen. Estimators of the number of objects per area unbiased by edge effects. *Microscopica Acta*, 81:107–117, 1978.

[217] H.J.G. Gundersen. Stereology — or how figures for spatial shape and content are obtained by observation of structures in sections. *Microscopica Acta*, 83:409–426, 1980.

[218] H.J.G. Gundersen. *Stereologi — Eller hvordan tal for rumlig form og indhold*

opnås ved iagttagelse af strukturer på snitplaner. D. Sc. thesis, København, 1981.

[219] H.J.G. Gundersen. Stereology of arbitrary particles. A review of unbiased number and size estimators and the presentation of some new ones, in memory of William R Thompson. *Journal of Microscopy*, 143:3–45, 1986.

[220] H.J.G. Gundersen. The nucleator. *Journal of Microscopy*, 151:3–21, 1988.

[221] H.J.G. Gundersen. Stereology: The fast lane between neuroanatomy and brain function or still only a tightrope? *Acta Neurologica Scandinavica*, 137:8–13, 1992.

[222] H.J.G. Gundersen. Personal communication, 2004.

[223] H.J.G. Gundersen and A.J. Baddeley. *Stereology: general principles and biological applications.* In preparation.

[224] H.J.G. Gundersen, P. Bagger, T.F. Bendtsen, S.M. Evans, L. Korbo, N. Marcussen, A Møller, K. Nielsen, J.R. Nyengaard, B. Pakkenberg, F.B. Sørensen, A. Vesterby, and M.J. West. The new stereological tools: disector, fractionator, nucleator and point sampled intercepts and their use in pathological research and diagnosis. *Acta Pathologica Microbiologica et Immunologica Scandinavica*, 96:857–881, 1988.

[225] H.J.G. Gundersen, T.F. Bendtsen, L. Korbo, N. Marcussen, A. Moller, K. Nielsen, J.R. Nyengaard, B. Pakkenberg, F.B. Sorensen, A. Vesterby, and M.J. West. Some new, simple and efficient stereological methods and their use in pathological research and diagnosis. *Acta Pathologica Microbiologica et Immunologica Scandinavica*, 96:379–394, 1988.

[226] H.J.G. Gundersen, M. Boysen, and A. Reith. Comparison of semiautomatic digitizer-tablet and simple point counting performance in morphometry. *Virchows Archiv B (Cell Pathology)*, 37:317–325, 1981.

[227] H.J.G. Gundersen and E.B. Jensen. The efficiency of systematic sampling in stereology and its prediction. *Journal of Microscopy*, 147:229–263, 1987.

[228] H.J.G. Gundersen and E.B.V. Jensen. Stereological estimation of the volume-weighted mean volume of arbitrary particles observed on random sections. *Journal of Microscopy*, 138:127–142, 1985.

[229] H.J.G. Gundersen, E.B.V. Jensen, K. Kiêu, and J. Nielsen. The efficiency of systematic sampling - reconsidered. *Journal of Microscopy*, 193:199–211, 1999.

[230] P.M. Gy. *Sampling of particulate materials: theory and practice.* Elsevier, Amsterdam/ New York, 1979.

[231] A. Haas, G. Matheron, and J. Serra. Morphologie mathématique et granulométries en place. *Annales des Mines*, 11:736–753, 1967.

[232] A. Haas, G. Matheron, and J. Serra. Morphologie mathématique et granulométries en place, II. *Annales des Mines*, 12:767–782, 1967.

[233] H. Hadwiger. *Vorlesungen über Inhalt, Oberfläche und Isoperimetrie.* Springer Verlag, Berlin, 1957.

[234] U. Hahn, E.B.V. Jensen, M.-C. van Lieshout, and L.S. Nielsen. Inhomogeneous spatial point processes by location-dependent scaling. *Adv. Appl. Prob. (SGSA)*, 35:319–336, 2003.

[235] U. Hahn, A. Micheletti, R. Pohlink, D. Stoyan, and H. Wendrock. Stereological analysis and modelling of gradient structures. *Journal of Microscopy*, 195:113–124, 1999.

[236] U. Hahn and D. Stoyan. Unbiased stereological estimation of the surface area of

gradient surface processes. *Advances in Applied Probability*, 30:904–920, 1998.

[237] J. Hájek. Representativní výběr skupin metodou dvou fází (Representative sampling by the method of two phases). *Statisticky Obzor*, 29:384–395, 1949. In Czech.

[238] J. Hájek. *Sampling from a finite population.* Marcel Dekker, New York and Basel, 1981.

[239] P. Hall. Correcting segment counts for edge effects when estimating intensity. *Biometrika*, 72:459–463, 1985.

[240] P. Hall. *An introduction to the theory of coverage processes.* John Wiley and Sons, New York, 1988.

[241] P. Hall and R.L. Smith. The kernel method for unfolding sphere size distributions. *Journal of Computational Physics*, 74:409–421, 1988.

[242] P. Halmos. *Measure theory.* Van Nostrand, 1950.

[243] K.-H. Hanisch, K.-D. Oberstedt, V. Reinsch, and D. Stoyan. Genauigkeitsfragen bei der Vorratsberechnung von Lagerstätten. *Neue Bergbautechnik*, 11:618–620, 1978.

[244] K.H. Hanisch. On stereological estimation of second-order characteristics and of the hard-core distance of systems of sphere centres. *Biometrical Journal*, 25:731–743, 1983.

[245] K.H. Hanisch and D. Stoyan. Stereological estimation of the radial distribution function of centres of spheres. *Journal of Microscopy*, 122:131–141, 1981.

[246] M.-L. Hannila, E.B.V. Jensen, A. Hobolth, and J.R. Nyengaard. Shape modelling of spatial particles from planar central sections - a case study. *Journal of Microscopy*, to appear, 2004.

[247] L.J. Hansen. Rekonstruktion og geometriske simulationer i en population af neuroner. Master's thesis, Department of Mathematical Sciences, University of Aarhus, 2000.

[248] H.O. Hartley and A. Ross. Unbiased ratio estimates. *Nature*, 174:841–851, 1954.

[249] H. Haug, editor. *First International Congress for Stereology*, Vienna, 1963. Congressprint.

[250] H. Haug. Gibt es Nervenzellverluste während der Alterung in der menschlichen Hirnrinde? Ein morphometrischer Beitrag zu dieser Frage. *Nervenheilkunde*, 4:103–109, 1985.

[251] H. Haug. History of neuromorphometry. *Journal of Neuroscience Methods*, 18:1–17, 1988.

[252] J.C. Hedreen. Lost caps in histological counting methods. *Anatomical Record*, 250:366–372, 1998.

[253] J.C. Hedreen. What was wrong with the Abercrombie and empirical cell counting methods? A review. *Anatomical Record*, 250(3):373–380, March 1998.

[254] J.C. Hedreen and J.P. Vonsattel. Centenary of Gaule and Lewin, pioneers in cell counting methodology. *Journal of Microscopy*, 187:201–203, 1997.

[255] A. Hennig. A critical summary of volume and surface measurements in microscopy. *Zeiss-Werkz.*, 30, 1959.

[256] A. Hennig. Länge eines räumlichen Linienzuges. *Zeitschrift für wissenschaftliche Mikroskopie*, 65:193–194, 1963.

[257] A. Hennig. Length of a three-dimensional linear tract. In *Proceedings 1st International Congress on Stereology*, pages 44/1–44/8. Congressprint, Vienna, 1963.

[258] H. Hermann and J. Ohser. Determination of microstructural parameters of random spatial surface fractals by measuring chord length distributions. *Journal of Microscopy*, 170:87–93, 1993.

[259] J.E. Hilliard. Specification and measurement of structural anisotropy. *Trans. Metallurg. Soc. AIME*, 224:1201–1211, 1962.

[260] J.E. Hilliard. Determination of structural anisotropy. In H. Elias, editor, *Second International Congress for Stereology*, pages 219–227, New York, 1967. Springer Verlag.

[261] J.E. Hilliard and J.W. Cahn. An evaluation of procedures in quantitative metallography for volume-fraction analysis. *Transactions of the Metallurgical Society of the AIME*, 221:344–351, 1961.

[262] J.E. Hilliard and L.R. Lawson. *Stereology and stochastic geometry.* Number 28 in Computational Imaging and Vision. Kluwer, 2003.

[263] M. Hiwatashi. A study of the comparison between the efficiency of the dot-grid method and planimeter method in the area measurement. *For. Exp. Stn. Tokyo, Japan, Bull.*, 214:111–125, 1968. (In Japanese).

[264] D. Hlubinka. Stereology of extremes; shape factor of spheroids. *Extremes*, 6:5–24, 2003.

[265] A. Hobolth. The spherical deformation model. *Biostatistics*, 4:583–595, 2003.

[266] A. Hobolth and E.B.V. Jensen. A note on design-based versus model-based variance estimation in stereology. *Advances in Applied Probability (SGSA)*, 34:484–490, 2002.

[267] A. Hobolth and E.B.V. Jensen. Stereological analysis of shape. *Image Analysis and Stereology*, 21:23–29, 2002.

[268] A.W. Hoogendoorn. Estimating the weight undersized distribution for the wicksell problem. *Statistica Neerlandica*, 4:259–282, 1992.

[269] E. Horikawa. Tetsu-no-haganė. *Journal of the Iron and Steel Institute of Japan*, 40(10):991, 1954. Cited by Saltykov [495].

[270] D.G. Horvitz and D.J. Thompson. A generalization of sampling without replacement from a finite universe. *Journal of the American Statistical Association*, 47:663–685, 1952.

[271] C.V. Howard, L.M. Cruz-Orive, and H. Yaegashi. Estimating neuron dendritic length in 3D from total vertical projections and from vertical slices. *Acta Neurologica Scandinavica*, Suppl.137:14–19, 1992.

[272] C.V. Howard and M.G. Reed. *Unbiased stereology: three dimensional measurement in microscopy.* Bios Scientific Publishers, Oxford, 1998.

[273] C.V. Howard, S. Reid, A.J. Baddeley, and A. Boyde. Unbiased estimation of particle density in the tandem-scanning reflected light microscope. *Journal of Microscopy*, 138:203–212, 1985.

[274] R.T. Howard and M. Cohen. Quantitative metallography by point-counting and lineal analysis. *Transactions of the AIME*, 172:413–426, 1947.

[275] P.P. Hujoel. A metaanalysis of normal ranges for root surface-areas of the permanent dentition. *Journal of Clinical Periodontology*, 21:225–229, 1994.

[276] M. Janicki, J. Mizera, B. Ralph, J.J. Bucki, and K.J. Kurzydłowski. Anisotropy of grain growth in polycrystals of α-Fe. *Materials Science and Engineering A– Structural Materials Properties Microstructure and Processing*, 298:187–192, 2001.

[277] A.M. Janson and A. Møller. Chronic nicotine treatment counteracts nigral cell loss induced by a partial mesodiencephalic hemitransection: an analysis of the total number and mean volume of neurons and glia in substantia nigra of the male rat. *Neuroscience*, 57:931–941, 1993.

[278] E.B. Jensen. A design-based proof of Wicksell's integral equation. *Journal of Microscopy*, 136:345–348, 1984.

[279] E.B. Jensen. On the use of point-sampling in the stereological analysis of arbitrarily shaped particles. In W. Nagel, editor, *Proceedings of Geobild '85*, pages 105–111. Friedrich-Schiller-Universität, Jena, 1985.

[280] E.B. Jensen. Design- and model-based stereological analysis of arbitrarily shaped particles. *Scandinavian Journal of Statistics*, 14:161–180, 1987.

[281] E.B. Jensen, A.J. Baddeley, H.J.G. Gundersen, and R. Sundberg. Recent trends in stereology. *International Statistical Review*, 53:99–108, 1985.

[282] E.B. Jensen and H.J.G. Gundersen. Stereological ratio estimation based on counts from integral test systems. *Journal of Microscopy*, 125:51–66, 1982.

[283] E.B. Jensen and H.J.G. Gundersen. On the estimation of moments of the volume-weighted distribution of particle sizes. In E.B. Jensen and H.J.G. Gundersen, editors, *Proceedings of the Second International Workshop on Stereology and Stochastic Geometry*, Memoir no. 6, pages 81–103, Aarhus, Denmark, 1983. Department of Theoretical Statistics, University of Aarhus.

[284] E.B. Jensen and H.J.G. Gundersen. The stereological estimation of moments of particle volume. *Journal of Applied Probability*, 22:82–98, 1985.

[285] E.B. Jensen and H.J.G. Gundersen. The corpuscle problem: reevaluation using the disector. *Acta Stereologica*, 6/II:105–122, 1987.

[286] E.B. Jensen and H.J.G. Gundersen. Stereological estimation of surface area of arbitrary particles. *Acta Stereologica*, 6(Suppl. III):25–30, 1987. Proceedings of the Seventh International Congress for Stereology, Caen (ed. J.L. Chermant).

[287] E.B. Jensen and H.J.G. Gundersen. Fundamental stereological formulae based on isotropically orientated probes through fixed points with applications to particle analysis. *Journal of Microscopy*, 153:249–267, 1989.

[288] E.B. Jensen, K. Kiêu, and H.J.G. Gundersen. Second-order stereology. In *Proceedings of the 5th European Congress for Stereology, Freiburg 1989*, pages 15–35. Acta Stereologica, 9/I, 1990.

[289] E.B. Jensen and F.B. Sørensen. A note on stereological estimation of the volume-weighted second moment of particle volume. *Journal of Microscopy*, 164:21–27, 1991.

[290] E.B. Jensen and R. Sundberg. Generalized associated point methods for sampling planar objects. *Journal of Microscopy*, 144:55–70, 1986.

[291] E.B. Jensen and R. Sundberg. Statistical models for stereological inference about spatial structures: on the applicability of best linear unbiased estimators in stereology. *Biometrics*, 42:735–751, 1986.

[292] E.B.V. Jensen. Recent developments in the stereological analysis of particles. *Annals of the Institute of Statistical Mathematics*, 43:455–468, 1991.

[293] E.B.V. Jensen. *Local Stereology*. Singapore. World Scientific Publishing, 1998.

[294] E.B.V. Jensen. On the variance of local stereological estimators. *Image Analysis and Stereology*, 19:15–18, 2000.

[295] E.B.V. Jensen and H.J.G. Gundersen. The rotator. *Journal of Microscopy*,

170:35–44, 1993.

[296] E.B.V. Jensen and M. Kiderlen. Directional analysis of digitized planar sets by configuration counts. *Journal of Microscopy*, 212:158–168, 2003.

[297] E.B.V. Jensen and K. Kiêu. A new integral geometric formula of Blaschke-Petkantschin type. *Mathematische Nachrichten*, 156:57–74, 1992.

[298] G. Badsberg Jensen and B. Pakkenberg. Do alcoholics drink their neurons away? *The Lancet*, 342:1201–1204, 1993.

[299] D. Jeulin, editor. *Advances in Theory and Applications of Random Sets.* World Scientific Publishing, Singapore, 1997.

[300] D. Jeulin. Dead leaves models: From space tessellation to random functions. In D. Jeulin, editor, *Advances in Theory and Applications of Random Sets,* pages 137–156, Singapore, 1997. World Scientific Publishing.

[301] D. Jeulin. Modelling random media. *Image Analysis and Stereology*, 21 (Supplt 1):S31–S40, 2002.

[302] A.E. Jones. Systematic sampling of continuous parameter populations. *Biometrika*, 35:283–296, 1948.

[303] L.M. Karlsson and L.M. Cruz-Orive. The new stereological tools in metallography: estimation of pore size in aluminium. *Journal of Microscopy*, 165:391–415, 1992.

[304] L.M. Karlsson and L.M. Cruz-Orive. Estimation of mean particle size from single sections. *Journal of Microscopy*, 186:121–132, 1997.

[305] A.M. Kellerer. Exact formulae for the precision of systematic sampling. *Journal of Microscopy*, 153:285–300, 1989.

[306] D.G. Kendall. On the number of lattice points inside a random oval. *Quarterly Journal of Mathematics (Oxford)*, 19:1–26, 1948.

[307] D.G. Kendall. Foundations of a theory of random sets. In E F Harding and D G Kendall, editors, *Stochastic geometry*, chapter 6.2, pages 322–376. John Wiley and Sons, Chichester, 1974.

[308] D.G. Kendall and R.A. Rankin. On the number of points of a given lattice in a random hypersphere. *Quarterly Journal of Mathematics (Oxford)*, 4:178–189, 1953.

[309] M.G. Kendall and P.A.P. Moran. *Geometrical Probability*. Charles Griffin, London, 1963. Griffin's Statistical Monographs and Courses no. 10.

[310] M. Kiderlen. Non-parametric estimation of the directional distribution. *Adv. Appl. Prob. (SGSA)*, 33:6–24, 2001.

[311] M. Kiderlen and E.B.V. Jensen. Estimation of the directional measure of planar random sets by digitization. *Advances in Applied Probability (SGSA)*, 35:583–602, 2003.

[312] K. Kiêu. Three lectures on systematic geometric sampling. Memoirs 13, Department of Theoretical Statistics, University of Aarhus, 1997.

[313] K. Kieu and E.B.V. Jensen. Stereological estimation based on isotropic slices through fixed points. *Journal of Microscopy*, 170:45–51, 1993.

[314] K. Kiêu and M. Mora. Variance of planar area estimators based on systematic sampling. In *Supplementary Notes: Summer School on Stereology and Geometric Tomography*, pages 64–79. Department of Mathematical Sciences, University of Aarhus, Denmark, 2000.

[315] K. Kiêu, S. Souchet, and J. Istas. Precision of systematic sampling and transitive

methods. *Journal of Statistical Inference and Planning*, 77:263–279, 1999.

[316] J.F.C. Kingman. *Poisson processes*. Oxford University Press, 1993.

[317] J.A. Kittelson. The postnatal growth of the kidney of the albino rat, with observations on an adult human kidney. *Anatomical Record*, 13:385–408, 1916.

[318] A. Kolmogoroff. *Grundbegriffe der Wahrscheinlichkeitsrechnung*. Ergebnisse der Mathematik und ihrer Grenzgebiete. Schriftleitung Zentralblatt für Mathematik, Berlin, 1933. Reprinted by Chelsea, New York, 1946.

[319] M. Konig, S.E. Larsen, L.M. Cruz-Orive, and E.R. Weibel. Stereology of membrane loss in vertical sections: II. Application to mitochondria. *Acta Stereologica*, 11(1):123–128, 1992.

[320] B.W. Konigsmark. Methods for the counting of neurons. In W.J.H. Nauta and S.O.E. Ebbesson, editors, *Contemporary research methods in neuroanatomy*, pages 315–338. Springer, Berlin, 1970.

[321] L. Korbo, B. Pakkenberg, O. Ladefoged, H.J.G. Gundersen, P. Arien-Soborg, and H. Pakkenberg. An efficient method for estimating the total number of neurons in rat brain cortex. *Journal of Neuroscience Methods*, 31:93–100, 1990.

[322] A. Korkmaz and L. Tümkaya. Estimation of section thickness and optical disector height with a simple calibration method. *Journal of Microscopy*, 187:104–109, 1997.

[323] K. Krickeberg. Processus ponctuels en statistique. In P.L. Hennequin, editor, *Ecole d'Eté de Probabilités de Saint-Flour X*, volume 929 of *Lecture Notes in Mathematics*, pages 205–313. Springer, 1982.

[324] P.R. Krishnaiah and C.R. Rao, editors. *Sampling*. Number 6 in Handbook of Statistics. North-Holland, Amsterdam, 1988.

[325] W.C. Krumbein and F.J. Pettijohn. *Manual of sedimentary petrography*. D. Appleton-Century, New York, 1938.

[326] L. Kubínová and J. Janáček. Estimating surface area by the isotropic fakir method from thick slices cut in an arbitrary direction. *Journal of Microscopy*, 191:201–211, 1998.

[327] L. Kubínová and J. Janáček. Confocal microscope and stereology: estimating volume, number, surface area and length by virtual test probes applied to three-dimensional images. *Microscopy Research and Technique*, 53:425–435, 2001.

[328] K.J. Kurzydłowski and B. Ralph. *The quantitative description of the microstructure of materials*. CRC Press, Boca Raton, FL, 1995.

[329] S. Kwasniewska, F. Fankhauser, S.E. Larsen, and L.M. Cruz-Orive. The efficacy of cw Nd:YAG laser trabeculoplasty. *Ophthalmic Surgery*, 24:304–308, 1993.

[330] M. Ladekarl. Quantitative histopathology in ductal carcinoma of the breast: prognostic value of mean nuclear size and mitotic counts. *Cancer*, 75:2114–2122, 1995.

[331] M. Ladekarl, T. Bæk-Hansen, R.H. Nielsen, C. Mouritzen, U. Henriques, and F.B. Sørensen. Objective malignancy grading of squamous cell carcinoma of the lung. Stereologic estimates of mean nuclear size are of prognostic value, independent of clinical stage of disease. *Cancer*, 76:797–802, 1995.

[332] D.B. Lahiri. A method for sample selection providing unbiased ratio estimates. *Bulletin of the International Statistical Institute*, 33:133–140, 1951.

[333] G. Landini and J.-P. Rigaut. A method for estimating the dimension of asymptotic fractal sets. *Bioimaging*, 5:65–70, 1997.

[334] H. Lange, G. Thörner, A. Hopf, and K.F. Schröder. Morphometric studies of the neuropathological changes in choreatic disease. *J. Neurol. Science*, 28:401–425, 1976.

[335] C. Lantuéjoul. Computation of the histograms of the number of edges and neighbours of cells in a tessellation. In R.E. Miles and J. Serra, editors, *Geometrical Probability and Biological Structures: Buffon's 200th Anniversary*, Lecture Notes in Biomathematics, No 23, pages 323–329. Springer Verlag, Berlin-Heidelberg-New York, 1978.

[336] C. Lantuéjoul. *La squelettisation et son application aux mesures topologiques des mosaiques polycristallines.* Thesis, docteur-ingénieur en Sciences et Techniques Minières, Ecole Nationale Supérieure de Mines de Paris, Fontainebleau, 1978.

[337] C. Lantuéjoul. On the importance of the quadrat size when analysing a specimen. *Acta Stereologica*, 6(1):103–106, 1987.

[338] C. Lantuéjoul. Some stereological and statistical consequences derived from Cartier's formula. *Journal of Microscopy*, 151:265–276, 1988.

[339] C. Lantuéjoul. Ergodicity and integral range. *Journal of Microscopy*, 161:387–403, 1991.

[340] G.J. Laroye and K. Grant. A simple algorithm to measure the volume-weighted and number-weighted mean volume of particles. *Journal of Microscopy*, 175:70–83, 1994.

[341] E.S. Larsen and F.S. Miller. The Rosiwal method and the modal determination of rocks. *Am. Min.*, 20:260–273, 1935.

[342] J.O. Larsen, H.J.G. Gundersen, and J. Nielsen. Global spatial sampling with isotropic virtual planes: estimators of length density and total length in thick, arbitrarily orientated sections. *Journal of Microscopy*, 191:238–248, 1998.

[343] F. Lincoln and H.L. Rietz. The determination of the relative volume of the components of rocks by mensuration methods. *Econ. Geol.*, 8:120–139, 1913.

[344] K. Linderstrøm-Llang, H. Holter, and A.S. Olsen. Studies on enzymatic histochemistry: XIII. The distribution of enzymes in the stomach of pigs as a function of its histological structure. *Comptes Rendus du Laboratoire Carlsberg*, 20:66–125, 1935.

[345] D.V. Little. A third note on recent results in geometrical probability. *Advances in Applied Probability*, 6:103–130, 1974.

[346] P.R. Lloyd. Quantisation error in area measurement. *Cartogr. J.*, 13:22–25, 1976.

[347] G.W. Lord and T.F. Willis. Calculation of air bubble size distribution from results of a Rosiwal traverse of aerated concrete. *ASTM Bulletin*, 177:56–61, 1951.

[348] W.E. Lorensen. Marching cubes: A high resolution 3-D surface construction algorithm. *ACM SIGGRAPH*, 21(4):38–44, 1987.

[349] S. Majumdar, R.S. Weinstein, and R.R. Prasad. Application of fractal geometry techniques to the study of trabecular bone. *Medical Physics*, 20:1611–1619, 1993. erratum appears in Med. Phys. 21 (3) (1994) 491.

[350] D.H. Maling. *Measurements from maps: principles and methods of cartometry.* Pergamon Press, Oxford, 1989.

[351] B.B. Mandelbrot. *The Fractal Geometry of Nature.* W.H. Freeman, New York, 1977.

[352] R. Marcos, E. Rocha, and R.A.F. Monterio. Stereological estimation of Ito cells from rat liver using the optical fractionator - a preliminary report. *Image Analysis*

and Stereology, 21:1–6, 2002.

[353] N. Marcussen. The double disector: unbiased stereological estimation of the number of particles inside other particles. *Journal of Microscopy*, 165:417–426, 1992.

[354] S. Mase. Stereological estimation of particle size distributions. *Advances in Applied Probability*, 27:350–366, 1995.

[355] A.V. Maslov. *Sposobi i tochnosti opredeleniya ploshchadey.* Geodezizdat, Moscow, 1955. (Methods and accuracies in the determination of area).

[356] B. Matérn. Methods of estimating the accuracy of line and sample plot surveys. *Meddelanden från Statens Skogsforskningsinstitut*, 36:1–138, 1947.

[357] B. Matérn. Spatial variation. *Meddelanden från Statens Skogsforskningsinstitut*, 49(5):1–114, 1960.

[358] B. Matérn. *Spatial Variation.* Number 36 in Lecture Notes in Statistics. Springer Verlag, New York, 1986.

[359] B. Matérn. Precision of area estimation: a numerical study. *Journal of Microscopy*, 153:269–284, 1989.

[360] G. Matheron. *Les variables régionalisées et leur estimation.* Masson, Paris, 1965.

[361] G. Matheron. *Éléments pour une Théorie des Milieux Poreux.* Masson, Paris, 1967.

[362] G. Matheron. La théorie des variables régionalisées et ses applications. Les Cahiers du Centre de Morphologie Mathématique de Fontainebleau 5, École Nationale Supérieure des Mines de Paris, Fontainebleau, France, 1970.

[363] G. Matheron. The theory of regionalized variables and its applications. Les Cahiers du Centre de Morphologie Mathématique de Fontainebleau 5, École Nationale Supérieure des Mines de Paris, Fontainebleau, France, 1970.

[364] G. Matheron. Random sets theory and its application to stereology. *Journal of Microscopy*, 95:15–23, 1972.

[365] G. Matheron. The intrinsic random functions and their applications. *Advances in Applied Probability*, 5:439–468, 1973.

[366] G. Matheron. *Random sets and integral geometry.* John Wiley and Sons, New York, 1975.

[367] G. Matheron. La formule de Crofton pour les sections épaisses. *Journal of Applied Probability*, 13:707–713, 1976.

[368] O. Mathieu, L.M. Cruz-Orive, H. Hoppeler, and E.R. Weibel. Measuring error and sampling variation in stereology: comparison of the efficiency of various methods for planar image analysis. *Journal of Microscopy*, 121:75–88, 1981.

[369] O. Mathieu, L.M. Cruz-Orive, H. Hoppeler, and E.R. Weibel. Estimating length density and quantifying anisotropy in skeletal muscle capillaries. *Journal of Microscopy*, 131:131–146, 1983.

[370] T. Mattfeldt. The accuracy of one-dimensional systematic sampling. *Journal of Microscopy*, 153:301–313, 1989.

[371] T. Mattfeldt, H. Frey, and C. Rose. Second-order stereology of benign and malignant alterations of human mammary gland. *Journal of Microscopy*, 171:143–151, 1993.

[372] T. Mattfeldt, H.-J. Möbius, and G. Mall. Orthogonal triplet probes: an efficient method for unbiased estimation of length and surface of objects with unknown orientation in space. *Journal of Microscopy*, 139:279–289, 1985.

[373] T. Mattfeldt, G. Mall, H. Gharehbaghi, and P. Möller. Estimation of surface area

and length with the orientator. *Journal of Microscopy*, 159:301–317, 1990.

[374] T.M. Mayhew. The new stereological methods for interpreting functional morphology from slices of cells and organs. *Experimental Physiology*, 76:639–665, 1991.

[375] T.M. Mayhew. Review article: The new stereological methods for interpreting functional morphology from slices of cells and organs. *Experimental Psysiology*, 76:639–665, 1991.

[376] T.M. Mayhew. A review of recent advances in stereology for quantifying neural structure. *Journal of Neurocytology*, 21:313–328, 1992.

[377] T.M. Mayhew. Invited Review: Adaptive remodelling of intestinal epithelium assessed using stereology: correlation of single cell and whole organ data with nutrient transport. *Histology and Histopathology*, 11:729–741, 1996.

[378] T.M. Mayhew. Recent applications of the new stereology have thrown fresh light on how the human placenta grows and develops its form. *Journal of Microscopy*, 186:153–163, 1997.

[379] T.M. Mayhew. Second-order stereology and ultrastructural examination of the spatial arrangements of tissue compartments within glomeruli of normal and diabetic kidneys. *Journal of Microscopy*, 195:87–95, 1999.

[380] T.M. Mayhew and L.M. Cruz-Orive. Caveat on the use of the Delesse principle of areal analysis for estimating component volume densities. *Journal of Microscopy*, 102:195–207, 1974.

[381] T.M. Mayhew, B. Huppertz, P. Kaufmann, and J.C. Kingdom. The 'reference trap' revisited: examples of the dangers in using ratios to describe fetoplacental angiogenesis and trophoblast turnover. *Placenta*, 24:1–7, 2003.

[382] T.M. Mayhew and D.R. Olsen. Magnetic resonance imaging (MRI) and model-free estimates of brain volume determined using the cavalieri principle. *Journal of Anatomy*, 178:133–144, 1991.

[383] T.M. Mayhew and E. Wadrop. Placental morphogenesis and the star volumes of villous trees and intervillous pores. *Placenta*, 15:209–217, 1994.

[384] M. McLean and J.W. Prothero. Three-dimensional reconstruction from serial sections. V. Calibration of dimensional changes incurred during tissue preparation and data processing. *Anal. and Quant. Cytol. and Hist.*, 13:269–278, 1991.

[385] A-M. McMillan and F.B. Sørensen. The efficient and unbiased estimation of nuclear size variability using the 'selector'. *Journal of Microscopy*, 165:433–437, 1992.

[386] V. McNulty, L.M. Cruz-Orive, N. Roberts, C.J. Holmes, and X. Gual-Arnau. Estimation of brain compartment volume from MR Cavalieri slices. *Journal of Computer Assisted Tomography*, 24:466–477, 2000.

[387] J. Mecke and W. Nagel. Stationäre räumliche Faserprozesse und ihre Schnittzahlrosen. *Elektronische Informationsverarbeitung und Kybernetik*, 475–483, 1980.

[388] J. Mecke, R.G. Schneider, D. Stoyan, and W.R.R. Weil. *Stochastische Geometrie*. DMV Seminar Band 16. Birkhäuser, Basel, 1990.

[389] J. Mecke and D. Stoyan. Formulas for stationary planar fibre processes I — general theory. *Mathematische Operationsforschung und Statistik, series Statistics*, 12:267–279, 1980.

[390] J. Mecke and D. Stoyan. Stereological problems for spherical particles. *Mathe-*

matische Nachrichten, 96:311–317, 1980.

[391] S.M.L.C. Mendis-Handagama and L.L. Ewing. Sources of error in the estimation of Leydig cell numbers in control and atrophied mammalian testes. *Journal of Microscopy*, 159(1):73–82, July 1990.

[392] W.A. Merz. Die Streckenmessung an gerichteten Strukturen im Mikroskop und ihre Anwendung zur Bestimmung von Oberfläche-Volumen-Relationen in kochengewebe. *Mikroskopie*, 22:132–142, 1967.

[393] R.P. Michel and L.M. Cruz-Orive. Application of the Cavalieri principle and vertical sections method to lung: estimation of volume and pleural surface area. *Journal of Microscopy*, 150:117–136, 1988.

[394] MicroBrightField, Inc. Software package 'Stereo Investigator'. http://www.microbrightfield.com/.

[395] H. Midzuno. On the sampling system with probability proportionate to sum of sizes. *Annals of the Institute of Statistical Mathematics*, 2:99–108, 1951.

[396] R.E. Miles. Multidimensional perspectives on stereology. *Journal of Microscopy*, 95:181–196, 1972.

[397] R.E. Miles. Stereological formulae based upon planar curve sections of surfaces in space. *Journal of Microscopy*, 121:211–237, 1972.

[398] R.E. Miles. The fundamental formula of Blaschke in integral geometry and its iteration, for domains with fixed orientation. *Australian Journal of Statistics*, 16(2):111–118, 1974.

[399] R.E. Miles. On the elimination of edge effects in planar sampling. In E.F. Harding and D.G. Kendall, editors, *Stochastic Geometry: a tribute to the memory of Rollo Davidson*, pages 228–247, London-New York-Sydney-Toronto, 1974. John Wiley and Sons.

[400] R.E. Miles. Estimating aggregate and overall characteristics from thick sections by transmission microscopy. *Journal of Microscopy*, 107:227–233, 1976.

[401] R.E. Miles. The importance of proper model specification in stereology. In R.E. Miles and J. Serra, editors, *Geometrical Probability and Biological Structures: Buffon's 200th Anniversary*, Lecture Notes in Biomathematics, No 23, pages 115–136. Springer Verlag, Berlin-Heidelberg-New York, 1978.

[402] R.E. Miles. The sampling, by quadrats, of planar aggregates. *Journal of Microscopy*, 113:257–267, 1978.

[403] R.E. Miles. Some new integral formulae, with stochastic applications. *Journal of Applied Probability*, 16:592–606, 1979.

[404] R.E. Miles. On the underlying relationship of plane section stereology to statistical sampling theory. *Mikoskopie*, 37 (Suppl):13–18, 1980.

[405] R.E. Miles. Contributed discussion (session on Stochastic Geometry). *Bulletin of the International Statistical Institute*, 44 (3):392, 1983.

[406] R.E. Miles. Stereology for embedded aggregates of not-necessarily-convex particles. In E.B. Jensen and H.J.G. Gundersen, editors, *Proceedings of the Second International Workshop on Stereology and Stochastic Geometry*, Memoir no. 6, pages 127–147, Aarhus, Denmark, 1983. Department of Theoretical Statistics, University of Aarhus.

[407] R.E. Miles. A comprehensive set of stereological formulae for embedded aggregates of not-necessarily-convex particles. *Journal of Microscopy*, 138:115–125, 1985.

[408] R.E. Miles and P.J. Davy. Precise and general conditions for the validity of a comprehensive set of stereological fundamental formulae. *Journal of Microscopy*, 107:211–226, 1976.

[409] R.E. Miles and P.J. Davy. Probabilistic foundations of stereology. In E Underwood, R de Wit, and G A Moore, editors, *Fourth International Congress on Stereology*, Washington, 1976. US Government Printing Office. National Bureau of Standards special publication number 431.

[410] R.E. Miles and P.J. Davy. On the choice of quadrats in stereology. *Journal of Microscopy*, 110:27–44, 1977.

[411] R.E. Miles and P.J. Davy. Particle number of density can be stereologically estimated by wedge sections. *Journal of Microscopy*, 113:45–53, 1978.

[412] R.E. Miles and J. Serra, editors. *Geometrical Probability and Biological Structures: Buffon's 200th Anniversary*, Lecture Notes in Biomathematics, No 23, Berlin-Heidelberg-New York, 1978. Springer Verlag.

[413] W.S. Miller and E.P. Carlton. The relation of the cortex of the cat's kidney to the volume of the kidney, and an estimation of the number of glomeruli. *Transactions of the Wisconsin Academy of Sciences*, 10:525–538, 1895.

[414] A. Milne. The centric systematic area sample treated as a random sample. *Biometrics*, 15:270–297, 1959.

[415] I. Molchanov. *Statistics of the Boolean Model for practitioners and mathematicians*. Wiley, 1997.

[416] A. Møller, P. Strange, and H.J.G. Gundersen. Efficient estimation of cell volume and number using the nucleator and disector. *Journal of Microscopy*, 159:61–71, 1990.

[417] P.A.P. Moran. Measuring the surface area of a convex body. *Annals of Mathematics*, 45:793–799, 1944.

[418] P.A.P. Moran. Surface area of small objects. *Nature*, 154:490–491, 1944.

[419] P.A.P. Moran. Numerical integration by systematic sampling. *Proceedings of the Cambridge Philosophical Society*, 46:111–115, 1950.

[420] P.A.P. Moran. Measuring the length of a curve. *Biometrika*, 53:359–364, 1966.

[421] P.A.P. Moran. A note on recent research in geometrical probability. *Journal of Applied Probability*, 3:453–463, 1966.

[422] P.A.P. Moran. Statistical theory of a high-speed photoelectric planimeter. *Biometrika*, 55:419–422, 1968.

[423] P.A.P. Moran. A second note on recent research in geometrical probability. *Advances in Applied Probability*, 1:73–89, 1969.

[424] P.A.P. Moran. The probabilistic basis of stereology. *Advances in Applied Probability*, 4 (Supplement):69–71, 1972.

[425] P.A.P. Moran. Quaternions, Haar measure and the estimation of a palaeomagnetic rotation. In J. Gani, editor, *Perspectives in Probability and Statistics: Papers in honour of M.S. Bartlett*, pages 295–302. Applied Probability Trust/Academic Press, London, 1975.

[426] B.R. Morris, A.M. Gokhale, and G.F. Vander Voort. Grain size estimation in anisotropic materials. *Metallurgical and Materials Transactions A-Physical Metallurgy and Materials Science*, 29:237–244, 1998.

[427] P.R. Mouton. *Principles And Practices Of Unbiased Stereology: An Introduction For Bioscientists*. The Johns Hopkins University Press, Baltimore and London,

2002.

[428] P.R. Mouton, A.M. Gokhale, N.L. Ward, and M.J. West. Stereological length estimation using spherical probes. *Journal of Microscopy*, 206(1):54–64, 2002.

[429] A. Müller, L.M. Cruz-Orive, P. Gehr, and E.R. Weibel. Comparison of two subsampling methods for electron microscopic morphometry. *Journal of Microscopy*, 123:35–49, 1981.

[430] M.J. Mulvany, U. Baandrup, and H.J.G. Gundersen. Evidence for hyperplasia in mesenteric resistance vessels of spontaneously hypertensive rats using a three-dimensional disector. *Circulation Research*, 57:794–800, 1985.

[431] A.O. Myking and L.M. Cruz-Orive. Rapid volume ratio estimation for 'star specimens'. *Mikroskopie*, 37 (Supplement):42–45, 1980.

[432] S. Nagel. A best unbiased ratio-estimator in stereology. In W. Nagel, editor, *Geobild '85*, pages 147–153, Jena, DDR, 1985. Friedrich-Schiller-Universität Jena.

[433] W. Nagel. Dünne Schnitte von stationäre räumlichen Faserprozessen. *Mathematische Operationsforschung und Statistik, series Statistics*, 14:569–576, 1983.

[434] W. Nagel and J. Ohser. On the stereological estimation of weighted sphere diameter distributions. *1988*, 7:17–31, Acta Stereologica.

[435] W. Näther. *Effective observation of random fields.* Number 72 in Teubner-Texte zur Mathematik. Teubner, Leipzig, 1985.

[436] F. Natterer. *The mathematics of computerized tomography.* Society for Industrial and Applied Mathematics (SIAM), Philadelphia, PA, 2001. Reprint of the 1986 original.

[437] R.C. Nester. Grain size standard. Letter to the Editor. *ASTM Standardization News*, March 2003. American Society for Testing and Materials.

[438] W.L. Nicholson, editor. *Symposium on Statistical and Probabilistic Problems in Metallurgy, Seattle, Washington.* Applied Probability Trust, 1972. Special Supplement to Advances in Applied Probability, 1972.

[439] W.L. Nicholson. Estimation of linear functionals by maximum likelihood. *Journal of Microscopy*, 107:323–336, 1976.

[440] D. Nychka, G. Wahba, S. Goldfarb, and T. Pugh. Cross-validated spline methods for the estimation of three-dimensional tumor size distributions from observations on two-dimensional cross sections. *Journal of the American Statistical Association*, 79:832–846, 1984.

[441] J.R. Nyengaard and H.J.G. Gundersen. The isector: a simple and direct method for generating isotropic, uniform random sections from small specimens. *Journal of Microscopy*, 165:427–431, 1992.

[442] J. Ohser. *Grundlagen und praktische Möglichkeiten der Charakterisierung struktureller Inhomogenitäten von Werkstoffen.* Dr. sc. techn. thesis, Bergakademie Freiberg, Freiberg (Sachsen), Germany, 1990.

[443] J. Ohser. Variances for different estimators of specific line length, December 1991. Lecture at the meeting on *Stochastic Geometry, Geometric Statistics, Stereology*, Oberwolfach.

[444] J. Ohser and F. Mücklich. *Statistical analysis of microstructures in materials science.* John Wiley and Sons, Chichester, 2000.

[445] J. Ohser, B. Steinbach, and C. Lang. Efficient texture analysis of binary images. *Journal of Microscopy*, 192:20–28, 1998.

[446] Olympus Denmark A/S, Ballerup, Denmark. Stereology hardware/software pack-

age 'CAST-Grid'. http://www.cast-grid.com/.

[447] D.E. Oorschot. Are you using neuronal densities, synaptic densities or neurochemical densities as your definitive data? There is a better way to go. *Progress in Neurobiology*, 44:233–274, 1994.

[448] J. G. Osborne. Sampling errors of systematic and random surveys of cover-type areas. *Journal of the American Statistical Association*, 37:256–264, 1942.

[449] S. Oster, P. Christoffersen, H.J.G. Gundersen, J.O. Nielsen, C. Pedersen, and B. Pakkenberg. Six billion neurons lost in AIDS — a stereological study of the neocortex. *APMIS*, 103:525–529, 1995.

[450] F. O'Sullivan. A statistical perspective on ill-posed problems. *Statistical Science*, 1:502–518, 1986.

[451] J.C. Pache, N. Roberts, P. Vock, A. Zimmermann, and L.M. Cruz-Orive. Vertical LM sectioning and parallel CT scanning designs for stereology: application to human lung. *Journal of Microscopy*, 170:3–24, 1993.

[452] B. Pakkenberg, J. Boesen, M. Albeck, and F. Gjerris. Unbiased and efficient estimation of total ventricular volume of the brain obtained from CT-scans by a stereological method. *Neuroradiology*, 31:413–417, 1989.

[453] B. Pakkenberg and H.J.G. Gundersen. Total number of neurons and glial cells in human brain nuclei estimated by the disector and the fractionator. *Journal of Microscopy*, 150:1–20, 1988.

[454] B. Pakkenberg and H.J.G. Gundersen. Neucortical neuron number in humans: Effect of sex and age. *Journal of Comparative Neurology*, 384:312–320, 1997.

[455] B. Pakkenberg, A. Møller, H.J.G. Gundersen, A. Mouritzen Dam, and H. Pakkenberg. The absolute number of nerve cells in substantia nigra in normal and in patients with Parkinson's disease. *Journal of Neurology, Neurosurgery and Psychiatry*, 54:30–33, 1991.

[456] H. Pakkenberg, B.B. Andersen, R.S. Burns, and B. Pakkenberg. A stereological study of substantia nigra in young and old rhesus monkeys. *Brain Research*, 693:201–206, 1995.

[457] I.H. Parkinson and N.L. Fazzalari. Methodological principles for fractal analysis of trabecular bone. *Journal of Microscopy*, 198:134, 2000.

[458] E. Parzen. *Stochastic processes.* Holden-Day, San Francisco, 1962.

[459] H.D. Patterson. The errors of lattice sampling. *Journal of the Royal Statistical Society, Series B*, 16:140–149, 1954.

[460] D. Paumgartner, G. Losa, and E.R. Weibel. Resolution effect on the stereological estimation of surface and volume and its interpretation in terms of fractal dimensions. *J. Microscopy*, 121:51–63, 1981.

[461] J.B. Pawley. *Handbook of biological confocal microscopy.* Plenum Press, New York, 1995.

[462] A. Peters, J.H. Morrison, D.L. Rosene, and B.T. Hyman. Feature article: are neurons lost from the primate cerebral cortex during normal aging? *Cerebral Cortex*, 8:295–300, 1998.

[463] J.R. Philip. Some integral equations in geometrical probability. *Biometrika*, 53:365–374, 1966.

[464] E. Philofsky and J. Hilliard. On the measurement of the orientation distribution of lineal and areal arrays. *Quarterly Journal of Applied Mathematics (Oxford)*, 27:79–86, 1969.

[465] F.F. Pitard. *Pierre Gy's sampling theory and sampling practice: heterogeneity, sampling correctness, and statistical process control.* CRC Press, Boca Raton, Florida, USA, 1993.

[466] E.P. Polushkin. Determination of structural composition of alloys by a metallographic planimeter. *Trans. AIMME*, 71:669–690, 1925.

[467] C.M. Pover and R.E. Coggeshall. Verification of the disector method for counting neurons, with comments on the empirical method. *Anatomical Record*, 231:573–578, 1991.

[468] M. Prokešová. Bayesian MCMC estimation of the rose of directions. *Kybernetika*, 39:701–718, 2003.

[469] J.S. Prothero and J.W. Prothero. Three-dimensional reconstruction from serial sections. IV: The reassembly problem. *Computers and Biomed. Res.*, pages 361–373, 1986.

[470] M.H. Quenouille. Problems in plane sampling. *Annals of Mathematical Statistics*, 20:335–375, 1949.

[471] B. Ralph and K.J. Kurzydłowski. The philosophy of microscopic quantification. *Materials Characterization*, 38:217–227, 1997.

[472] C.R. Rao. Information and accuracy obtainable in estimation of statistical parameters. *Bulletin of the Calcutta Mathematical Society*, 37:81–91, 1945.

[473] T. Rasmussen, T. Schliemann, J.C. Sorensen, J. Zimmer, and M.J. West. Memory impaired aged rats: No loss of principal hippocampal and subicular neurons. *Neurobiology of Ageing*, 17:143–147, 1996.

[474] L. Regeur, G. Badsberg Jensen, H. Pakkenberg, S.M. Evans, and B. Pakkenberg. No global neocortical nerve cell loss in brains from patients with senile dementia of Alzheimer's type. *Neurobiology of Aging*, 15:347–352, 1994.

[475] W.P. Reid. Distribution of sizes of spheres in a solid from a study of slices of the solid. *J Math Phys*, 34:95–102, 1955.

[476] R.-D. Reiss. *A course on point processes.* Springer, 1993.

[477] F.N. Rhines. Measurement of topological parameters. In H. Elias, editor, *Stereology. Proceedings of the Second International Congress for Stereology*, pages 235–250. Springer-Verlag, New York, 1967.

[478] J.-P. Rigaut. An empirical formulation relating boundary lengths to resolution in specimens showing 'non-ideally fractal' dimensions. *J. Microscopy*, 133:41–54, 1984.

[479] B.D. Ripley. *Spatial statistics.* John Wiley and Sons, New York, 1981.

[480] H.E. Robbins. On the measure of a random set. *Annals of Mathematical Statistics*, 15:70–74, 1944.

[481] H.E. Robbins. On the measure of a random set II. *Annals of Mathematical Statistics*, 16:342–347, 1945.

[482] H.E. Robbins. Acknowledgement of priority. *Annals of Mathematical Statistics*, 18:297, 1947.

[483] N. Roberts, A.S. Garden, L.M. Cruz-Orive, G.H. Whitehouse, and R.H.T. Edwards. Estimation of fetal volume by MRI and stereology. *The British Journal of Radiology*, 67:1067–1077, 1994.

[484] N. Roberts, C.V. Howard, L.M. Cruz-Orive, and R.H.T. Edwards. The application of total vertical projections for the unbiased estimation of the length of blood vessels and other structures by magnetic resonance imaging. *Magnetic Resonance*

Imaging, 9:917–925, 1991.

[485] N. Roberts, G. Nesbitt, and F. Fens. Visualisation and quantification of oil and water phases in a synthetic porous medium using NMRM and stereology. *Nondestructive Testing and Evaluation*, 11:273–291, 1994.

[486] A. Rosiwal. Über geometrische Gesteinsanalysen. Ein einfacher Weg zur ziffermäßigen Feststellung des Quantitätsverhältnisses der Mineralbestandteile gemengter Gesteine. *Verhandlungen der Kaiserlich-Königlichen Geologischen Reichsanstalt Wien*, pages 143–175, 1898. (On geometric rock analysis. A simple surface measurement to determine the quantitative content of the mineral constituents of a stony aggregate).

[487] H. Röthlisberger. An adequate method of grain-size determination in sections. *Journal of Geology*, 63:579–584, 1955.

[488] R.M. Royall. On finite population sampling theory under certain linear regression models. *Biometrika*, 57:377–387, 1970.

[489] J.C. Russ and R.T. DeHoff. *Practical stereology*. Plenum Press, second edition, 1999.

[490] S.A. Saltykov. Sposub geometričeskogo količestvennogo analiza metallov (The method of sections in metallography). Patent 72704 USSR 28, 18 may 1945. Announced in *Bjul. izobr.* 10 (1948) 27.

[491] S.A. Saltykov. The method of intersections in metallography. *Zavodskaya Laboratoriya*, 12(9–10):816–825, 1946. (In Russian).

[492] S.A. Saltykov. Determination of surface area of grains in a deformed structure. *Zavodskaya Laboratoriya*, 18(5):575–579, 1952. (In Russian).

[493] S.A. Saltykov. *Stereometric metallography*. State Publishing House for Metals and Sciences, Moscow, 2nd edition, 1958. 3rd edition published 1970.

[494] S.A. Saltykov. The determination of the size distribution of particles in an opaque material from measurement of the size distribution of their sections. In H. Elias, editor, *Second International Congress for Stereology (Proceedings)*, pages 163–173, New York, 1967. Springer-Verlag.

[495] S.A. Saltykov. *Stereometrische Metallographie*. VEB Deutscher Verlag für Grundstoffindustrie, Leipzig, DDR, 1974. German translation, edited by H.-J. Eckstein.

[496] P.D. Sampson and P. Guttorp. Nonparametric estimation of nonstationary spatial covariance structure. *J. Amer. Statist. Assoc.*, 87:108–119, 1992.

[497] K. Sandau. How to estimate the area of a surface using a spatial grid. *Acta Stereologica*, 6(3):31–36, 1987.

[498] K. Sandau and U. Hahn. Some remarks on the accuracy of surface area estimation using the spatial grid. *Journal of Microscopy*, 173:67–72, 1994.

[499] L.A. Santaló. Sobre la distribución de las tamaños de corpúsculas contenidos en un cuerpo a partir de la distribución en sus secciones a projecciones. *Trabajos de Estadística*, 6:191–196, 1955.

[500] L.A. Santaló. *Integral Geometry and Geometric Probability*. Encyclopedia of Mathematics and Its Applications, vol. 1. Addison-Wesley, 1976.

[501] C.B. Saper. Any way you cut it: A new journal policy for the use of unbiased counting methods. *Journal of Comparative Neurology*, 364:5, 1996.

[502] C.B. Saper. Reply. *Trends in Neurosciences*, 22(347), 1999.

[503] I. Saxl. *Stereology of objects with internal structure*. Number 50 in Materials

Science Monographs. Elsevier, 1989.

[504] E. Scheil. Die Berechnung der Anzahl und Grossenverteilung kugelformiger Kristalle in undurchsichtigen Körpern mit Hilfe der durch einen ebenen Schnitt erhalten Schnittkreise. *Zeitschr. anorg. allgem. Chem.*, 201:259–264, 1931.

[505] E. Scheil. Statistische Gefügeuntersuchungen I. *Z. Metallk.*, 27:199–??, 1935.

[506] K. Schladitz. Surprising optimal estimators for the area fraction. *Advances in Applied Probability*, 31:995–1001, 1999.

[507] C. Schmitz, H. Korr, and H. Heinsen. Design-based counting techniques: the real problems. *Trends in Neurosciences*, 22:345, 1999.

[508] C. Schmitz, H. Korr, D.P. Perl, and P.R. Hof. Advanced use of 3-D methods for counting neurons. *Trends in Neurosciences*, 24:377–378, 2001.

[509] R. Schneider and W. Weil. *Integralgeometrie.* Teubner Skripten zur Mathematischen Stochastik. [Teubner Texts on Mathematical Stochastics]. B. G. Teubner, Stuttgart, 1992.

[510] R. Schneider and W. Weil. *Stochastische Geometrie.* Teubner, Stuttgart, 2000.

[511] A. Schwandtke, J. Ohser, and D. Stoyan. Improved estimation in planar sampling. *Acta Stereologica*, 6(2):325–334, 1987.

[512] A.R. Sen. Present status of probability sampling and its use in estimation of farm characteristics (abstract). *Econometrica*, 20:103, 1952.

[513] J. Serra. Stereology and structuring elements. *Journal of Microscopy*, 95:93–103, 1972.

[514] J. Serra. *Image analysis and mathematical morphology.* Academic Press, London, 1982.

[515] J. Serra, editor. *Image analysis and mathematical morphology, volume 2: Theoretical advances.* Academic Press, London, 1988.

[516] B. W. Silverman, M. C. Jones, J. D. Wilson, and D. W. Nychka. A smoothed EM approach to indirect estimation problems, with particular reference to stereology and emission tomography (with discussion). *Journal of the Royal Statistical Society, series B*, 52:271–324, 1990.

[517] H. Smalbruch. The number of neurons in dorsal root ganglia L4–L6 of the rat. *Anatomical Record*, 219:315–322, 1987.

[518] J.C. Smit. Estimation of the mean of a stationary stochastic process by equidistant observations. *Trabajos de Estadistica*, 12:35–45, 1961.

[519] C.S. Smith and L. Guttman. One and two dimensional sections of three dimensional structures. *Journal of Metals*, 4(2):150, 1952.

[520] C.S. Smith and L. Guttman. Measurement of internal boundaries in three-dimensional structures by random sectioning. *Transactions of the AIME (Journal of Metals)*, 197:81–87, 1953.

[521] A.J. Smolen, L.L. Wright, and T.J. Cunningham. Neuron numbers in the superior cervical sympathetic ganglion of the rat: a critical comparison of methods for cell counting. *Journal of Neurocytology*, 12:729–750, 1983.

[522] G.W. Snedecor and W.G. Cochran. *Statistical methods.* Iowa State University Press, seventh edition, 1980.

[523] R.R. Sokal and F.J. Rohlf. *Biometry*. W.H. Freeman and Co., second edition, 1981.

[524] H. Solomon. A coverage distribution. *Annals of Mathematical Statistics*, 21:139–140, 1950.

[525] H. Solomon. Distribution of the measure of a random two-dimensional set. *Annals of Mathematical Statistics*, 24:650–656, 1953.

[526] H. Solomon. *Geometric Probability*. Number 28 in CBMS-NSF Regional Conference Series in Applied Mathematics. Society for Industrial and Applied Mathematics, Philadelphia, Pennsylvania, 1978.

[527] F.B. Sørensen. Objective histopathological grading of cutaneous malignant melanomas by stereological estimation of nuclear volume: prediction of survival and disease-free period. *Cancer*, 63:1784–1798, 1989.

[528] S. Souchet. Précision de l'estimation de Cavalieri. Rapport de stage, D.E.A. de Statistiques et Modèles Aléatoires appliqués à la Finance, Université Paris-VII, Laboratoire de Biométrie, INRA, Versailles, 1995.

[529] A.G. Spektor. Analysis of distribution of spherical particles in non-transparent structures. *Zavodskaja Laboratoria*, 16:173–177, 1950.

[530] A.G. Spektor. Opredelenie orientatsii poverkhnosti elementov mikrostrukturei (Determination of the orientation of surface elements of microstructures). *Zavodskaya Laboratoriya*, 21(8):955–960, 1955.

[531] A.G. Spektor. Izmerenije poverkhnosti strukturi sostavlyushchikh metodom sekushcheii tsikloidy (Measuring the surface of structural components using the method of intersecting cycloids). *Zavodskaya Laboratoriya*, 26(3):303–305, 1960.

[532] M. Spivak. *Differential Geometry*, volume II. Publish or Perish, Inc., Berkeley, 1979.

[533] A. Steiner, M. Binder, U. Mossbacher, K. Wolff, and H. Pehamberger. Estimation of the volume-weighted mean nuclear volume discriminates Spitzs nevi from nodular malignant melanomas. *Laboratory Investigation*, 70:381–385, 1994.

[534] H. Steinhaus. Sur la portée pratique et théorique de quelques theéorèmes sur la mesure des ensembles de droites. In *Comptes Rendus 1er Congr. Mathématiciens des Pays Slaves*, pages 348–354, Warszawa, 1929.

[535] H. Steinhaus. Praxis der Rectifikation und zur Längenbegriff. *Berichten Akademie der Wissenschaften Leipzig*, pages 120–130, 1930.

[536] H. Steinhaus. Length, shape and area. *Colloq. Math*, 3:1–13, 1954.

[537] Stereology Resource Center, Chester, Maryland, USA. Software package 'Stereologer'. `http://www.disector.com/`.

[538] D.C. Sterio. The unbiased estimation of number and size of arbitrary particles using the disector. *Journal of Microscopy*, 134:127–136, 1984.

[539] D. Stoyan. On the accuracy of lineal analysis. *Biometrical Journal*, 21:439–449, 1979.

[540] D. Stoyan. Proofs of some fundamental formulae of stereology for non-Poisson grain models. *Mathematische Operationsforschung und Statistik, series Optimization*, 10:573–581, 1979.

[541] D. Stoyan. Stereological formulae for size distributions through marked point processes. *Probability and Mathematical Statistics*, 2:161–166, 1982.

[542] D. Stoyan. Stereology and stochastic geometry. *International Statistical Review*, 58:227–242, 1990.

[543] D. Stoyan, W.S. Kendall, and J. Mecke. *Stochastic Geometry and its Applications*. John Wiley and Sons, Chichester, second edition, 1995.

[544] D. Stoyan and J. Mecke. *Stochastische Geometrie: eine Einführung*. Akademie-

Verlag, Berlin, 1983.

[545] D. Stoyan and H. Stoyan. *Fractals, random shapes and point fields.* Wiley, 1995.

[546] D. Stoyan and H. Stoyan. Improving ratio estimators of second order point process characteristics. *Scandinavian Journal of Statistics*, 27:641–656, 2000.

[547] D.J. Struik. *Lectures on classical differential geometry.* Addison-Wesley, 1961.

[548] A. Stuart. *The Ideas of Sampling.* Griffin, London, 1984.

[549] R. Takahashi and M. Sibuya. The maximum size of the planar sections of random spheres and its application to metallurgy. *Ann. Inst. Statist. Math.*, 48:361–377, 1996.

[550] T. Tandrup. A method for unbiased and efficient estimation of number and mean volume of specified neuron subtypes in rat dorsal root ganglion. *Journal of Comparative Neurology*, 329:269–276, 1993.

[551] T. Tandrup and H. Brændgaard. Number and volume of rat dorsal root ganglion cells in acrylamide intoxication. *Journal of Neurocytology*, 23:242–248, 1994.

[552] T. Tandrup, H.J.G. Gundersen, and E.B.V. Jensen. The optical rotator. *Journal of Microscopy*, 186:108–120, 1997.

[553] C.C. Taylor. A new method for unfolding sphere size distributions. *Journal of Microscopy*, 132:57–66, 1983.

[554] J. Thioulouse, J.P. Royet, H. Ploye, and F. Houllier. Evaluation of the precision of systematic sampling: nugget effect and covariogram modelling. *Journal of Microscopy*, 172:249–256, 1993.

[555] S.K. Thompson. *Sampling.* John Wiley and Sons, 1992.

[556] S.K. Thompson and G.A.F. Seber. *Adaptive Sampling.* Wiley, New York, 1996.

[557] W.R. Thompson. The geometric properties of microscopic configurations. I. General aspects of projectometry. *Biometrika*, 24:21–26, 1932.

[558] W.R. Thompson, R. Hussey, J.T. Matteis, W.C. Meredith, G.C. Wilson, and F.E. Tracy. The geometric properties of microscopic configurations. II. Incidence and volume of islands of Langerhans in the pancreas of a monkey. *Biometrika*, 24:27–38, 1932.

[559] E. Thomson. Quantitative microscopic analysis. *Journal of Geology*, 38:193–222, 1930.

[560] S.T. Tomkeieff. Linear intercepts, areas and volumes. *Nature*, 155:24, 1945. correction on page 107.

[561] E. Underwood, R. de Wit, and G.A. Moore, editors. *Fourth International Congress on Stereology*, Washington, 1976. US Government Printing Office. National Bureau of Standards special publication number 431.

[562] E.E. Underwood. *Quantitative Stereology.* Addison-Wesley, Reading, Mass, 1970.

[563] E.E. Underwood. The stereology of projected images. *Journal of Microscopy*, 95:25–44, 1972.

[564] A. van der Vaart. *Asymptotic statistics.* Cambridge University Press, 1998.

[565] F. Valladares, L.G. Sancho, and C. Ascaso. Water storage in the Lichen family Umbilicariaceae. *Botanica Acta*, 111:99–107, 1998.

[566] B. van Es and A. Hoogendoorn. Kernel estimation in Wicksell's corpuscle problem. *Biometrika*, 77:139–145, 1990.

[567] A. Vesterby. Star volume of marrow space and trabeculae in iliac crest: sampling

procedure and correlation to star volume of first lumbar vertebra. *Bone*, 11:149–155, 1990.

[568] A. Vesterby, H.J.G. Gundersen, and F. Melsen. Star volume of marrow space and trabeculae of first lumbar vertebra: sampling efficiency and biological variation. *Bone*, 10:7–13, 1989.

[569] A. Vesterby, J. Kragstrup, H.J.G. Gundersen, and F. Melsen. Unbiased stereologic estimation of surface-density in bone using vertical sections. *Bone*, 8:13–17, 1987.

[570] S. Ya. Vilenkin. On the estimation of the mean in stationary processes. *Theory of Probability and its Applications*, 4:415–416, 1959.

[571] Voltaire. *Micromégas: Histoire Philosophique.* Oeuvres de Voltaire, Tome XXXIII. Imprimerie A. Firmin Didot, Paris, 1829. Original publication date approx 1752.

[572] G. Vander Voort. Committee E-4 and grain size measurements: 75 years of progress. *ASTM Standardization News,*, May 1991.

[573] G.S. Watson. Estimating functionals of particle size distributions. *Biometrika*, 58:483–490, 1971.

[574] G.S. Watson. Characteristic statistical problems of stochastic geometry. In *Geometrical Probability and Biological Structures: Buffon's 200th Anniversary*, Lecture Notes in Biomathematics, No 23. Springer Verlag, Berlin-Heidelberg-New York, 1978.

[575] E.R. Weibel. *Morphometry of the human lung.* Springer, Berlin etc, 1963.

[576] E.R. Weibel. Structure in space and its appearance on sections. In H. Elias, editor, *Second International Congress for Stereology*, pages 15–, New York, 1967. Springer Verlag.

[577] E.R. Weibel. A stereological method for estimating volume and surface of endoplasmic reticulum. *Journal of Microscopy*, 95:229–242, 1972.

[578] E.R. Weibel. The value of stereology in analysing structure and function of cells and organs. *Journal of Microscopy*, 95:3–13, 1972.

[579] E.R. Weibel. The non-statistical nature of biological structure and its implications on sampling for stereology. In R.E. Miles and J. Serra, editors, *Geometrical Probability and Biological Structures: Buffon's 200th Anniversary*, Lecture Notes in Biomathematics, No 23. Springer Verlag, Berlin-Heidelberg-New York, 1978.

[580] E.R. Weibel. *Stereological Methods, 1. Practical Methods for Biological Morphometry.* Academic Press, London, 1979.

[581] E.R. Weibel. *Stereological Methods, 2. Theoretical Foundations.* Academic Press, London, 1980.

[582] E.R. Weibel and L.M. Cruz-Orive. Morphometric methods. In R.G. Crystal, J.B. West, et al., editors, *The Lung*, chapter 22, pages 333–344. Lippincott-Raven Publishers, Philadelphia, PA, 1997.

[583] E.R. Weibel and H. Elias. Introduction to stereologic principles. In E.R. Weibel and H. Elias, editors, *Quantitative methods in morphology*, pages 89–98. Springer-Verlag, 1966.

[584] E.R. Weibel and H. Elias. Introduction to stereology and morphometry. In E.R. Weibel and H. Elias, editors, *Quantitative methods in morphology*, pages 3–19. Springer-Verlag, 1966.

[585] E.R. Weibel and H. Elias, editors. *Quantitative methods in morphology.* Springer-

Verlag, 1966.

[586] E.R. Weibel, C. Fisher, J. Gahm, and A. Schaefer. Current capabilities and limitations of available stereological techniques. *Journal of Microscopy*, 95:367–392, 1972.

[587] E.R. Weibel and D.M. Gomez. A principle for counting tissue structures on random sections. *Journal of Applied Physiology*, 17:343–348, 1962.

[588] E.R. Weibel, G.S. Kistler, and W.R. Scherle. Practical stereological methods for morphometric cytology. *Journal of Cell Biology*, 30:23–38, 1966.

[589] E.R. Weibel and B.W. Knight. A morphometric study on the thickness of the pulmonary air-blood barrier. *Journal of Cell Biology*, 21:367–384, 1964.

[590] E.R. Weibel, G. Meek, B. Ralph, P. Echlin, and R Ross, editors. *Third International Congress for Stereology*, Oxford–London–Edinburgh–Melbourne, 1972. Blackwell. reprinted from Journal of Microscopy, vol 95, parts 1 and 2.

[591] E.R. Weibel, W. Stäubli, H.R. Gnägi, and F.A. Hess. Correlated morphometric and biochemical studies on the liver cell. I. Morphometric model, stereologic methods, and normal morphometric data for rat liver. *Journal of Cell Biology*, 42:68–91, 1969.

[592] W. Weil. Stereology: a survey for geometers. In P.M. Gruber and J.M. Wills, editors, *Convexity and its applications*, pages 360–412. Birkhauser, Basel, Boston, Stuttgart, 1983.

[593] R.S. Weinstein and S. Majumdar. Fractal geometry and vertebral compression fractures. *J. Bone Miner. Res.*, 4:1797–1802, 1994.

[594] M.J. West. New stereological methods for counting neurons. *Neurobiology Of Aging*, 14:275–285, 1993.

[595] M.J. West. Advances in the study of age-related neuron loss. *Seminars in Neuroscience*, 28:403–411, 1994.

[596] M.J. West. Reply. *Trends in Neurosciences*, 22:345–346, 1999.

[597] M.J. West. Stereological methods for estimating the total number of neurons and synapses: issues of precision and bias,. *Trends in Neurosciences*, 22:51–61, 1999.

[598] M.J. West, K. Æstergaard, O.A. Andreassen, and B. Finsen. Estimation of the number of somatostatin neurons in the striatum: An in situ hybrizidation study using the optical fractionator method. *Journal of Comparative Neurology*, 370:11–22, 1996.

[599] M.J. West and H.J.G. Gundersen. Unbiased stereological estimation of the number of neurons in the human hippocampus. *Journal of Compararative Neurology*, 296:1–22, 1990.

[600] M.J. West and L. Slomianka. What is an optical disector? *Trends in Neurosciences*, 24:374, 2001.

[601] M.J. West, L. Slomianka, and H.J.G. Gundersen. Unbiased stereological estimation of the total number of neurons in the subdivisions of the rat hippocampus using the optical fractionator. *Anatomical Record*, 231:482–497, 1991.

[602] S.D. Wicksell. Et iagttagelsesteoretisk problem. *Assurandøren*, 24:243, 1923. (In Danish).

[603] S.D. Wicksell. The corpuscle problem. A mathematical study of a biometric problem. *Biometrika*, 17:84–89, 1925.

[604] S.D. Wicksell. The corpuscle problem. Second memoir. Case of ellipsoidal corpuscles. *Biometrika*, 18:152–172, 1926.

[605] B.M. Wiebe and H. Laursen. Lung morphometry by unbiased methods in emphysema: bronchial and blood vessel volume, alveolar surface area and capillary length. *APMIS*, 106:651–656, 1998.

[606] R.W. Williams and P. Rakic. Three-dimensional counting: an accurate and direct method to estimate numbers of cells in sectioned material. *Journal of Comparative Neurology*, 278:344–352, 1988.

[607] R.W. Williams and P. Rakic. Erratum and addendum. *Journal of Comparative Neurology*, 281:335, 1989.

[608] T Wilson, editor. *Confocal Microscopy*. Academic Press, London, 1990.

[609] J.R. Wilton. The lattice points of a circle: a historical account of the problem. *Messenger of Mathematics*, 58:67–80, 1929.

[610] L. Wojnar. Unbiased estimation of fracture surface area. *Acta Stereologica*, 11:651–656, 1992.

[611] B. Wondimu, F.P. Reinholt, and T. Modeer. Stereologic study of Cyclosporine-A-induced gingival overgrowth in renal-transplant patients. *European Journal of Oral Sciences*, 103:199–206, 1995.

[612] D. Wulfsohn, J.R. Nyengaard, H.J.G. Gundersen, A.J. Cutler, and T.M. Squires. Non-destructive, stereological estimation of plant root length, branching pattern and diameter distribution. *Plant & Soil*, 214:15–26, 1999.

[613] Z.-W. Yang, Y.-H. Qin, and S.-R. Su. Use of star volume to measure the size of the alveolar space in the asthmatic guinea-pig lung. *Respirology*, 7:117–121, 2002.

[614] D. Ylvisaker. Designs on random fields. In J. Srivastava, editor, *A survey of statistical design and linear models*, pages 593–607. North-Holland, Amsterdam, 1975.

[615] R.S. Yuill. Areal measurement error with a dot planimeter: some experimental estimates. Interagency Report 213, U.S. Geological Survey, Washington, 1971. 9 pages.

[616] M. Zähle. A kinematic formula and moment measures of random sets. *Mathematische Nachrichten*, 149:325–340, 1990.

[617] J.H. Zar. *Biostatistical analysis*. Prentice Hall, Upper Saddle River, N.J., 4th ed edition, 1999.

[618] S. Zubrzycki. Remarks on random, stratified and systematic sampling in a plane. *Colloquium Mathematicum*, 6:251–264, 1958.

Index

Abercrombie's method, 39, 81, 256
  bias, 39, 41
  contrasted with empirical method, 247
  experimental studies, 41, 256
  ill-defined, 47
  method 1, 40, 256
  method 2, 40
  ratio estimation in, 47
  weaknesses, 40, 47
Abercrombie's relation, 39, 47
Abercrombie, M., 29, 31, 39, 40, 48, 53
absolute quantities, 140, 279
  definition, 140
  distinct from relative parameters, *see* parameters
  estimation by stratified sampling, 281
  estimation from IUR designs, 155–173
  estimation from vertical designs, 179, 182
  estimation via ratios, 213–216
  require knowledge of section thickness, 280
accuracy, *see* bias, *see* variance
  deterministic view, 13
  estimate of volume fraction, 10, 13, 297
  of stereology, 80
  point counting estimators, 13, 17, 327
  point counting vs other methods, 13, 327
  sampling methods, 13
adaptive cluster sampling, 71, 246
Aitkin, M., 329
allocation of sampling effort, 80, 318
  Do More Less Well, 321
  pilot experiments, 292
  stratification, 281
  subsampling, 282
analysis of variance
  nested ANOVA, 284, 293, 318–321
  non-Normal, 329
Andersen, B.B., 41
Anderssen, R.S., 36, 53
anisotropy, 330
Antoniadis, A., 53
Archimedes, 2, 172
area fraction, 11, 22
  estimation by length fraction, 11
  estimation by point counting, 12, 219
  units, 22
area of plane region
  by point-counting, 13, 158
  estimation, 13, 158–159
  in cartography, 12, 13
  systematic subsampling, 161
  using line probes, 88, 90
area of surface in space, *see* surface area
area-weighted mean profile area
  spherical particles, 83
area-weighted sampling, 64
  for unbiased estimation of ratios, 65, 224
  sampling designs, 224
associated point rule for counting profiles, *see* counting rules
'assumption-free' stereology, 82
assumptions
  about covariance, 299, 301
  about particles
    ellipsoidal particles, 37
    particle shape, 29, 34
    size-shape distribution, 140
    spherical particles, 34–37
  for calculating variances, 322
  for variance estimation, 298
  geometry of boundaries, 308

homogeneity, 14, 20, 26, 28, 30, 32, 58, 59, 61, 139, 153
in Extended case, 137, 149
in Random case, 138, 144
in Restricted case, 137, 147
independence of replicates, 325
isotropy, 20, 61, 297
isotropy about the vertical axis, 175
Poisson process, 58, 299
random effects model, 320
regularity conditions, 108
stationarity, 140
ASTM, 41
axial sections, 175, 199–202
are not vertical sections, 177

Bach, G., 35, 53, 327
Baddeley, A.J., 5, 17, 53, 71, 81, 83, 142, 173, 175, 181, 206, 253, 299, 315, 328
Barbier, E., 53, 133
Basu, D., 339
Bendtsen, T.F., 84
Benes, F., 81
Benes, F.M., 5, 58
Beneš, V., 83, 133, 331
Bertram, J., iii, 54, 83, 253
best linear unbiased estimation
conditional, 225
for random field, 300
of a ratio, 212–213
bias, 50, 78–79, 228
and variance, 16
correction of, 244
difference of two biased estimates, 79
due to design errors, 129
due to thick sections, 39
eliminating, 338
estimation bias
definition, 235
external sources of, 289
in Abercrombie's method, 39
in ASTM standard E112, 41
in estimating population mean, 336
in ratio estimation, 210
in ratio estimator
minimising, 218
methodological, 289
most dangerous source of error, 78
sampling bias, 30–32, 234–235
definition, 235
leads to estimation bias, 235
sampling bias and estimation bias, 235
two meanings, 235
bias-variance tradeoff, 79
binomial approximation
to variance of point count, 301, 327
binomial regression, 226
biological variability, 80, 318, 320
overestimated by sample variance, 325
bivariate binomial model, 301
Blaschke, W. von, 54, 108
Blaschke-Petkantschin formulae, 202
bone, 188
biopsy, 152
fractals in, 331
star volume, 207
vertical sections, 176
vertical slices, 189
Boolean model, 45, 299
botany, 54, 188
Brændgaard, H., 276
brain
age-related neuron loss, 5, 62
cerebellum, 38
definition of reference space, 136
hippocampus, 41
importance of counting neurons, 25
number of neurons, 41
sampling design for, 282
thick section, 38
vertical slices, 189
breadcrust "theorem", 73, 96
is false, 97
Bronowski, J., 327
Buckland, S.T., 84, 252, 332
Buffon's needle problem, 133
Buffon, Comte de, 53, 133, 173

Cabo, A.J., 268
caliper diameter, 114, 116
mean, 116, 309
for convex sets, 116

capillaries, 22
Carlton, E.P., 252
Cartier's formula, 328
cartography
  cluster sampling, 232, 252
  estimation of area, 12, 13, 173, 327
  explanation for Jacobian, 93
  spherical coordinates, 93
Casley-Smith, J.R., 62, 143
Cavalieri estimator of volume, 65, 83, 156, 252
  applications, 66
  ratios, 214, 220
  requires section thickness, 214
  tissue shrinkage, 214
  variance, 157, 307, 308
Cavalieri's principle, 15, 157
Cavalieri, F.B., 157, 172
Chalkley, H., 19
Chia, J.L.C., 226, 228, 301, 314, 323, 328, 329
circus elephant example, 339
classical stereology, 9–54
  assumptions, 14
  bias corrections, 80
  conflates sources of variability, 59
  experimental studies of bias, 253
  history, 52
  homogeneity assumption ill-defined, 55
  homogeneity assumption inappropriate, 55, 59
  inference inappropriate, 59
  methodological bias, 289
  statistical critique of, 46
  vs modern stereology, 55
  was model-based, 5
cluster sampling, 230, 236, 237, 247, 252, 337
  adaptive, *see* adaptive cluster sampling
  in cartography, 232, 252
  in spatial sampling, 252
Cochran, W.G., 56, 78, 83, 209, 210, 228, 319, 328, 333, 335
Coggeshall, R., 5, 41, 71, 81, 84, 246, 253
Coleman, R., 53, 84, 153, 328
colon
  vertical sections, 177
concrete
  sand grains in, 25
  stereology of, 4
  vertical slices, 189
conditional BLUE, 225
conditional distribution
  of spherical profile diameter, 35
conditional property
  for FUR planes, 118
  for FUR quadrats, 121
  for IUR lines in the plane, 116
  for uniformly random points, 113
confocal microscopy, 156, 278, 284
  application of disector, 259
  problems with quantitation, 81
consistent estimation, 59, 145, 153
  of ratios, 211
containing space, 142, 155, 213, 229, 231
  definition, 142
  particles in, 233
  partition of, 232, 236
  stratification of, 281
contrast, 50, 289, 290
core samples, 3, 176
Cornfield, J., 19
Corrsin, S.J., 52
counting rules, 41, 68–71
  associated point rule, 69, 241
  guard region, 69
  minus-sampling, 68
    Miles-Lantuéjoul correction, 70
  multiple associated points, 242
  plus-sampling, 68, 234
  sampling bias, 68
  three dimensions, 70–71
  tiling rule, 69, 239
  unbiased, 68
  unbiased counting frame, 69
covariance
  between terms in a sum, 295
  explains variance paradoxes, 299
  of periodic processes, 317
  positive at small scales, 17
  spatial, *see* spatial covariance function
covariogram, 305

Cox, D.R., 84, 252
Cressie, N.A.C., 327, 330
Crofton formula, 72, 73, 87–96, 162
  for lines in 2D, 90
  general form, 89
Crofton, M.W., 53, 74, 158, 173
Cruz-Orive Models I and II, 225
Cruz-Orive, L.M., iv, 17, 35, 36, 38, 52, 53, 71, 81, 83, 84, 135, 143, 153, 181, 192, 214, 216, 218, 225–226, 228, 252, 253, 257, 258, 268, 299, 302, 308, 309, 311, 313, 315–317, 322, 323, 327, 328, 330, 331
curvature, 104–107
  Gaussian, 106
    integral of, 106
  integral of, 244
  mean, *see* mean curvature, 105
    integral of, 106
  of plane curve, 104
  of surface, 105
  principal, 105
  total, 243
cycloid, 77
  history, 206
cycloid test curves, 78, 184–185, 187, 191, 192
cycloidal test surfaces, 191
cylinder
  principal curvatures, 105
cylindrical coordinates, 74, 199
  expression for volume, 76, 199
cylindrical particles, 256
cylindrical structures, 22, 53
Czuber, E., 53, 133

Daniels, H.E., 84, 252
Davy, P.J., 3, 5, 56, 60, 64, 84, 133, 153, 228, 252, 314, 316, 327, 328
DeHoff, R.T., iii, 27, 29, 46, 52–55, 83, 242, 253, 257
Delesse's principle, 2, 9–17, 144
  accuracy, 3, 17
  in Extended case, 149
  precision, 16
  proof in Random case, 144
  reduction of effort, 3
  statistical content, 46
  unbiased, 15
  unbiased estimation, 145
  validity, 14
Delesse, A.E., 2, 9, 52, 153, 327
delta method, 211, 216
Deltheil, R., 53, 133
dendrites, 189
density, 11, 140
  length density, 20
  numerical density N_V, *see* numerical density
  stereological notation, 11, 22, 23
  surface area density, 19
  volume density, 11
dentistry, 54
dermatology, 188
design-based sampling inference, 57, 137, 148
design-based stereology, 5, 59–61, 137, 148
  applicability, 61
  controversy, 253
  depends on adherence to protocol, 61
  estimation of volume, 60
  first mention of, 153
  no assumptions about material, 60
  of particle populations, 257–269
  ratio parameters, 61
  unbiased estimation, 60
  variance estimation
    problems, 329
  variances, 302–314
  vertical sections, 175, 188
Dhani, A., 228, 328
Diggle, P.J., 327, 332
digital stereology, 330
dilation, 120
direction
  fixed, *see* FUR
  of line in 3D space, 95
  sin-weighted, 267
  vertical, 176, *see* vertical
directional statistics, 330
directions in 3D space, 124
  coordinates, 92, 124

isotropic random, 124, 285
directions in the plane, 123
coordinates, 123
isotropic random, 123
disector, 70, 82, 237, 257
basic principle, 238
comparison with empirical method, 253
history, 252
optical, 252, 259, 280
physical, 70, 252, 280, 287
basic principle, 238
systematic disectors, 259
three-dimensional observation, 280
disector count, 238
example data, 323
goodness-of-fit of Poisson, 323
statistical models, 321
Do More Less Well, 321
Dorph-Petersen, K.A., 188, 253, 276, 282, 291
Downie, A.S., 228, 328

ecological sampling, 97, 252
edema, 62
edge effects, 41, 122, 123, 253
for quadrats, 122, 123
efficiency
definition, 17
estimator of volume fraction, 17, 228
experimental studies, 327
gain using Rao-Blackwell, 268
improved by data modelling, 224
improved by PPS sampling, 338
improved by sub-sampling, 249
multi-stage ratio estimation, 216
of stereological techniques, 3
of subsampling, 282
of systematic sampling
optimization, 291
optimization of, 277
partition of material, 248
point counting vs line intercept measurement, 300
ratio estimator, 211, 223
superefficiency, 157, 304
explanation of, 304
of point count, 309
systematic sampling vs simple random sampling, 328, 339
theory, 327
egg slicer, 65, 155
Elias, H., 4, 18, 29, 52, 54, 55, 83
empirical method, 71, 245–247, 253, 257
comparison with disector, 253
environmental surveys, 252
erosion, 122, 298
estimation, 10, 46
bias and variance in, 16
choice of parameters, 49, 280
consistent, 37
efficiency of, 17
not always possible, 23
of particle size distribution, 58
of stereological ratios, 23, 216–224
fixed sample size, 218
general theory, 217
systematic sampling, 219
weighted sampling, 224
unbiased, *see* unbiased estimation
unidentifiability, 37
via ratios, 213–216
multi-stage, 214
estimation bias, *see* bias
caused by sampling bias, 235
estimation of ratios, 63
biased, 64
unbiased, 64
Evans, S.M., 265, 276, 331
Ewing, L.L., 41, 256, 289
Exner, H.E., iii
Extended case of stereological inference, 137, 149–151, 217–219, 223
extension term, 307, 308

Fan, J., 53
feature of interest
generic term, 135
operational definition, 290
Federer, H., 72
first-time rule, 237
fixed orientation, *see* FUR

geometrical identities, 87
lines, 96
planes, 96
fixed sample size
in ratio estimation, 218
in stereology, 143
sample mean is unbiased, 336
Flatland, 18
Floderus method, 41
experimental studies, 41, 256
Floderus, S., 41, 53
Flook, A., 290
fluorescence, 278
fractals, 151, 290, 331
fracture surfaces
vertical sections, 176
Fullman, R.L., 53
Fundamental Formulae of Stereology, 22–23, 54, 58, 83
cannot be extended, 73
depend on geometrical identities, 72
in Extended case, 150
in Random case, 146
in Restricted case, 148
under weighted sampling, 65
FUR (fixed orientation, uniformly random), 111, 117
FUR lines in the plane, 114–115
definition, 115
mean values, 115
FUR planes, 117–119, 179
conditional property, 118
definition, 117
estimation of volume, 119, 147
hitting probabilities, 118
normalising constant, 117
sampling bias for particles, 118
simulation, 118
FUR quadrats, 119, 149, 218
definition, 119–120
hitting probability, 121
mean values, 121
properties, 121
FUR serial sections, 155, 179

Garcia-Fiñana, M., 81, 188, 311, 328
Gardner, R., 43, 72, 83, 108, 194
Gaussian curvature, 106
integral of, 106
geometric measure theory, 72, 108
geometric tomography, 72
geometrical identities, 87–109
breadcrust "theorem", 96
is false, 97
Crofton formula, 87–96
general form, 89
fixed orientation, 87
for integral of mean curvature, 106
for lines in 2D, 89
for lines in 3D, 95
for planes in 3D, 92
for planes of fixed orientation, 96
for quadrats, 97–104
general form, 87
isotropic orientation, 87
kinematic formula, 87, 97–104
section formula
for integral of mean curvature, 88, 106
stereology depends on, 71
vertical sections, 179–182
vertical slices, 190
geometrical probability, 19, 75
literature, 133
Geuna, S., 5, 81
Giger, H., 46, 52–54, 73, 83, 84, 153
Gijbels, I., 53
Glagolev, A.A., 12, 52, 173
Glagolev-Thomson technique, 12
Gokhale, A.M., 188, 189, 192, 207
Gomez, D.M., 53
grain size
controversy about measuring, 5
influences variance of point counting, 301
granuloblastic rock, 1
granulosa cells, 25, 192
grid constant, 223
Groeneboom, P., 36, 53
Guillery, R.W., 84
Gundersen tiling rule, 69, 239
Gundersen, H.J.G., iv, 17, 41, 54, 62, 68, 69, 71, 76, 77, 83, 84, 128, 142, 173,

175, 181, 187, 198, 200, 203, 206, 207, 223, 228, 239, 249, 252, 253, 258, 264, 265, 268, 269, 273, 276, 285, 291, 299, 311, 316, 318–320, 327–329, 331
Guttman, L., 19
Guttorp, P., 330
Gy, P., 54, 328, 329, 332

Haar measures, 108
Haas, A., 175
Hadwiger characterization theorem, 73
Hadwiger, H., 46, 53, 54, 73, 108, 153
Hahn, U., 173, 330
Hájek, J., 228
Hall, P.G., 36, 53, 242, 253, 327
Haug, H., 5, 53, 62, 84
Hedreen, J.C., 53, 84, 253, 328
Hennig, A., 53, 299, 316
Herrup, K., 84
Hilliard, J.E., iii, 54, 77, 83, 175, 188, 206, 299
Hlubinka, D., 331
Hobolth, A., 317, 332
homogeneity assumption, 55, 58, 59, 82
Hoogendoorn, A.W., 36, 53
Hoppeler, H., 228, 330
Horvitz-Thompson estimator, 70, 71, 252, 253
  definition, 339
  poor performance, 339
Hyde, D., 54, 83

ice
  vertical sections, 176
  vertical slices, 189
image analysis, 54
immunolabelling, 278
inference
  design-based, 57
  hybrid, 61
  model-based, 57
integral equations
  in Wicksell's problem, 35, 36
integral geometry, 19, 72, 87, 108
  literature, 83, 108
  stereology depends on, 72
  stochastic interpretations, 75
integral of Gaussian curvature, 106
  for polyhedral surface, 106
integral of mean curvature, 22, 88, 106
  for polyhedral surface, 106
integral of plane curvature, 244
integral range, 297
integral test system, 223
intercept count, 22
  units, 22
International Society for Stereology, iv, 4, 54
invariant density
  for 2-quadrats in 3D, 103
  for lines in 2D, 89, 107
  for lines in 3D, 95, 107
    factorisation, 180–182
  for planes in 3D, 92, 190
  for rotations, 99
  for vertical planes, 178
invariant measure
  of lines hitting a plane set, 116
  of planes hitting a solid, 117
invariant measures, 89, 108
isector, 128, 278, 285
islets of Langerhans, 83
isotropic orientation
  estimation of length without, 164
  geometrical identities, 87
  need for, 92, 96
  needed for surface area and length, 162, 280
  not needed for volume, 280
isotropic random
  directions, 123
  lines in 2D, 123
  lines in 3D, 124
  planes
    orientator, 124–126
  planes in 3D, 124
Ito index, 30
IUR (isotropic, uniformly random), 111
IUR lines in the plane, 114–117
  conditional property, 116
  coordinates, 115
  definition, 115

examples, 115
hitting probability, 116
mean values, 117
normalising constant, 116
rejection method, 116
IUR planes, 126–129, 148
are not isotropic, 129, 132
are not marginally isotropic, 127
conditional property, 127
definition, 126
distribution of coordinates, 127
hitting probability, 127
normalising constant, 126
section formulae, 128
virtual, 285
IUR quadrats, 129, 150, 217, 218
IUR sections, 148
point-sampled intercepts, 193
serial sections, 164, 220
of blocks, 221
sampling designs, 278, 285
subselection of local sections from, 193
virtual, 285

Jakeman, A.J., 36, 53
Janáček, J., 168, 173, 285
Jensen, E.B.V., 17, 53, 54, 71, 75, 76, 83, 84, 175, 192, 200, 202–204, 207, 223, 226, 228, 242, 253, 255, 257, 262, 264, 265, 268, 269, 273, 291, 299, 311, 316, 317, 327–332
Jensen-Gundersen paradox, 299
Jeulin, D., 54, 58, 133, 153
joint distribution
area-weighted plane coordinates, 227
cartesian coordinates of UR point, 112
fallacies, 129
IUR line coordinates, 115
IUR plane coordinates, 126
numerator and denominator of ratio, 211
particle size and shape, 37
polar coordinates of UR point, 130
VUR plane coordinates, 178
Jones, M.C., 53
Jongbloed, G., 36, 53
Journal of Microscopy, 54

Karlsson, L.M., 84, 258
Kellerer, A.M., 328
Kendall, D.G., 54, 309, 327
Kendall, M.G., 19, 53, 75, 111, 133, 327
Kendall, W.S., iii, 35, 45, 53, 54, 75, 77, 83, 133, 146, 153, 255, 265, 271, 297, 299, 301, 302
Kiderlen, M., 54, 330
kidney
biopsy, 152
cortex volume fraction, 213
counting in, 253
disector, 70
glomeruli
total filtration surface, 59, 141
total length, 138
volume-weighted mean volume, 140
inefficient sampling in, 80
inhomogeneous, 138
sampling inference, 59, 137, 138
second-order stereology, 331
section of brush border, 21
serial sections, 66
sharing tissue from, 152
Kiêu, K., 76, 264, 265, 306, 307, 328, 331
kinematic density, 98
for 1-quadrats in the plane, 100
for 2-quadrats in 3D, 103
in d dimensions, 99
in the plane, 98
kinematic formula, 87, 97–104
for 1-quadrats in the plane, 101
for 2-quadrats in 3D, 103
for objects of full dimension, 99
for quadrats of full dimension, 99
Kolmogorov, A.N., 327
Kubínová, L., 168, 173, 285
Kurzydłowski, K., iii, 54, 83, 188, 258, 331

Lange, H., 62

Lange, N., 5, 58, 81
Lantuéjoul, Ch., 69, 84, 253, 314–316, 327, 328
Larsen, J.O., 173, 285
Lawson, L.R., iii, 54, 83
length
  of curve in space
    from section planes, 88, 94
  of plane curve
    estimation, 163
    estimation without isotropic orientation, 164
    using line probes, 88, 90, 312
    variance of estimator, 163
length density, 20, 22
  definition, 20
  depends on magnification, 290
  estimation, 21
  estimation of, 17, 61, 146, 192, 207, 219, 279, 280, 301
  history, 52
  interpretation, 141
  kidney microvilli, 21
  thick slices, 44
  units, 22
length fraction, 22
  units, 22
lines in 2D
  coordinates, 89
  Crofton formula, 90
  geometrical identities, 89
  invariant density, 89, 107
  IUR, 115–117
  random, 114
lines in 3D
  coordinates, 95
  Crofton formula, 95
  determine surface area, 95
  determine volume, 95
  geometrical identities, 95
  invariant density, 107
liver, 18, 30
local curvature
  of plane curve, 104
local stereology, 76, 192–207, 331
Losa, G., 290, 331
lost caps, 48
lung
  air-blood barrier thickness, 143
  sampling inference, 61
  scales of organisation, 214
  star volume of alveoli, 207
  vertical sections, 188

magnification
  determined by scale of features, 216, 278
  entails subsampling, 166, 282
  fixed, 218
  fractal effects, 151, 290
  high, 132
  influences definition of features, 290
  low, 288
  lung, 214
  multiple, 228
  must be known, 280
  not needed for estimating particle number, 280
  not needed for estimating volume fraction, 214, 280
Mall, G., 124, 168, 285, 331
Mandelbrot, B.B., 331
Martensite, 18
Mase, S., 36, 53
Matérn, B., 327, 328
mathematical morphology, 54, 58, 175
  dilation, 120
  erosion, 122, 298
  Minkowski subtraction, 122
  Minkowski sum, 120
  star volume, 207
Matheron, G., 16, 45, 53, 54, 58, 84, 153, 175, 207, 297, 303, 309, 328
Mathieu, O., 228, 330
Mattfeldt, T., 124, 168, 173, 285, 328, 331
Mayhew, T.M., 62, 81, 83, 143, 207, 228, 331
mean caliper diameter, 116
mean cross-sectional area
  of tubular structure, 143
mean curvature, 105
  integral of, *see* integral of mean curvature, 88, 106

mean particle size
  estimation, 32
mean sphere diameter
  estimation
    sensitivity to assumptions, 48
  estimation unstable, 35
mean thickness
  of air-blood barrier, 143
  of skin layer, 143
Mecke, J., iii, 35, 45, 53, 54, 58, 75, 77, 83, 133, 146, 153, 173, 255, 265, 271, 297, 299, 301, 302
Mendis-Handagama, S.M.L.C., 41, 256, 289
Merz grid, 83, 164
Merz, W.A., 83
methodological bias, 289
methodological errors
  breadcrust "theorem", 96
  in cell biology, 29, 31, 39
  interpretation of ratios, 79, 141
  interpretation of sections, 18
  isotropy, 96, 280
  liver, 4, 18
  Merz grid, 280
  particles
    counting in thick sections, 39
    numerical density, 26
    numerical density changes, 28
    relative abundance, 30
    sampling bias, 4
  quenched steel, 4, 18
  reference trap, *see* reference trap
  sampling design, 79, 129
  variance, 325
microfilaments, 22
microscopy
  acoustic, 278
  confocal, *see* confocal
  electron, 21, 42, 214, 259, 278, 288
  modes of, 4, 278
  optical, 1, 10, 20, 25, 160, 167, 168, 207, 278, 284, 289
  preparation of material, 278
microtome, 20, 48, 65, 156
microvilli, 22
Miles
  generic scheme for stereological experiments, 135, 143
    modifications, 142
  three cases of stereological inference, 137
Miles, R.E., iv, 3, 5, 16, 46, 54, 56, 60, 64, 68, 69, 83, 84, 133, 135, 153, 207, 228, 242, 252, 253, 255, 314, 316, 327, 328
Miles-Lantuéjoul correction, 69
Miller, W.S., 252
mining, 54
Minkowski sum, 120
minus-sampling, 68
  Miles-Lantuéjoul correction, 70
'Miraculous case', 139
Model I, 143, 225
Model II, 143, 225
model-based sampling inference, 57
model-based stereology, 5, 58–59
  applicability, 59
  estimation, 144–147
  Miles' Random case, 138, 139
  of particle populations, 269–275
  parameters, 58, 140
  rigorous foundations, 58
model-unbiased, 212
modern stereology, 55–85
  development of, 56
  history, 84
Molchanov, I.S., 45, 299
moments
  circle diameter, 37
  of area fraction, 53
  of random sets, 58
  particle size, 75, 76, 140, 257, 262, 265
  particle volume, 74, 193, 207
  point-sampled intercept length, 193
  second-order, 47
  sphere diameter, 34, 37
    weighted, 36
  stochastic processes, 58
  weighted, 37
    sphere diameter, 36
  Wicksell's problem, 37

Moran, P.A.P., 19, 53, 75, 111, 133, 153, 163, 173, 305, 309, 312, 327
mouse
  ovarian follicle, 25, 192
Mouton, P., iii, 54, 83, 84, 168, 173
Mücklich, F., iii, 16, 33, 35, 36, 38, 53, 54, 83, 298, 330
multi-stage ratio estimation, 214
  efficiency, 216
muscle
  vertical sections, 176
Myking, A.O., 84, 228

Nagel, S., 228
Nagel, W., 53, 153, 173
nested ANOVA, 293, 318–321
nested sampling designs, 282–284
  analysis of variance, 318–321
neurons
  age-related loss of, 62
  controversy about counting methods, 5
  definition of reference space, 136
  discrepancies in estimates of number, 41
  importance of counting, 25
  in brain, 25, 41
  in dorsal root ganglia, 41
  in hippocampus, 41
  loss of, *see* brain
  number per unit area, 49
  numerical density, 63
  second-order stereology, 331
  volume estimation, 202
neuroscience, 62, 188
Neyman, J., 327
Nielsen, J., 173, 285, 311, 328
nonparametric methods, 330
nucleolus, 76, 192, 202, 241
nucleus, 76, 241
numerical density N_V, 23, 62, 136
  across species, 63
  estimation, 29, 36, 255, 259
    Abercrombie's method, *see* Abercrombie
    empirical studies, 256
    methodological errors, 26, 39
    shape assumptions, 256
    variance, 301, 321
  interpretation, 62, 141
  of neurons in brain, 62
  reciprocal, 141
  under atrophy, 63
  under oedema, 62
  under shrinkage, 63
Nychka, D., 36, 53
Nyengaard, J.R., 38, 53, 66, 84, 128, 252, 253, 276, 285, 332

O'Sullivan, F., 36, 53
observation artefacts, 290
oedema, 62
Ohser, J., iii, 16, 17, 33, 35, 36, 38, 53, 54, 83, 298, 316, 328, 330, 331
oocyte, 192
Oorschot, D., 62
operator effects, 290
optical disector, *see* disector
optical sections, 76, 160, 242, 269, 331
optimal sampling design, 320
orientator, 124–126, 278, 285
Østerby, R., 318–320, 329
ovarian follicle, 25, 192

Pakkenberg, B., 41, 62, 81, 258, 276
paradox
  Simpson's, 141
paradoxes
  Jensen-Gundersen, 299
  of uniform distribution, 132
  Ohser, 316
  probability, 133, 285
  Simpson's, 151
  superefficiency of point counting, 289, 299–301
  superefficiency of subsampling, 17
  variance, 316
parameters, 61, 139–142
  absolute, 63
  distinction between absolute and relative, 61, 139
  interpretation of, 140
  methodological errors, 141

some cannot be estimated, 73
superpopulation, 58
types of, 139
Park, H., 19
particles, 25, 255–276
counting, 70–71
disector, *see* disector
empirical method, 71
Horvitz-Thompson weights, 71
cylindrical, 256
design-based stereology, 257
disector, *see* disector
ellipsoidal, 37
empirical method, 245–247
general shape, 75, 255
geometrical models, 256
"lost caps", 41
mean size, 32, 53, 262–268, 272–274
unweighted mean, 263, 272
volume-weighted mean, 266, 273
mean volume, 74, 140
methodological errors for, 26, 28, 30
model-based stereology, 269–275
moments of size, 140
moments of volume, 74
number
does not require knowledge of magnification, 280
variance of estimators, 321
numerical density, 26, 271
difficulty of estimating, 29
from thick sections, 38
population parameters, 47
profile counts, 26
relative numbers, 30, 280
sampling bias, 30, 255
strategies for eliminating, 257
sampling bias in plane sections, 32, 118
sampling effects, 31
sampling of, 75
size distribution, 34, 53, 268, 274
estimation, 34
spherical, 34–37
size distribution, 34
tomato salad analogy, 28, 31, 32, 53, 70
volume-weighted mean volume, 140
volume-weighted sampling, 75
weighted sampling, 75
particulate materials, 54
pathology, 62
Paumgartner, D., 290, 331
physical disector, 70, 259
pilot experiments, 292
placenta, 62
star volume, 207
planes in 3D
coordinates for, 92
Crofton formula, 94
determine area of curved surface, 94
determine length of space curve, 94
determine mean curvature, 106
determine volume of solid, 94
geometrical identities, 92
invariant density, 92
IUR, *see* IUR planes
plus-sampling, 68, 234
point counting, 158
bivariate binomial model, 301
estimator of area, 13, 158
variance of, 159, 309
estimator of relative area, 13, 219
experimental studies, 301
Poisson distribution
for disector counts, 321
goodness-of-fit, 323
Poisson generalised linear model, 323
Poisson process assumption, 58, 299
Poisson regression, 226
Poisson variance contribution, 323
polar coordinates, 74, 195
expression for volume, 74, 76, 195
for IUR lines in 2D, 116
for IUR planes in 3D, 126
for lines in 2D, 89, 114, 163
for planes in 3D, 92, 155, 164
for vertical planes, 178
of UR point, 130
pps sampling, 64
precision, 80, 295, 327
principal curvatures, 105
probe, 64, 71, 87, 217
arbitrary, 138

dimension, 22, 314, 328
fixed orientation, 87
fixed size and shape, 218
FUR, *see* FUR
generic term, 56, 136
included in reference space, *see* Model I
inference depends on, 138
isotropic random encounters with structure, 175
isotropic random orientation, 87
IUR, *see* IUR
line in 2D, 88
line in 3D, 88
local, 76, 192
notation, 178
plane, 88
quadrat, 97, 129
random, 111, 137
uniformly random, 75
virtual, 192
weighted random, 224
profile curve, 18, 19
profiles, 18, 21
and particle size, 33
and traces, 26
area-weighted mean area, 83
area-weighted random selection, 76, 267
counting rules, *see* counting rules, 70
definition, 25
diameter, 73
identifiability, 255
lost caps, 48
number of, 26, 41
misinterpretation, 28, 30
of ellipsoids, 37
of spheres, 34
of tubes, 49
realities, 48
relative numbers
misinterpretation, 30
sampling bias, 41, 68
unbiased sampling, 68
projection, 42–46
effect on dimension, 43
opaque, 44
total, 44
Prokešová, M., 330
proportional regression, 212, 225, 226

quadrat sampling, 119
quadrats, 97, 119
edge effects, 122, 123
FUR, *see* FUR quadrats
IUR, *see* IUR
ratio estimation, 219
strictly inside a region, 121
with uniform reference point, 123

Rakic, P., 41, 253
Ralph, B., iii, 54, 83
Random case of stereological inference, 138, 144–147, 217–219, 223
random directions, 123
isotropic, 123
random effects model, 319
random lines
fixed orientation, *see* FUR lines
in the plane, 114–117
isotropic orientation, *see* IUR lines
random planes
fixed orientation, *see* FUR planes
IUR, *see* IUR planes
random sampling, 63, 333–340
adaptive cluster sampling, 246
cluster sampling, 230, 247, 252, 337
nested, 249
discrete designs, 67, 229–253
estimation from, 335
fractionator, 249
nested cluster sampling, 249
nonuniform sampling designs, 338–339
PPS sampling, 64, 338
sampling bias, 335
sampling probability, 334
simple, 333
stratified sampling, 252, 336
systematic sampling, 64, 65, 333
uniform sampling probability, 334
variance, 339
weighted sampling, 64

random sets, 54, 58, 75, 138, 140, 144, 205, 296, 297, 299, 302, 317
Rankin, R.A., 309, 327
Rao-Blackwell theorem, 17, 204, 314–317, 328
Rataj, J., 83, 133
ratio estimation, 209–228
  conditional BLUE, 225
  does not require section thickness, 214
  estimation of stereological ratios, 63, 216–224
    fixed sample size, 218
    general theory, 217
    systematic sampling, 219
    weighted sampling, 224
  estimation via ratios, 213–216
  in Abercrombie's method, 47
  in finite population survey sampling, 209
  in stereology, 213–224
  kidney cortex volume fraction, 213
  Model I, 225
  Model II, 225
  multi-stage, 214
    efficiency, 216
  parametric model, 226
  practical advantages, 214
  rationale, 210
  robust against tissue shrinkage, 214
  semiparametric model, 225
  single stage, 213
ratio estimator, 210
  best linear unbiased estimator, 212–213
  bias
    depends on population, 211
    tends to be unimportant, 210
  biased, 210
  conditions for optimality, 212
  efficiency, 223
  of population ratio, 210
  of population total, 210
  unbiased in weighted sampling design, 211
  variance, 211–212
ratios
  ratio parameters, 61, 140
    interpretation, 79, 140
    misinterpretation, 141
reference space, 11, 44, 56, 60–64, 141–144, 148, 149, 151, 155
  and containing space, 142
  generic term, 136
reference trap, 62–63, 141–142, 151
  foetal development, 141
  interspecies comparison, 63
  neoprene, 142
  neurons in brain, 62
  neuroscience, 62
  particle means, 142
  pathology, 62
  placenta, 62
rejection method
  for FUR planes, 118
  for IUR lines in the plane, 116
  for uniformly random points, 113
relative area, *see* area fraction
relative proportions, 25
  alternatives, 49, 280
  methodological errors, 30
relative quantities, 140
  definition, 140
  do not require section thickness, 280
replication, 266, 321
  assumption of independence, 325
  desirability, 325
  rock samples, 327
  source of variability, 80
Restricted case of stereological inference, 137, 147–148, 155, 217, 220, 223, 224, 255, 266, 275, 281
Rhines, F.N., 27, 29, 52, 55, 252
Rhines-DeHoff relation, 27
Rigaut, J.-P., 331
Ripley, B.D., 35, 327
Robbins, H., 327
Roberts, N., 81, 311, 328
rock
  homogeneity assumption, 9, 14, 61, 153
  isotropy assumption, 61
  microstructure, 17
  mineral grains, 25
  polished plane sections, 1, 9

reference space, 56
replicated samples, 327
scale of variability, 16
star volume of pores, 175
test lines, 11
test points, 12
vertical sections, 176
Röthlisberger, H., 84, 175
Rosiwal's technique, 11
Rosiwal, A., 11, 52, 173, 327
Russ, J., iii, 54, 83

Saltykov, S.A., 4, 18, 19, 35, 52, 53, 188
sampling, 333–340
allocation of effort, 80
area-weighted, 64
cluster sampling, 337
discrete designs, 67
estimation from, 335
PPS sampling, 64, 338
sampling bias, 335
sampling probability, 334
simple random sampling, 333
stratified random sampling, 336
systematic sampling, 64, 65, 333
variance, 339
weighted, 64
sampling bias, iii, 4, 30–32, 234–235, 255
causes, 335
definition, 335
due to preferential orientation, 175
eliminating, 257
in counting profiles, 41, 68
in plane sections, 118
in vertical sections, 179
leads to estimation bias, 235
not measurable, 256
orientation-dependent, 179
strategies for eliminating, 229
sampling design
optimal, 320
sampling design errors
IUR planes are not isotropic, 129, 132
uniform random locations, 129
sampling rules
unbiased, 233
sampling techniques
cluster sampling, *see* cluster sampling
spatial, *see* spatial sampling
sampling theory
and stereology, 56
kinds of inference, 57
sampling variability, 80
due to spatial inhomogeneity, 80
Sampson, P.D., 330
Sandau, K., 167, 173
Santaló, L.A., 19, 54, 72, 83, 108, 153
Saper, C.B., 84
Saxl, I., iii, 54, 83
scale changes, 292
scale families of particles, 34
scale of variability, 16, 297
scales of organisation, 278
determine magnifications, 216
liver, 215
lung, 214
Schladitz, K., 316
Schmitz, C., 81, 253
Schneider, R., 54, 72, 83, 108, 133
second-order stereology, 331
section thickness, 20, 38, 287
measurement of, 287
not needed in ratio estimation, 214
sections
optical, 76, 160, 242, 269, 331
projected thick, 42–46
thick, 38
thin, 38
virtual, 284, 285
selector, 252
serial sections, 65, 155
kidney, 66
ratio estimation, 220
Serra, J., 16, 35, 45, 53, 54, 84, 175, 207, 253, 327, 328, 330
sharing tissue, 152
shrinkage, 62, 63, 276, 290
age-dependent, 62
and ratio estimation, 209, 214
in Abercrombie's method, 40
in atrophy, 62
Silverman, B.W., 36, 53

simple random sampling, 333
Simpson's paradox, 151
size distribution
  of spheres, 34
size-shape distribution
  ellipsoids, 37
  estimation, 37
skin
  layer thickness, 143
  mitotic fraction, 30
  stereological parameters, 143
  vertical sections, 77, 176, 267
Slomianka, L., 41, 81
Smith, C.S., 19, 52
Smith, R.L., 36, 53
Solomon, H., 53, 111, 133, 327
sources of variability, 59, 80
spatial covariance function, 16, 58, 297, 330
  derivative, 297
  estimation, 16, 298
  exponential model, 298
  integral range, 297
  isotropic, 297
  modelling, 329, 330
  of random field, 300
spatial grid, 167
spatial grid of test points, 159
spatial sampling, 252, 332
spatial statistics, 332
specimen
  generic term, 136
Spektor, A.G., 35, 53, 77, 175, 188, 206
sphere size distribution, 34
spherical coordinates, 93
staining, 278, 290
star, 195
star volume, 175, 198, 205, 207
  applications, 207
star-shaped sets, 194–198, 341
statistical formulation of stereology, 135–153
  Extended case, 137, 149–151
    example, 139
  generic models, 137–139
  implications of, 135
  Miles' generic scheme, 135–139
  'Miraculous case', 139
  Model I, 143
  Model II, 143
  need for, 135
  parameters, *see* parameters
  Random case, 138, 144–147
    example, 139
  Restricted case, 137, 147–148
    example, 138
  sample size, 143
  terminology, 135–136
steel
  grain boundaries, 187
  quenched, 18
  rolled, 187
  stainless, 16, 59
  stochastic model, 138
  vertical section, 187
Steinhaus estimator of curve length, 163
  relative error, 312
  variance, 312
Steinhaus, H., 19, 52, 163, 173, 312
stereological parameters, *see* parameters
stereological thinking, 3–5
stereology
  advantages and disadvantages, 3
  and sampling inference, 1
  applications, 4
  'assumption-free', 82
  bias corrections, 80
  classical, 9–54
    conflates sources of variability, 59
    history, 52
    homogeneity assumption ill-defined, 55
    homogeneity assumption inappropriate, 55, 59
    inference inappropriate, 59
    statistical critique of, 46, 47
    was model-based, 5
    weaknesses, 55
  classical vs modern, 55
  connection with sampling theory, 56
  depends on geometrical identities, 71
  depends on integral geometry, 72
  design-based, 5, 59–61
    applicability, 61

depends on adherence to protocol, 61
no assumptions about material, 60
ratio parameters, 61
unbiased estimation, 60
digital, 330
Fundamental Formulae of, 22–23
hybrids of design-based and model-based, 61
in bone, 188
in botany, 54, 188
in dentistry, 54, 188
in dermatology, 188
in pathology, 62
in placenta, 62
interdisciplinary, 4
International Society for, *see* International Society for Stereology
is not tomography, 2
key problems, 2
kinds of inference, 57
literature, 83
local, 76, 331
model-based, 5, 58–59
applicability, 59
parameters, 58
rigorous foundations, 58
modern, 55–85
development of, 56
history, 84
modern definition, 3
notation, 11, 23
of the lung, 188
parameters, *see* parameters
second-order, 331
'unbiased', 57
Sterio, D.C., 75, 84, 252, 255
stochastic geometry, 58, 75, 133
literature, 83
Stoyan, D., iii, 35, 45, 53, 54, 58, 75, 77, 83, 133, 146, 153, 255, 265, 271, 297, 299, 301, 302, 327, 330
stratified random sampling, 336
Stuart, A., 56, 333
Sundberg, R., 83, 226, 228, 242, 253
superefficiency, *see* efficiency
superpopulation parameters, 58
surface area
"fakir's bed", 166
estimation from IUR serial sections, 164
estimation requires isotropic orientation, 162
from line probes, 88, 95
from section planes, 94
surface area density, 19, 22
estimation of, 17, 219
units, 22
surfactor, 203
systematic sampling, 64, 155–173, 333
of orientations, 312
Cavalieri estimator of volume, 156
"egg slicer", 155
in ratio estimation, 219
serial sections, 155
superefficiency, 304
variance, 328
variance reduction, 290
systematic subsampling, 160

Takahashi, R., 331
Tandrup, T., 258, 276
Taylor, C.C., 36, 53
test lines, 11
IUR in 2D, 115–117
IUR in 3D, 132
sin-weighted, 181, 267
through fixed point, 192, 197
test points, 12, 158
estimation of area, 158
spatial grid, 159
test system, 11
integral test system, 223
ratio estimation, 219, 221
thick sections, 38
Thompson, S.K., 83, 209, 210, 228, 252, 319, 332, 333
Thompson, W.R., 71, 83, 153, 252
Thomson, E., 12, 52, 173
three-dimensional context, 278
three-dimensional imaging, 156
weaknesses, 81
three-dimensional interpretation, 48, 73

three-dimensional observation, 278, 280
three-dimensional parameters, 49, 72
three-dimensional reconstruction, 2, 15, 81
tiling rule for counting profiles, 69, 239
tissue shrinkage
  and ratio estimation, 214
tomato salad analogy, 28, 31, 32, 53, 70
Tomkeieff, S.T., 19, 52
total curvature, 243
  of plane curve, 104
trisector, 188
tubular organ
  vertical section designs, 286
tubular structure
  average cross-sectional area, 143

unbiased
  model-unbiased, 212
  sampling rule, 229, 233
unbiased brick, 253
unbiased counting frame, 69, 239
unbiased estimation, 10, 16
  confused with consistent estimation, 84
  in design-based stereology, 60
  of absolute volume, 157
  of area in the plane, 158
  of length
    without isotropy, 164
  of length of space curve
    from IUR serial sections, 164
  of surface area
    "fakir's bed", 166
    from IUR serial sections, 164
    spatial grid of lines, 167
  superpopulation sense, 58, 59
'unbiased stereology', 57, 81–82
  'assumption-free', 82
  misnomer, 81
Underwood, E., 35, 52, 53, 173, 188
uniform distribution, 112
  for random points, *see* uniformly random points
uniformly random
  probe, 111
uniformly random points, 112–114
  cannot be generated using FUR planes, 118
  Cartesian coordinates
    are conditionally uniform, 113
    marginal distributions, 113
  cartesian coordinates
    are jointly uniform, 112
    are not independent, 112
    are not marginally uniform, 112
  conditional property, 113
  non-cartesian coordinates, 114
  sampling probability, 112
  simulation, 113
urban legends
  breadcrust "theorem", 96
  Merz grid ensures isotropy, 280

van Es, B., 53
variability
  biological, 80, 318, 320, 325
  due to spatial inhomogeneity, 80
  sampling variability, 80
  sources of, 80
variable sample size, 143
variance, 80, 295–328
  binomial approximation, 301, 327
  estimation of, 310
  extension term, 307, 308
  nested ANOVA, 318–321
  of Cavalieri estimator, 157, 307, 308
  of estimator of curve length, 163, 312
  of estimators of particle number, 321
  of estimators of volume fraction, 296
    approximation, 299
  of point-count estimator of area, 159
  of point-counting estimator of area, 309
  of systematic sampling, 302–314, 328
    associated with boundary, 304
    estimation, 304
    in one dimension, 305–308
  prediction variance, 317
  Zitterbewegung, *see* Zitterbewegung
variance components, 319
variance paradoxes, 17, 299, 316

Jensen-Gundersen, 299
variance reduction
in systematic sampling, 290
Vedel, E., *see* Jensen, E.B.V.
vertical direction, 176
aligned with structure, 176
arbitrary choice, 177
test system must align with, 185
vertical projections, 188–192, 279
estimation of curve length, 191
vertical rotator, 200
vertical sections, 77, 175–188
applications, 176, 187, 188
choice of vertical direction, 176, 286
cycloid test curves, 184–185
estimation of surface area, 182–187
examples, 176–177, 185, 187
geometrical identities, 179–182
in botany, 188
in dentistry, 188
in dermatology, 188
in neuroscience, 188
isotropic test lines in, 279
model-based estimation, 187
of fracture surfaces, 188
of lung, 188
random (VUR), 178–179
sampling designs, 286
bars design, 286
for tubular organ, 286
slabs design, 286
serial, 182
stratified, 282
vertical slices, 188–192, 279
estimation of curve length, 192
examples, 188
geometrical identities, 190
vertical spatial grid, 192
villi, 177
virtual IUR sections, 285
volume
cylindrical coordinates, 74, 199
from FUR plane, 119
from section planes, 119
of solid
estimation, 155–160
from line probes, 88, 95
from section planes, 94
polar coordinates, 74, 195
volume fraction, 22
comparison of estimators, 13
Delesse estimator, 10, 15
does not require knowledge of magnification, 280
estimation of, 219
units, 22
volume-weighted sampling of particles, 75
VUR planes, 179
VUR serial sections, 179

Wahba, G., 53
Watson, G.S., 53
Weibel, E.R., iii, 29, 35, 52, 53, 55, 65, 83, 84, 135, 143, 153, 214, 216, 224, 228, 290, 321, 330, 331
weighted distribution
of sphere diameter, 34
estimation, 36
weighted moments
of particle size, 266, 273
of sphere diameter, 36
estimation, 37
weighted sampling, 64
in ratio estimation, 224
of particles, 75
Weil, W., 54, 72, 83, 108, 133
West, M., 41, 81, 253, 276, 328
Wicksell's problem, 34
design-based formulation, 268
for thick sections, 53
integral equation, 35
methods of solution, 36
moment relationships, 37
statistical theory, 53
Wicksell, S.D., 34, 37, 53
Williams, R.W., 41, 253
Wilson, J.D., 53
Wojnar, L., 188
Wreford, N., iii, 54, 83
Wulfsohn, D., 54, 189

Zitterbewegung, 303, 307, 323